Principles of plasma diagnos

Plasma physics is currently one of the most active subdisciplines of physics. Measurements of the parameters of laboratory plasmas, termed plasma diagnostics, are based on a wide variety of characteristic plasma phenomena. Understanding these phenomena allows standard techniques to be applied and interpreted correctly and also forms the basis of innovation. This book provides a detailed derivation and discussion of the principles of plasma physics upon which diagnostics are based. These include magnetic measurements, electric probes, refractive index, radiation emission and scattering, and ionic processes.

Principles of Plasma Diagnostics is the most modern and comprehensive book in its field to date. It gathers for the first time a body of knowledge previously scattered throughout scientific literature. The text is based on first-principles development of the required concepts, so it is accessible to students and researchers with little plasma physics background. Nevertheless, even seasoned plasma physicists should appreciate the work as a valuable reference and find insight in the lucid development of the fundamentals as they apply to diagnostics.

Although most of the examples of diagnostics in action are taken from fusion research, the focus on principles will make it useful to all experimental and theoretical plasma physicists, including those interested in space and astrophysical applications as well as laboratory plasmas.

Ian H. Hutchinson has been engaged in research in experimental physics and controlled fusion since 1973, first at the Australian National University as Commonwealth Scholar and subsequently at M.I.T. and the U.K. Atomic Energy Authority. He has published numerous journal articles on a wide variety of plasma experiments. Currently Associate Professor in the Department of Nuclear Engineering at M.I.T., Dr. Hutchinson teaches graduate level plasma physics.

I. H. HUTCHINSON

Massachusetts Institute of Technology

Principles of plasma diagnostics

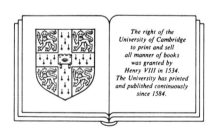

The right of the
University of Cambridge
to print and sell
all manner of books
was granted by
Henry VIII in 1534.
The University has printed
and published continuously
since 1584.

CAMBRIDGE UNIVERSITY PRESS

Cambridge
New York Port Chester Melbourne Sydney

HUTCHINSON

Published by the Press Syndicate of the University of Cambridge
The Pitt Building, Trumpington Street, Cambridge CB2 1RP
40 West 20th Street, New York, NY 10011, USA
10 Stamford Road, Oakleigh, Melbourne 3166, Australia

© Cambridge University Press 1987

First published 1987
Reprinted 1990

Printed in the United States of America

Library of Congress cataloguing in publication data
Hutchinson, I. H. (Ian H.)
Principles of plasma diagnostics.
Bibliography: p.
Includes index.
1. Plasma diagnostics.   I. Title.
QC718.5.D5H88   1987   530.4'4   87-11690

British Library cataloguing in publication data applied for

ISBN 0-521-32622-2

S. D. G.

Contents

Preface

The practice of plasma diagnostics is a vast and diverse subject, far beyond the span of a single volume, such as this, to cover in all its detail. Therefore, some limitations on the objectives adopted here have to be accepted. The title *Principles of Plasma Diagnostics* refers to the fact that the physical principles used for plasma measurements are to be our main concern. In brief, this book seeks to give a treatment of the fundamental physics of plasma diagnostics, and thus to provide a sound conceptual foundation upon which to base any more detailed study of applications. I hope, therefore, to bring the reader to the point where he or she may, with confidence and understanding, study the details of any diagnostic discussed in the literature.

Most journal articles and reviews on plasma diagnostics tend, of necessity, to begin from a mere citing of the required equations governing the principles employed. For all but the experienced specialist, this means that the reader must accept the equations without much justification or else pursue a deeper understanding through references to original papers. One of my main objectives here is to overcome this difficulty by a systematic presentation from first principles. Therefore, if in some cases it may seem that the development stops just as we approach the point of practicality, I can only plead that, in bringing the reader to the point of being able comfortably to understand the basis of any application, I have fulfilled a major part of my task.

Some justification of the fact that I provide very little detailed discussion of instruments and techniques may be appropriate, since they are by no means uninteresting or irrelevant.

First, to describe the various experimental technologies in a way accessible to the uninitiated, at anything other than a pure "cookbook" level, would require so much space as to be overwhelming for a single volume. Second, instruments and technology are not really unique to the plasma field, in most cases, even though the needs of plasma diagnostics are sometimes the driving force behind their development. Third, the technology is developing so rapidly that any extensive treatment tends to become outdated almost immediately. Fourth, there are several recent journal article reviews and specialist book series that emphasize instrumentation.

My philosophy, then, is to include only sufficient description of the technology to provide a fundamental understanding of the applications, rather than a detailed analysis of the instruments themselves. Only when the plasma is virtually part of the instrumental configuration, such as in an interferometer, is more detailed discussion given. As compensation, brief summaries of some of the present technological capabilities are given in an appendix.

By concentrating on the physical principles, my intention has been to produce a book of interest to plasma physicists as a whole, whatever the area of their major specialization. However, it is necessary in a work such as this to have a fairly clear perspective. Otherwise, one is forever qualifying statements in a way that ought to be implicit. My perspective is that of *laboratory* plasma diagnostics. What is more, most of the examples are taken from controlled fusion research applications, partly because fusion is the area in which by far the most study of plasma diagnostics has been done. I trust, nevertheless, that the material may be useful also to experimentalists and theoreticians in other plasma fields, such as space or astrophysical research, since it is a discussion mostly of general principles, applicable to these very different plasmas as well as those in the laboratory.

The level of the treatment may perhaps best be described as intermediate graduate. This means that a good basic undergraduate physics background should be sufficient to enable the reader to follow the material, even though the approach may be more demanding than in an undergraduate text. Very little detailed prior knowledge of plasma physics is assumed; therefore, researchers entering the plasma field, should find most of the material accessible. There is, however, no pretence at a systematic introduction to plasma physics, and the presumption is that basic plasma physics concepts, at least, are familiar. The more senior researcher I hope will also find useful material here for reference and to gain a broader perspective, although length restrictions prevent discussion of many important practical details.

The literature and references cited are intended to serve two limited purposes: to provide representative examples of the principles in action and to provide starting points for more detailed study of the scientific literature in any specialized area. There is no attempt to provide exhaustive references and I apologize to anyone who feels their own work to have been unjustly omitted.

I also thank all my teachers, friends, students, and colleagues who have provided information, figures, criticism, suggestions, corrections, and so on. Like all of science, plasma research is a cooperative enterprise, and so the material here represents an overview of the work of a large number of people over many years. Special thanks also go to my colleagues at MIT who have taken time to read sections of the manuscript and make suggestions for improvements, especially Bruce Lipschultz, Earl Marmar, Steve McCool, Jim Terry, Reich Watterson, and Steve Wolfe. The shortcomings of the book are mine, though! Thanks to Cathy Lydon for managing so much difficult word processing.

Thank you, Fran, for making it all possible by your constant support and love.

1   Plasma diagnostics

1.1   Introduction

During the past few decades a great deal of research has been undertaken in plasma physics. As a result, the field includes a very substantial body of knowledge covering a wide variety of branches, from the most theoretical to the most practical, comparable to any other subdiscipline of physics. As with any other science, progress has been made most effectively when an early quantitative confrontation between theory and experiment has been possible. This confrontation places strong demands upon theory to do calculations in realistic configurations and circumstances, but it also requires that the properties of plasmas be measured experimentally as completely and accurately as possible. For this reason much of the effort in experimental plasma physics is devoted to devising, developing, and proving techniques for diagnosing the properties of plasmas: plasma diagnostics.

A major driving force behind the research on plasmas has been, and still is, the prospect of generating economically significant amounts of power from controlled thermonuclear fusion. Fusion has its own imperatives of temperature, density, confinement, and so on, which provide a stimulating and relevant environment in which plasma research is conducted. Moreover, the vitally important diagnosis of fusion plasmas poses problems that are often enhanced by the nature of the fusion goal. For example, the high temperatures sought for fusion frequently eliminate the possibility of internal diagnosis by material probes.

The overall objective of plasma diagnostics is to deduce information about the state of the plasma from practical observations of physical processes and their effects. This usually requires a rather elaborate chain of deduction based on an understanding of the physical processes involved. In more mundane situations the same is true of other diagnostic measurements; for example, an ordinary mercury/glass thermometer relies on the physical process of thermal expansion of mercury, which determines the height of the mercury column observed. However, since plasmas have properties that are often rather different from the more familiar states of matter met in everyday life, the train of reasoning is sometimes more specialized and may seem more obscure, especially since plasma diagnostics are rarely routine.

What is required, then, for an understanding of the principles of plasma diagnostics is a thorough knowledge of plasma physics. For that reason, the reader with a good basic background in plasma physics will find the going easier than someone with little background. It would be inappropriate to attempt to provide a broad introduction to plasma physics in a work such as this, not least because many excellent texts exist to which reference can be made. (A brief bibliography is given at the end of this and other chapters.) However, the intent here is to develop, essentially from first principles, those aspects of plasma physics that are necessary to diagnostics, and to include those areas of general physics (for example, electromagnetic theory) that are also important in plasma diagnostics. Therefore, a good basic physics background should be sufficient to enable the reader to follow the material.

To reach useful and accurate results requires rather complete quantitative mathematical analysis; more so sometimes, than in a general text where a qualitative treatment is sufficient. When the mathematical analysis required here is not too difficult or lengthy it is given, though occasionally with some details left as an exercise. If the results are expressible in reasonably compact analytic form they are given fairly completely for reference. Using these results does not absolutely require that the reader follow every step of the mathematics (although it may help); so if, at first reading, you feel in danger of getting bogged down in mathematics, plunge on to the end and see how it all works out.

Some diagnostics require substantial amounts of data (for example atomic data) that are not easily expressed analytically. Such data are mostly not included here except in the form of examples and where it needs to be invoked for general understanding. Therefore, reference must be made to the specialized literature and tables for detailed information. On various occasions, simple heuristic physical arguments are used to obtain rough estimates in order to understand the dominant processes. These are not intended to be a substitute for the detailed results, often obtained through painstaking efforts over years of research. Therefore, care should be exercised in applying these estimates when accurate calculations are essential.

1.2   Plasma properties

At its simplest, a plasma is a gaseous assembly of electrons, ions, and neutral molecules residing in electric and magnetic fields. We shall regard the electromagnetic fields as an essential ingredient of the plasma. Indeed, since the plasma particles usually generate or at least modify these fields we can regard them as properties of the plasma. A complete classical description of a plasma, then, would consist of a specification of the fields and the position and velocity of all the particles throughout the volume of interest.

Naturally, as with any discussion of gaseous media, such a description is an inaccessible and rather fruitless ideal and we must appeal to the concepts and methods of statistical mechanics to provide a useful description. Thus, we suppose that at each position $\mathbf{x}$, in an element of volume $d^3\mathbf{x}$, there are on average a number $f_j(\mathbf{x}, \mathbf{v}) \, d^3\mathbf{v} \, d^3\mathbf{x}$ of particles of type j that have velocity in the element $d^3\mathbf{v}$ at $\mathbf{v}$. Then f_j is the *distribution function* of particles of type j. It may, of course, vary as a function of time t and space $\mathbf{x}$. A complete statistical description of our plasma is thus provided by a knowledge of $f_j(\mathbf{x}, \mathbf{v}, t)$ for all appropriate $\mathbf{x}$, $\mathbf{v}$, and t together with the electromagnetic fields ensemble averaged over all possible realizations of the particle distributions.

To determine completely the distribution functions is still a task beyond our practical capabilities in most circumstances, although certain diagnostics can, in principle, determine f_j at a given position and time. Therefore, it is usually necessary to restrict our ambitions still further. We would like to be able to determine f_j, but instead we aim to determine a few of the most important facts about f_j.

1.2.1 *Moments of the distribution function*

Consider a situation homogeneous in space and constant in time so that for a single species of particles the distribution is only a function of velocity $f(\mathbf{v})$. The kth moment of the distribution is defined as

$$\mathbf{M}^k = \int f(\mathbf{v})(\mathbf{v})^k \, d^3\mathbf{v}, \qquad (1.2.1)$$

where in general $\mathbf{M}$ is a tensor of order k. If we know $\mathbf{M}^k$ for all $k = 0$ to ∞, that determines $f(\mathbf{v})$ completely. However, knowledge of only the lower order moments is often sufficient to provide us with the information we need because the equations of motion of a plasma can often be written with sufficient accuracy as moment equations involving only low order moments and closed (i.e., truncated) by using phenomenological transport coefficients.

Because of their importance, the lower order moments are called by specific names reflecting their physical significance. Order $k = 0$ is a scalar moment,

$$M^0 = \int f(\mathbf{v}) \, d^3\mathbf{v} \equiv n, \qquad (1.2.2)$$

which is simply the particle *density* (number of particles per unit volume). Order $k = 1$ is a vector moment that is usually normalized by n,

$$\frac{1}{n}\mathbf{M}^1 = \frac{1}{n}\int f(\mathbf{v})\mathbf{v} \, d^3\mathbf{v} \equiv \mathbf{V}. \qquad (1.2.3)$$

This is simply the mean particle *velocity*.

The second order moment usually appears in the form

$$m \int (\mathbf{v} - \mathbf{V})(\mathbf{v} - \mathbf{V}) f(\mathbf{v}) \, d^3 v = m[\mathbf{M}^2 - n\mathbf{V}\mathbf{V}] \equiv \mathbf{p} \qquad (1.2.4)$$

and is the *pressure tensor*.

Because $\mathbf{p}$ is symmetric, there exists a coordinate system in which it is diagonal [see, e.g., Morse and Feshbach (1953)]. In this coordinate system $\mathbf{p}$ is said to be "referred to principal axes." When a magnetic field is present, because of the gyration of the plasma particles about the field, the distribution function usually acquires rotational symmetry about the field direction. As a result, the field direction becomes one of the principal axes (usually taken as the z axis) and the pressure tensor has the form

$$\mathbf{p} = \begin{bmatrix} p_\perp & 0 & 0 \\ 0 & p_\perp & 0 \\ 0 & 0 & p_\parallel \end{bmatrix}. \qquad (1.2.5)$$

If the distribution function is fully isotropic then the two components of the pressure tensor become equal, $p_\perp = p_\parallel = p$, and one can define the *temperature* (in energy units) as

$$T \equiv p/n. \qquad (1.2.6)$$

If $p_\perp \neq p_\parallel$ one can similarly define perpendicular and parallel temperatures $T_\perp, T_\parallel$. If the distribution is Maxwellian then these definitions will yield the temperature of the Maxwellian distribution. Otherwise, for nonthermal distributions one can regard the definitions as providing the *effective* temperature.

The third order moment is usually written as

$$\mathbf{Q} = m \int (\mathbf{v} - \mathbf{V})(\mathbf{v} - \mathbf{V})(\mathbf{v} - \mathbf{V}) f(\mathbf{v}) \, d^3 v, \qquad (1.2.7)$$

a third order tensor (of which the explicit expansion is rather cumbersome in dyadic notation) called the *heat flux tensor*.

These are then the quantities upon which we shall focus in our discussions of diagnostic measurements: the density, velocity, pressure and temperature, and, less importantly from the viewpoint of direct measurement, heat flux. In a situation that is inhomogeneous we shall sometimes be able to deduce these properties as a function of space and sometimes have to be content with spatial averages, much as we have had to content ourselves with averages over the velocity distribution (i.e., the moments).

Of course, these moments and our treatment of them so far are precisely what appears in a development of the kinetic theory of gases [see, e.g., Chapman and Cowling (1970)] in relating the microscopic state of the gas to its macroscopically observable fluidlike parameters. In the case of a normal gas of neutral molecules, it seems intuitively obvious that measurements will focus on these parameters and that, for example, we shall determine the density of gas by measuring its mass, the pressure by measuring the force it exerts, and so on. In a plasma the low order moments of the distribution are very rarely measured in such familiar ways, but they nevertheless provide the fundamental parameters that define the state of the plasma insofar as it can be described in simple fluid terms.

In order for the low order moments to be sufficient to describe the plasma, it is necessary for the plasma to be close to local thermodynamic equilibrium so that locally the distribution function f_j is approximately Maxwellian. If this restriction holds then local measurement of density, velocity, pressure, etc., together with the electromagnetic fields is sufficient to tell us essentially everything we need to know about the plasma. If the plasma is not close to thermal equilibrium (and there are many cases when it is not), then the moments still provide valuable information, but a complete description requires us to return to the distribution function, in this case non-Maxwellian.

1.2.2 *Multiple species*

So far we have considered only a single particle species, but in practice plasmas have usually at least two species, electrons and ions. More often than not, it is necessary to consider also neutral molecules and various types of different impurity ions. Moments can be taken for each species separately, of course, giving rise to possibly different densities, temperatures, etc. for each species. This results in a multiple-fluid description. The charged species, electrons and ions, are strongly coupled together in some aspects by electromagnetic forces. This coupling reduces the independence of the moments.

Consider a case where the plasma consists of just two species, electrons and ions, of charge $-e$ and Ze. The charge density is $e(-n_e + Zn_i)$, where subscripts e and i refer to electrons and ions (we shall also use subscript 0 to denote neutrals). If n_e were different from Zn_i then a space charge would exist and the corresponding electric field would tend to oppose the buildup of this charge. The effects of such electric fields are so strong that for most plasmas there can never be much imbalance in the electron and ion charge densities, that is, $n_e - Zn_i$ is always much less than n_e. This is the condition of so-called quasineutrality: $n_e \approx Zn_i$, consisting of a relationship between the zero order moments of the distribution of electrons and ions.

The relationship between the mean velocities involves the electric current rather than charge. The total current is

$$\mathbf{j} = n_i Ze\mathbf{V}_i - n_e e\mathbf{V}_e \approx n_e e(\mathbf{V}_i - \mathbf{V}_e), \tag{1.2.8}$$

Table 1.1. *The moments of the distribution for fluid plasma description.*

Order	Quantity	Symbol	Multiple species combinations
0	Density	n_j	Charge density (~ 0)
			Mass density ρ
1	Mean velocity	$\mathbf{V}_j$	Current $\mathbf{j}$.
			Mass flow $\mathbf{V}$.
2	Pressure	$\mathbf{p}_j$	Total pressure
	Temperature	$T_j = p_j/n$	
3	Heat flux	$\mathbf{Q}_j$	Total heat flux

which shows that the first order moments are related via the electromagnetic fields, since $\mathbf{j}$ is related to $\mathbf{E}$ and $\mathbf{B}$ through Ampere's law.

The pressures and temperatures of the different species generally are not strongly coupled. The collisional processes that cause energy transfer between the species are often slow compared to other energy transport mechanisms. Thus the electron and ion temperatures are generally thought of as distinct quantities to be measured. However, it is often convenient for some purposes to treat a plasma as if it were a single conducting fluid, in which case the quantity of interest is the total pressure given by the sum of electron and ion (and possibly neutral) pressures. In the single-fluid description a single velocity, generally the mass flow velocity

$$\mathbf{V} = \frac{1}{n_e m_e + n_i m_i}(n_e m_e \mathbf{V}_e + n_i m_i \mathbf{V}_i), \qquad (1.2.9)$$

the current $\mathbf{j}$, and a single density, generally the mass density $n_e m_e + n_i m_i$, together with the pressure and electromagnetic fields complete the plasma description.

Table 1.1 summarizes the low order moments with which we shall be most concerned.

1.3   Categories of diagnostics

There are several different ways to group diagnostics when studying them. First, we might group them by the plasma parameter measured. Thus, we might consider separately ways of measuring density, temperature, and so on. This has a certain logical appeal but suffers the drawback that many diagnostics measure more than one parameter or a combination of parameters so that this grouping tends to lead to an artificial and repetitive division of material, a single diagnostic appearing under several different headings.

A second possible categorization is by experimental technique: that which can be learned using a certain measuring instrument, for example a certain detector or spectrometer. Such a division is more appropriate if the treatment emphasizes the details of technique, which we do not.

The treatment we shall adopt categorizes diagnostics by the physical process or property of the plasma that is directly measured; for example, the refractive index of the plasma, the electromagnetic waves emitted by free electrons, and so on. This differs from the first option in treating together the various plasma parameters that measurement of a specific physical process allows us to estimate, and from the second in treating together measurements of a specific process even by widely differing techniques, though sometimes separating measurements made with the same instruments.

None of these options, of course, provides an ideal universal solution; however, our choice seems most suitable for concentrating on an understanding of the underlying physical principles of the processes that enable us to make plasma measurements. Even having made this decision as to the guiding principle of categorizing the diagnostics, a certain degree of arbitrariness remains in dividing up the material, but we shall deal in the succeeding chapters with the following main topics:

Magnetic measurements (Chapter 2). Measurements made by sensing directly the magnetic fields in various places inside and outside the plasma using coils and probes of various types.

Plasma particle flux measurements (Chapter 3). Measurements based on directly sensing the flux of plasma particles using probes of various types in contact with the plasma.

Plasma refractive index (Chapter 4). Diagnostics based on measurement of the plasma's refractive index for electromagnetic waves of appropriate frequency by transmission of such waves through the plasma.

Electromagnetic emission from free electrons (Chapter 5). The deduction of plasma properties from observation of radiation emitted by free electrons including cyclotron (synchrotron) emission, bremsstrahlung, and Čerenkov processes.

Electromagnetic emission from bound electrons (Chapter 6). Diagnostics using observation of the line radiation from atoms and ions that are not fully ionized.

Scattering of electromagnetic waves (Chapter 7). Measurements of the radiation scattered by plasma particles when subjected to incident radiation.

Ion processes (Chapter 8). Measurements of phenomena occurring to heavy nonelectron species, for example charge-exchange reactions, nuclear reactions, and interactions with probing beams of heavy particles.

Which of these processes provides information upon which plasma parameter depends to some extent upon the ingenuity of the application. There are, however, some uses that may be regarded as reasonably well

Table 1.2. *The plasma parameters that can be diagnosed using measurements of different properties. An estimate of the quality of the data to be expected is indicated by the range 1 (good) to 3 (poor).*

Property measured	Parameter diagnosed										
	f_e	f_i	n_e	n_i	n_0	V_i	T_e	T_i	p	E	B
Magnetic measurements							2		1	1	1
Plasma particle flux	1	1	2	2		2	1	1		1	
Refractive index			1								1
Emission of EM waves											
by free electrons											
Cyclotron	3		2				1				
Bremsstrahlung	2		2		2		1				
Čerenkov	3		3								
Line radiation			2	2	2	1	2	1			3
EM wave scattering	3		2	3			1	3			3
Ion processes											
Charge exchange		2			1			1			
Nuclear reactions		3	2					1			
Heavy probe beams			2						1		3

established (even if not necessarily widely used). For those techniques that have been extensively explored, either theoretically or experimentally, the matrix in Table 1.2 is intended to indicate which parameter is measurable by which process. The extent to which reliable information is available is indicated on a scale of 1 to 3, 1 meaning reliable, direct, or high quality information, 3 meaning rather unreliable, indirect, or poor quality. Naturally, such a rating is somewhat subjective and also can only reflect the current state of development of the techniques used. In some cases, though possibly only a minority, further improvement is foreseeable.

Of course, not all process measurements are possible with all plasmas. An example already mentioned is that most hot fusion plasmas cannot be diagnosed internally with material probes, so that direct flux of plasma particles cannot be measured. Another is that cold plasmas may have negligible nuclear reactions. Similarly the quality of information may vary from plasma to plasma. Nevertheless, the matrix provides a guiding framework that broadly indicates the applicability of various types of measurement.

Further reading

Numerous texts on basic plasma physics exist. Some of the more comprehensive include:

Boyd, T. J. M. and Sanderson, J. J. (1969). *Plasma Dynamics*. London: Nelson.
Clemmow, P. C. and Dougherty, J. P. (1969). *Electrodynamics of Particles and Plasmas*. Reading, Mass.: Addison-Wesley.

Krall, N. A. and Trivelpiece, A. W. (1973). *Principles of Plasma Physics*. New York: McGraw-Hill.

Schmidt, G. (1979). *Physics of High Temperature Plasmas*. 2nd ed. New York: Academic.

The following are treatments specifically of general plasma diagnostics:

Huddlestone, R. H. and Leonard, S. L. (1965). *Plasma Diagnostic Techniques*. New York: Academic. An excellent, though rather old, collection of research reviews emphasizing techniques.

Lochte-Holtgreven, W. (1968). *Plasma Diagnostics*. Amsterdam: North-Holland. A collection of reviews with more emphasis on spectroscopic methods.

Podgornyi, I. M. (1971). *Topics in Plasma Diagnostics*. New York: Plenum.

Tolok, V. T. (1971). *Recent Advances in Plasma Diagnostics*. New York: Consultants Bureau. A less useful collection of Russian papers.

Lovberg, R. H. and Griem, H. R. (1971). *Methods of Experimental Physics*. Vols. 9A and 9B. New York: Academic.

Sindoni, E. and Wharton, C. eds. (1978). *Diagnostics for Fusion Experiments*, Proc. Int. School of Plasma Physics, Varenna. London: Pergamon.

Stott, P. E. et al., eds. (1982). *Diagnostics for Fusion Reactor Conditions*, Proc. Int. School of Plasma Physics, Varenna. Brussels: Commission of E.E.C.

Some general plasma physics books have sections on diagnostics. One of the more extensive is in:

Miyamoto, K. (1980). *Plasma Physics for Nuclear Fusion*. English edition. Cambridge, Mass.: MIT.

In addition to countless specialized articles there are also some general reviews of plasma diagnostics in scientific journals, for example:

Equipe TFR (1978). Tokamak plasma diagnostics. *Nucl. Fusion* 18:647.

Luhmann, N. C. and Peebles, W. A. (1984). Instrumentation for magnetically confined fusion plasma diagnostics. *Rev. Sci. Instrum.* 55:279. Both of these tend to emphasize techniques.

2 Magnetic diagnostics

In very many plasma experiments the main parameters of the experiment consist of the magnitude of currents and magnetic and electric fields inside and outside the plasma volume. Reliable measurement of these parameters is basic to performing and understanding the experiments. Moreover, in many cases, measurements of these global quantities can give considerable information about the microscopic properties of the plasma such as temperature, density, and composition. It is therefore logical to begin our consideration of the topic of plasma diagnostics by consideration of electric and magnetic techniques. These may not seem quite so exciting or to involve such interesting areas of physics as the more exotic techniques but there is little doubt that they are extremely productive and practical in routine use.

2.1 Magnetic field measurements

2.1.1 *The magnetic coil*

The simplest way to measure the magnetic field in the vicinity of a point in space is to use a small coil of wire. Such a magnetic coil, illustrated in Fig. 2.1, may be considered the archetype of magnetic measurements. In a uniform magnetic field, varying with time $B(t)$, the voltage induced in the coil is

$$V = NA\dot{B}, \tag{2.1.1}$$

where N is the number of turns in the coil of area A and the dot denotes time derivative. As indicated in the figure, because one is normally interested in B rather than $\dot{B}$, an analog integrating circuit, such as the (somewhat schematic) one shown, is generally used to obtain a signal proportional to the field

$$V_0 = \frac{NAB}{RC}, \tag{2.1.2}$$

where RC is the time constant of the integrator.

It is instructive to adopt a rather more formal approach to calculating the voltage appearing at the output of a magnetic coil, even though this may seem unnecessary for such a simple case as Fig. 2.1. This is because we shall be concerned later with more complicated situations in which how to obtain the solution may not be so self-evident. This general method is based simply on application of the integral form of Faraday's law (the Maxwell

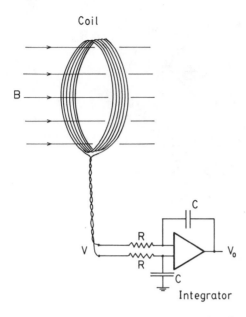

Fig. 2.1. Typical magnetic coil and integrating circuit.

equation relating electric field and the magnetic field time derivative) to an appropriately chosen closed contour C. The equation is

$$\oint_C \mathbf{E} \cdot \mathbf{dl} = - \int_S \dot{\mathbf{B}} \cdot \mathbf{ds}, \tag{2.1.3}$$

where S is a surface spanning the contour C.

In the case of a coil we choose the contour to lie within the conducting material of the coil itself, as illustrated in Fig. 2.2. To the ends of the coil are attached some kind of electronics, for example, an integrator or an oscilloscope that may have some nontrivial impedance. It is the voltage across the coil ends that this electronics senses.

In the most general case we must take into account the measuring electronics' impedance, but to begin with let us assume this to be so large that the coil can be taken as an open circuit. There is then no current flowing in the coil (ignoring any capacitive effects) so, within the electrically conducting material of the coil, the electric field must be zero. The contour integral of the electric field may then be written as

$$\int_C \mathbf{E} \cdot \mathbf{dl} = \int_{coil} \mathbf{E} \cdot \mathbf{dl} + \int_{ends} \mathbf{E} \cdot \mathbf{dl} = 0 + V = - \int_S \dot{\mathbf{B}} \cdot \mathbf{ds}, \tag{2.1.4}$$

the two integral parts here being around the coil and across the ends.

Naturally this has given us essentially the result we had before [Eq. (2.1.1)] except that it is now more explicit that we measure only the

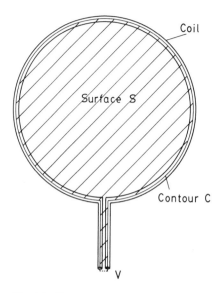

Fig. 2.2. Contour to be used in Faraday's law to calculate the output voltage of a coil.

component of $\dot{\mathbf{B}}$ normal to the plane of the coil and that, if $\dot{B}$ is nonuniform, it is the mean value over the surface that appears. The surface integral strictly spans the space between the leads to the coil as well. This space is kept as small as possible and usually the leads are twisted in order to make this contribution negligible.

It is worth commenting that Eq. (2.1.4) shows that we could equally interpret a coil as measuring the inductive electric field via the equality of the first and third quantities. From this viewpoint the effect of the wire is to convert the inductive electric field, which would have appeared as a nonzero integral around the coil part if we had chosen our contour just *outside* the coil, into an electrostatic field giving rise to the voltage V across the coil ends. Thus, as we shall see, we often use magnetic coils to measure electric fields. The designation magnetic measurements remains appropriate, though, because these electric fields are always inductive.

If current flows in the coil, because of finite impedance in the measurement electronics, then there may be some finite electric field within the coil (i.e., some potential drop) due to its resistivity. Also there may be some modification of the field at the coil due to currents flowing in it (i.e., its self-inductance). Exercises 2.1 and 2.2 explore these probabilities.

2.1.2   *Hall effect and Faraday effect measurements*

The most serious handicap of magnetic coils in measuring the magnetic field is that they respond to the rate of change of field $\dot{B}$, not the field itself. For steady fields this means that a magnetic coil becomes ineffective unless it can be physically moved within the field in a controlled way, a process that is usually very cumbersome. For these time invariant fields it therefore becomes attractive to use a different physical process to sense the magnetic field: the Hall effect.

The Hall effect is, in essence, a plasma phenomenon, though for practical measurements the plasma used is virtually always a solid state plasma within a semiconductor. Figure 2.3 illustrates how a Hall probe works. A slab of semiconductor resides within the field B. A current (j) is passed through it and the current carriers (electrons or holes, depending on the semiconductor) experience a Lorentz force, due to their motion, tending to deviate them perpendicular to j and B. The resulting charge buildup on the faces of the slab gives rise to an additional electric field that cancels the magnetic force. This additional field is sensed by electrodes on the semiconductor faces.

The combination of Hall probe and accompanying electronics, sometimes called a Gaussmeter (or Teslameter – S.I. units!), thus gives a way of measuring (with appreciable accuracy when appropriately calibrated) a local value of magnetic field. Nevertheless in many, if not most, experiments with magnetically confined plasmas the majority of measurements are made with coils rather than Hall probes. This is because Hall probes are inherently more complex, because they tend to be sensitive to stray pickup in the electrically noisy environment of a plasma experiment, and because they become nonlinear at high magnetic fields; also the majority of high power experiments are pulsed so integrators can be used with coils.

A third physical process that has attracted interest for magnetic measurements is the Faraday effect of magnetic field upon light propagating in an optical fiber. This offers, in principle, a measurement based upon polariza-

Fig. 2.3. Schematic illustration of the operation of a Hall probe.

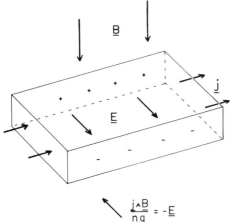

tion rotation proportional to the longitudinal magnetic field. It avoids the integration problem and lends itself naturally to the construction of the equivalent of a Rogowski coil (see Sec. 2.1.3). It suffers from numerous practical difficulties associated with nonideal optical behavior of the fibers due to residual strains created in the manufacturing process. However, recent research has shown that these difficulties can be overcome [see, for example, Chandler et al. (1986)]. It may be that this method will find practical application in the near future.

2.1.3   *Rogowski coils*

Many different kinds of magnetic coil configuration can be used. One which is very widely used is the Rogowski coil. This is a solenoidal coil whose ends are brought around together to form a torus as illustrated in Fig. 2.4. Consider a coil of uniform cross-sectional area A, with constant turns per unit length n. Provided the magnetic field varies little over one turn spacing, that is, if

$$|\nabla B|/B \ll n,\qquad\qquad (2.1.5)$$

the total flux linkage by the coil can be written as an integral rather than a sum over individual turns:

$$\Phi = n \oint_l \int_A dA\, \mathbf{B} \cdot \mathbf{dl}\qquad\qquad (2.1.6)$$

Fig. 2.4. The Rogowski coil.

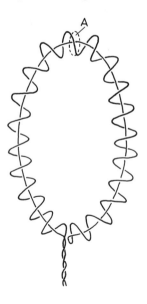

where **dl** is the line element along the solenoidal axis as illustrated in Fig. 2.5. Note that it is important to have the return wire back down the coil as shown in Fig. 2.4 or else to "back wind" the coil; otherwise, Eq. (2.1.6) also includes a term arising from the flux passing through the torus center. Now we note that the order of integration may be changed in Eq. (2.1.6) and that Ampere's law is quite generally

$$\oint_l \mathbf{B} \cdot \mathbf{dl} = \mu I, \qquad (2.1.7)$$

where I is the total current encircled by l and μ is the magnetic permeability of the medium in the solenoid. Thus

$$\Phi = nA\mu I \qquad (2.1.8)$$

and the voltage out of the Rogowski coil is

$$V = \dot{\Phi} = nA\mu \dot{I}, \qquad (2.1.9)$$

which again is usually integrated electronically to give a signal proportional to I.

The Rogowski coil thus provides a direct measurement of the total current flowing through its center. Note particularly that it is independent of the distribution of that current within the loop provided that Eq. (2.1.5) is satisfied. This principle is used in many different types of electrical circuits since it has the merit of requiring no circuit contact at all with the current being measured. The typical situation in plasma diagnostics in which it is used is to measure the total current flowing in the plasma, particularly for toroidal plasmas such as tokamak or pinch. For this purpose the Rogowski coil links the toroidal plasma as illustrated in Fig. 2.6.

Fig. 2.5. Equivalent geometry for the integral form of flux through a Rogowski coil.

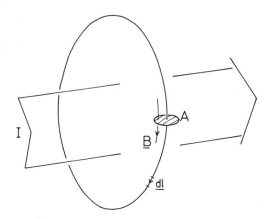

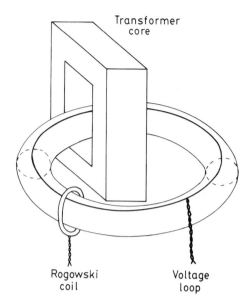

Fig. 2.6. Typical use of a Rogowski coil and a voltage loop to measure current and voltage in a toroidal plasma.

2.1.4   *Ohmic power and conductivity*

In a toroidal experiment the plasma current is driven by a voltage induced by transformer action, the plasma being the secondary. The toroidal loop voltage is usually measured by a so-called voltage loop, which is simply a single wire encircling the machine in the toroidal direction as illustrated in Fig. 2.6. If the plasma current is not varying with time, then the resistance of the plasma is evidently V_ϕ/I_ϕ, where V_ϕ and I_ϕ are the toroidal loop voltage and plasma current, respectively. In the following discussion much of what is said applies equally as well to linear plasmas as to toroidal ones. It is inconvenient to maintain a terminological distinction. So it should be noted that to apply the concepts to a linear plasma, where appropriate, one need only identify the toroidal (ϕ) direction as the axial (z) direction of a linear system. Thus, for example, the voltage and current between the ends of a linear discharge V_z and I_z give the resistance of the plasma. These parallel applications will henceforth remain implicit. The plasma resistance is important because it determines the ohmic heating input to the plasma and also because it may be used to estimate electron temperature. However, before moving on to these matters we must consider the more general situation in which the currents are not constant and the inductance makes a significant contribution.

We write Poynting's theorem as applied to a volume V bounded by the toroidal surface ∂V (generally outside the plasma) on which the measuring loops lie:

$$\int_V \mathbf{E} \cdot \mathbf{j} + \frac{1}{2\mu_0} \frac{\partial}{\partial t} (B^2) \, d^3 \mathbf{x} = -\frac{1}{\mu_0} \int_{\partial V} (\mathbf{E} \wedge \mathbf{B}) \cdot d\mathbf{S}, \qquad (2.1.10)$$

where $\mathbf{j}$ is the current density and $d\mathbf{S}$ is the outward pointing surface element. The first term is the total ohmic dissipation within the volume, the second is the rate of change of stored magnetic field energy ($\varepsilon_0 E^2/2$ is negligible), and the right hand side is the Poynting flux, which is the rate of input of energy from the external circuits. The Poynting flux may be written as $V_\phi I_\phi + V_\theta I_\theta$, where θ refers to the poloidal direction and $I_\theta = B_\phi 2\pi R/\mu_0$ is the total poloidal current linking the torus, R and a being the major and minor radii of ∂V. In the case (usually valid for tokamaks) where $\partial B_\phi/\partial t$ and hence V_θ are negligible, the energy equation (2.1.10) may be written as

$$P \equiv \int_V \mathbf{E} \cdot \mathbf{j} \, d^3 \mathbf{x} = V_\phi I_\phi - \frac{\partial}{\partial t} \left(\frac{1}{2} L I_\phi^2 \right), \qquad (2.1.11)$$

where the inductance,

$$L \equiv \frac{1}{\mu_0 I_\phi^2} \int_V B_\theta^2 \, d^3 \mathbf{x}, \qquad (2.1.12)$$

is determined by the distribution of the (toroidal) current density j_ϕ. If the toroidal field terms are not negligible (e.g., for a pinch plasma), the full energy equation (2.1.10) must be retained, but in either case one often has an estimate of the effective inductance L from other knowledge of the field profiles. If that is so, the inductive corrections may be estimated and the ohmic power P deduced. A word of caution should be mentioned concerning the treatment of the plasma in terms of lumped circuit parameters such as L. When inductive corrections are important, very often the effective inductance is changing as well as the current; therefore, the full derivative must be retained. Also, for a distributed current such as flows in a typical plasma, more than one effective inductance can be defined. The inductance we introduced in Eq. (2.1.12) is the *energy* inductance, which is different from, for example, the stored flux inductance defined by $\Phi = LI$. When any doubt arises, it is always best to return to the appropriate integral form of Maxwell's equations to be sure of using the correct definition.

Regardless of whether the plasma is effectively steady (so that its resistance is V_ϕ/I_ϕ) or varying [so that inductive (stored magnetic) energy must be accounted for], we normally wish to relate global quantities such as resistance or ohmic power to the local plasma properties; in particular, the resistivity or, equivalently, the conductivity. In order to determine the conductivity of the plasma, Ohm's law is used, usually in the form $\mathbf{j} = \sigma \mathbf{E}$ because the anisotropy of conductivity is usually irrelevant for these estimates. Then if we know, from V_ϕ and I_ϕ measurements, etcetera, the power

$$P = \int_V j^2/\sigma\, dV, \tag{2.1.13}$$

we can deduce a typical value of σ. For example, a simple way to define a kind of averaged conductivity $\bar{\sigma}$ is to write the plasma conductance as

$$\frac{\pi a^2}{2\pi R}\bar{\sigma} = \frac{I_\phi^2}{P} \quad \left(= \frac{I_\phi}{V_\phi} \quad \text{if} \quad \frac{\partial}{\partial t} = 0 \right), \tag{2.1.14}$$

where a and R are the minor and major radii of the plasma, respectively. Of course, if we have additional information about the spatial variation of j and σ it may be possible to deduce a local value of σ (see Exercise 2.3). Otherwise, we shall have to be content with some average value such as is given by Eq. (2.1.14).

The usefulness of having a measurement of σ, apart from determining the ohmic heating power density, is that it gives us an estimate of the electron temperature. This estimate is based on the equation for the conductivity of a fully ionized plasma,

$$\sigma = 1.9 \times 10^4 \frac{T_e^{3/2}}{Z_\sigma \ln \Lambda} \ \Omega^{-1}\, \mathrm{m}^{-1}, \tag{2.1.15}$$

where T_e is the electron temperature in electron volts, Z_σ is the resistance anomaly determined by the ion charge, and $\ln \Lambda$ is the Coulomb logarithm. It would take us too far out of our way to derive this expression (called the Spitzer conductivity) properly, but we can gain a general understanding of its form from the following considerations.

The conductivity is determined by a balance between the acceleration of charge carriers (electrons primarily) in an applied electric field and their deceleration by collisions. In a plasma in which neutral collisions can be ignored, the collisions of interest are Coulomb collisions with the ions. These are discussed in more detail in Chapter 5; for now we simply note that an electron of velocity v colliding with an ion of charge Ze will be scattered though 90° if its kinetic energy is equal to half the potential energy at a distance equal to the impact parameter in the Coulomb field. If the electron has impact parameter b_{90} for 90° scattering, this means

$$m_e v^2 = \frac{Ze^2}{4\pi\varepsilon_0 b_{90}}. \tag{2.1.16}$$

If we suppose a collision to occur if an electron approaches an ion within the impact parameter of a 90° collision and not otherwise, the resulting estimate for the collision frequency (i.e., number of collisions per unit time) is

$$\nu_{ei} \approx \upsilon n_i \pi b_{90}^2 \propto \frac{n_e Z}{\upsilon^3}. \tag{2.1.17}$$

The conductivity will then be given by substituting in this expression a typical thermal electron velocity ($\propto T_e^{1/2}$) and writing

$$\sigma \approx \frac{n_e e^2}{m_e \nu_{ei}} \propto \frac{T_e^{3/2}}{Z}. \tag{2.1.18}$$

This rough estimate of how the conductivity scales agrees with the accurate expression (2.1.15) in giving σ as independent of n_e. This is because the increase of collision frequency with n_e is exactly compensated by the increase in the number of charge carriers. There are two important corrections to σ in the accurate expression. The first is the Coulomb logarithm $\ln \Lambda$, which accounts for the fact that glancing collisions are important and in fact increase the collision frequency substantially above our crude estimate. The quantity Λ is approximately the ratio of the Debye length (see Chapter 3) to b_{90} and $\ln \Lambda$ is a very slowly varying function of plasma parameters, typically ~ 15 for hot plasma experiments. A convenient approximate expression for Λ, valid when $T_e > 10$ eV, is

$$\ln \Lambda = 31 - \ln\left(n_e^{1/2}/T_e\right), \tag{2.1.19}$$

where T_e is in electron volts. The second correction is that Z_σ and Z are not exactly the same (nor is Z_σ equal to Z_{eff}, discussed in Chapter 5) except in hydrogen plasmas when $Z_\sigma = Z = 1$. The reason is that the coefficient of Eq. (2.1.15) includes corrections for electron–electron collisions, which are important for small Z. These make the conductivity scale not exactly proportional to $1/Z$. Figure 2.7 shows how Z_σ, the resistivity anomaly, varies with the ion charge Z. This and other details of the proper derivation of plasma conductivity were originally worked out by Spitzer (1962) and others.

Knowledge of σ, Z_σ, and $\ln \Lambda$ allows us to deduce T_e. The most common approach is to take an appropriate value for $\ln \Lambda$, take $Z_\sigma = 1$ for hydrogenic plasmas, and then deduce a temperature from Eq. (2.1.15) using the measured value of σ. This temperature is then called the conductivity temperature T_σ. Uncertainties in Z_σ due to the presence of unknown

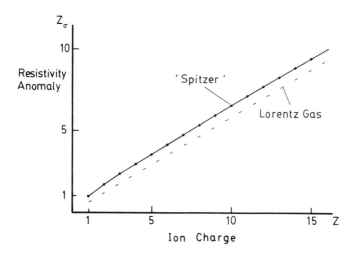

Fig. 2.7. The resistivity anomaly. Z_σ as a function of ion charge Z. The Lorentz gas value is what is obtained from a calculation ignoring electron–electron collisions. The Spitzer value is the correct one when these are included.

quantities of impurities often make T_σ a relatively poor quantitative measure of T_e. Nevertheless, T_σ is still useful as a first estimate of temperature particularly because it can be deduced from the relatively simple measurements of the gross parameters I_ϕ, V_ϕ, etcetera, of the plasma.

2.2   Magnetohydrodynamic equilibrium

The electric current flowing in a plasma can give us an estimate of electron temperature via the conductivity, as we have seen. However, in many plasmas, notably those that are magnetically confined, a potentially more powerful diagnostic is available, based on the role played by the current in balancing the plasma pressure. This can be investigated using magnetic measurements. In a sense, we can speak of this method as a way of determining the plasma pressure by measuring the force it exerts on the magnetic field by which it is contained.

We adopt a single-fluid description of the plasma and suppose, for the moment, that the pressure is isotropic. The condition for the plasma to be in equilibrium is then that the sum of kinetic pressure force density and electromagnetic force density should be zero:

$$-\nabla p + \mathbf{j} \wedge \mathbf{B} = 0. \tag{2.2.1}$$

Ampere's law and straightforward vector manipulation allow us to write this also as

$$-\nabla\left(p + \frac{B^2}{2\mu_0}\right) + \frac{1}{\mu_0}(\mathbf{B}\cdot\nabla)\mathbf{B} = 0 \qquad (2.2.2)$$

or equivalently

$$\nabla\cdot\mathbf{T} = 0, \quad \text{where} \quad \mathbf{T} = \left(p + \frac{B^2}{2\mu_0}\right)\mathbf{1} - \frac{\mathbf{BB}}{\mu_0} \qquad (2.2.3)$$

is the Maxwell stress tensor.

To solve this equilibrium equation in general geometries is an extremely difficult task. However, many plasmas have approximate cylindrical symmetry and it is then possible to obtain convenient results by appropriate Fourier analysis. In a cylindrical polar coordinate system (r, θ, z) appropriate to a truly cylindrical plasma or, for example, a large aspect ratio torus in which r is the minor radius and $z = R\phi$, we express components of the field at radius r as a sum of poloidal Fourier harmonics:

$$B(\theta) = \frac{C_0}{2} + \sum_{m=1}^{\infty} C_m\cos m\theta + S_m\sin m\theta, \qquad (2.2.4)$$

where

$$C_m = \frac{1}{\pi}\int_0^{2\pi} B(\theta)\cos m\theta\, d\theta,$$

$$S_m = \frac{1}{\pi}\int_0^{2\pi} B(\theta)\sin m\theta\, d\theta. \qquad (2.2.5)$$

(See Appendix 1 for a brief review of Fourier analysis.)

Insofar as the plasma is approximately cylindrical, the higher Fourier components will be smaller. Thus, we regard Eq. (2.2.4) as a perturbation expansion for B in which the zeroth order is independent of θ, the first order is proportional to the cosine or sine of θ, and so on. One can thus, by equating appropriate orders in the equilibrium equation, obtain a solution as a perturbation expansion. For the purposes of pressure diagnosis it is the zeroth and first order terms that are most useful. It is to these we now turn explicitly.

2.2.1 *Diamagnetism (m = 0 term)*

For the purposes of the zeroth order we can suppose the fields all to be cylindrically symmetric. The only component of the equilibrium equation of interest is then the radial one:

$$(-\nabla p + \mathbf{j}\wedge\mathbf{B})\cdot\hat{\mathbf{r}} = -\left[\frac{dp}{dr} + \frac{dB_\phi}{dr}\frac{B_\phi}{\mu_0} + \frac{1}{r}\frac{d(rB_\theta)}{dr}\frac{B_\theta}{\mu_0}\right] = 0$$

$$(2.2.6)$$

22   *Magnetic diagnostics*

(again ϕ is the axial coordinate equivalent to z). Multiplying this equation by r^2 and integrating from 0 to a, one obtains (see Exercise 2.4)

$$\beta_\theta \equiv \frac{2\mu_0 \langle p \rangle}{B_{\theta a}^2} = 1 + \frac{B_{\phi a}^2 - \langle B_\phi^2 \rangle}{B_{\theta a}^2}, \tag{2.2.7}$$

where $\langle \ \rangle$ indicates average over the plasma cross section, the subscript a means quantities evaluated at $r = a$, and we have assumed $p_a = 0$. This is an expression for the ratio of kinetic pressure to (poloidal) magnetic field pressure, the plasma beta β_θ. In its present form the equation is not very useful because it is not clear how to measure $\langle B_\phi^2 \rangle$, but if B_ϕ varies only weakly across the plasma, which will be the case if $\beta_\phi \equiv 2\mu_0 \langle p \rangle / B_\phi^2 \ll 1$ and $B_\theta \ll B_\phi$ (as occurs in tokamaks, stellarators, and some mirror machines but not, for example, in a reversed field pinch), then

$$B_{\phi a}^2 - \langle B_\phi^2 \rangle \approx 2B_{\phi a}\left(B_{\phi a} - \langle B_\phi \rangle\right) \tag{2.2.8}$$

and so

$$\beta_\theta \approx 1 + \frac{2B_{\phi a}\left(B_{\phi a} - \langle B_\phi \rangle\right)}{B_{\theta a}^2}. \tag{2.2.9}$$

In this equation $B_{\phi a}$ and $B_{\theta a}$ may be measured by magnetic coils outside the plasma at $r = a$ (or by less direct techniques), while $\langle B_\phi \rangle$ is proportional to the total magnetic flux in the toroidal direction and so may be measured by a poloidal loop around the plasma as shown in Fig. 2.8. Such a loop measures V_θ so it might be called a poloidal voltage loop, but because,

Fig. 2.8. The flux loop.

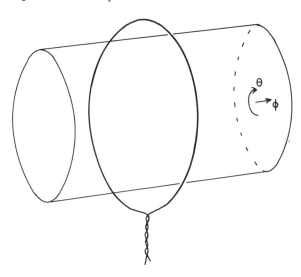

when integrated, it measures toroidal flux, it is more often called a flux loop.

If I_ϕ (and hence B_θ) is negligible, Eq. (2.2.9) should be written

$$\beta_p = \frac{2(B_{\phi a} - \langle B_\phi \rangle)}{B_{\phi a}}.$$

(2.2.10)

Since β_ϕ is always positive by definition, it is clear that in this case the mean $\langle B_\phi \rangle$ must be less than the edge field $B_{\phi a}$. Thus the plasma acts to decrease the magnetic field within it. The plasma is then said to be diamagnetic and, as a recognition of this, the flux loop is sometimes called a diamagnetic loop. It is possible to demonstrate the diamagnetic nature of the plasma by consideration of the effective magnetic dipole moments of electrons and ions gyrating around the field lines. This approach is equivalent to the magnetohydrodynamic (MHD) equilibrium calculation we have done but is much more cumbersome to evaluate in nontrivial geometries, which is why we adopt the MHD model. In cases where the pressure is anisotropic, the preceding derivation is still valid provided we interpret p as the perpendicular pressure $p_\perp$, which acts in the r direction.

Equation (2.2.9) shows that when I_ϕ is not negligible the plasma may have $\langle B_\phi \rangle$ greater than $B_{\phi a}$, that is, it may be paramagnetic. This is due to the extra pinching force of the poloidal field. In general, the plasma is diamagnetic or paramagnetic according to whether β_θ is greater or less than 1. However, the measurement of $\langle p \rangle$ from $B_{\phi a} - \langle B_\phi \rangle$ is called the diamagnetic measurement in recognition that the plasma kinetic pressure always acts to decrease the field.

The measurement of the diamagnetic effect is an important measurement of total plasma kinetic energy W because energy density is proportional to p and so $\langle p \rangle \propto W$. However, on tokamaks, for example, it is in practice quite a difficult measurement to achieve because, although $\beta_\theta \sim 1$, typical tokamaks have $B_{\theta a}/B_{\phi a} \sim 10^{-1}$, so that the diamagnetic effect gives only a very small fractional change in B_ϕ. To obtain even 20% accuracy in measuring β_θ when it is of order unity requires $(B_{\phi a} - \langle B_\phi \rangle)/B_{\phi a}$ to be measured accurately to about 1 part/10^3; clearly, considerably better accuracy is desirable. Despite this exacting requirement, reliable measurements of plasma energy have been achieved on many tokamaks. An example of the results possible is shown in Fig. 2.9.

The striking fact is that the external magnetic measurements of the diamagnetic effect and the plasma resistance are sufficient to calculate the energy confinement time τ_E of a quasistationary ohmically heated plasma. This arises from the relationship

$$\tau_E = \frac{W}{P} = \frac{3}{8}\mu_0 \beta_\theta (R/R_p)$$

(2.2.11)

Fig. 2.9. Measurement of the evolution with time of β_θ and hence τ_E as measured by diamagnetism on the small tokamak LT-3 (Hutchinson 1976a). Two different discharges are shown.

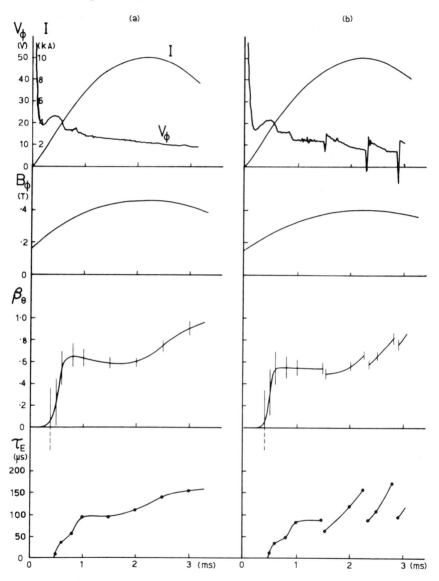

(where R_p is the plasma resistance $P = I_\phi^2 R_p$), which may be demonstrated directly from the definitions of β_θ, W, and P (see Exercise 2.5). This shows that β_θ and R_p are sufficient to determine the energy confinement time τ_E. Figure 2.9 includes this parameter.

2.2.2 *Position and asymmetry measurements (m = 1)*

Measurement of the $m = 1$ components of the magnetic field serves primarily to determine the position of a current carrying plasma. Consider the situation, illustrated in Fig. 2.10, of a straight plasma with cylindrical symmetry about an axis at position $x = \Delta \ll a$, $y = 0$. We suppose the azimuthal field (B_θ) to be measured at radius $r = a$. The field there (due to the plasma alone, any applied external field being subtracted off) is

$$
B_\theta(\theta) = \frac{\mu_0 I}{2\pi a} \frac{1}{\left[\sin^2\theta + (\cos\theta - \Delta/a)^2\right]^{1/2}}
$$

$$
\approx \frac{\mu_0 I}{2\pi a}\left(1 + \frac{\Delta}{a}\cos\theta\right) \tag{2.2.12}
$$

to first order in Δ/a. Thus the cosine Fourier component of the field measures the horizontal displacement; specifically

$$
\Delta = 2a(C_1/C_0). \tag{2.2.13}
$$

In just the same way the sine component S_1/C_0 would give any vertical displacement.

It is relatively straightforward to perform the measurement of the $m = 1$ (and higher) components of the poloidal magnetic field. There are two main

Fig. 2.10. Measurement of the position of a cylindrical current carrying plasma from field asymmetry.

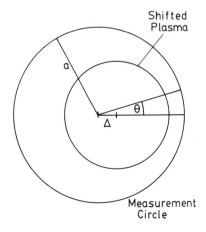

approaches. The first is to use a kind of Rogowski coil but with an effective winding density that varies like $\cos m\theta$ or $\sin m\theta$ such as is illustrated in Fig. 2.11(a). This gives an output directly proportional to the (time derivative of the) required Fourier component. The other method commonly adopted is to use a set of discrete coils ranged around the plasma at different values of θ [Fig. 2.11(b)]. By using an appropriate weighted sum of the signals from these coils, discrete approximations to the Fourier integrals can be synthesized. These types of measurement are routinely used in tokamaks, for example, to monitor, and hence control by feedback, the plasma position.

In toroidal plasmas, although the sine component still gives the vertical position in just the same way as for a cylinder, the cosine component and the horizontal position are related in a more complicated way. The reason for this is that even if the magnetic surfaces (in which the field lines lie) are circular in cross section, they are not necessarily concentric circles, but generally show an outward shift of the inner surfaces relative to the outer ones, as illustrated in Fig. 2.12. To calculate this effect requires a solution of the $m = 1$ component of the MHD equilibrium. We shall follow approximately the original derivation of Shafranov (1963), whose name has

Fig. 2.11. The measurement of a specific Fourier component of the field can be done using, for example, (a) cosine coil with appropriately varying winding density or (b) an array of discrete coils.

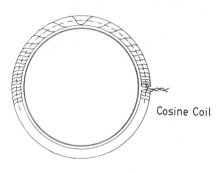

Cosine Coil

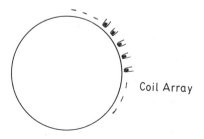

Coil Array

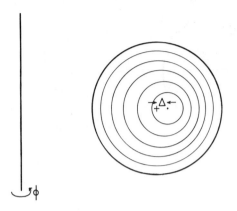

Fig. 2.12. The Shafranov shift of magnetic surface makes them not concentric in a poloidal cross section.

become attached to the relative displacement of the magnetic surfaces: the Shafranov shift.

We wish to solve the MHD equilibrium for a toroidal plasma with circular poloidal cross section. To do this we consider a thin toroidal "slice" of plasma of angle $d\phi$, having a major radius R_0, as shown in Fig. 2.13. We take the fields, etcetera, to be given just by their first two Fourier harmonics:

$$B_\phi = B_{\phi 0} + B_{\phi 1}\cos\theta,$$
$$B_\theta = B_{\theta 0} + B_{\theta 1}\cos\theta,$$
$$B_r = \qquad B_{r1}\cos\theta,$$
$$p = p_0 + p_1\cos\theta. \tag{2.2.14}$$

Ignoring higher Fourier modes amounts to an expansion in inverse aspect ratio:

$$\frac{r}{R_0} \sim \frac{B_1}{B_0} \ll 1. \tag{2.2.15}$$

Sine components are zero by symmetry (for zero vertical displacement). We set the total force in the major radial direction acting on this plasma slice to zero for equilibrium. In terms of the stress tensor this may be written

$$\int_S \mathbf{T} \cdot \mathbf{dS} = 0, \tag{2.2.16}$$

where S is the surface of the slice. Thus the total force is composed of the total stress integrated over the surface of the slice. The surface consists of two disks and a round outer surface. The stress in general consists of normal (pressurelike) stress and tangential stress. However, one can show by evaluation that the tangential stress integrated over the surfaces is zero

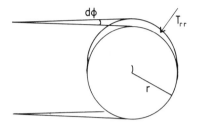

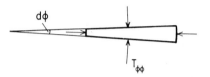

Fig. 2.13. The "slice" of plasma whose equilibrium is to be calculated.

to the order of approximation we are employing. Therefore, we shall concentrate only upon the normal stress.

The force in the R direction on the circular disk faces is

$$F_1 = d\phi \int_0^r T_{\phi\phi} 2\pi r' \, dr' = d\phi \int_0^r \left[p_0 + \frac{B_0^2}{2\mu_0} - \frac{B_{\phi 0}^2}{\mu_0} \right] 2\pi r' \, dr',$$

$$(2.2.17)$$

the cosine components averaging to zero. The R direction force on the outer round surface is

$$F_2 = - \int_0^{2\pi} d\phi \, R \left[p + \frac{B^2}{2\mu_0} - \frac{B_r^2}{\mu_0} \right] \cos\theta \, r \, d\theta, \qquad (2.2.18)$$

where R is the major radius of the surface at angle θ, that is, $R = R_0 + r\cos\theta$. Substituting for R and for B in terms of the expansion (2.2.14), we perform the θ integral and retain only the terms of lowest order in r/R; then

$$F_2 = -d\phi \, \pi \left[R_0 r \left\{ p_1 + \frac{1}{\mu_0} (B_{\phi 0} B_{\phi 1} + B_{\theta 0} B_{\theta 1}) \right\} + r^2 \left\{ p_0 + \frac{B_0^2}{2\mu_0} \right\} \right].$$

$$(2.2.19)$$

The equilibrium $F_1 + F_2 = 0$ then becomes

$$\langle p \rangle = \frac{1}{2\mu_0}\left(\langle B_{\theta 0}^2\rangle - \langle B_{\phi 0}^2\rangle\right) - \frac{R_0}{r}\left\{p_1 + \frac{1}{\mu_0}\left(B_{\phi 0}B_{\phi 1} + B_{\theta 0}B_{\theta 1}\right)\right\}$$

$$-\left\{p_0 + \frac{1}{2\mu_0}\left(B_{\theta 0}^2 + B_{\phi 0}^2\right)\right\} = 0, \tag{2.2.20}$$

where $\langle \ \rangle$ denotes, as before, an average over the poloidal cross section ($\int 2\pi r' \, dr'/\pi r^2$).

We now choose our coordinate system so that R_0 is measured to the center of a magnetic surface of minor radius r. This means that the outer rounded surface, on which the quantities in Eq. (2.2.20) are to be evaluated, is a magnetic surface, and implies that quite generally p is constant and $B_\phi \propto 1/R$ on the surface (see Exercise 2.6). Hence,

$$p_1 = 0, \qquad B_{\phi 1} = -\frac{r}{R_0}B_{\phi 0}, \tag{2.2.21}$$

and the equilibrium can be rewritten

$$\frac{B_{\theta 1}B_{\theta 0}}{\mu_0} = \frac{r}{R_0}\left[\langle p \rangle - p + \frac{\langle B_{\theta 0}^2\rangle - B_{\theta 0}^2}{2\mu_0} - \frac{\langle B_{\phi 0}^2\rangle - B_{\phi 0}^2}{2\mu_0}\right]. \tag{2.2.22}$$

The final term on the right hand side of this equation we have already evaluated. It is the diamagnetic term [Eq. (2.2.7)] and is equal to $\langle p \rangle - p - B_{\theta a}^2/2\mu_0$. We therefore substitute for it and rearrange slightly to get the form

$$B_{\theta 1} = B_{\theta 0}\frac{r}{R_0}\left[\frac{2\mu_0}{B_{\theta 0}^2}(\langle p \rangle - p) + \frac{\ell_i}{2} - 1\right], \tag{2.2.23}$$

where ℓ_i is a nondimensional form of the energy inductance of the poloidal field

$$\ell_i \equiv \langle B_{\theta 0}^2\rangle / B_{\theta 0}^2. \tag{2.2.24}$$

From a practical viewpoint, the measurement of $B_{\theta 1}$ is virtually always done outside the plasma, where $p = 0$, so that we can write Eq. (2.2.23) in its most common form:

$$B_{\theta 1} = B_{\theta 0}\frac{r}{R_0}\Lambda, \qquad \Lambda \equiv \left(\beta_\theta + \frac{\ell_i}{2} - 1\right). \tag{2.2.25}$$

The $m = 1$ component of the magnetic field outside the plasma thus provides a measurement of the combination $\beta_\theta + \ell_i/2$ via the asymmetry factor Λ. (This is not the same Λ as in the Coulomb logarithm.)

If β_θ is known from the diamagnetic measurement, for example, then measurement of Λ gives a value of ℓ_i, the plasma inductance. This is determined by the radial distribution of toroidal current density within the plasma and so Λ gives information essentially about the "width of the current channel." If on the other hand, ℓ_i is sufficiently well known from other measurements, we have two measurements of β_θ, one from diamagnetism and one from Λ: Call them β_{dia} and β_Λ. The potential value of these two measurements is to diagnose any anisotropy of the pressure. Although the preceding analysis assumed isotropic pressure, we could perfectly well have performed our equilibrium calculation with an anisotropic pressure tensor. In doing so we should have arrived at the same results except that the pressure average $\langle p \rangle$ from the force on the circular disks would be $\langle p_{\phi\phi} \rangle$, whilst that from the diamagnetism is $\langle p_{rr} \rangle$. For a tokamak, in which $B_\phi \gg B_\theta$, these would be essentially $\langle p_\parallel \rangle$ and $\langle p_\perp \rangle$, respectively. Recall that the $\langle p \rangle$ in Eq. (2.2.23) came half from the pressure on the disks and half from the diamagnetism. Hence

$$\beta_{\text{dia}} = \beta_{\theta\perp}, \qquad \beta_\Lambda = \tfrac{1}{2}\left(\beta_{\theta\perp} + \beta_{\theta\parallel}\right). \tag{2.2.26}$$

Thus it is possible in principle to distinguish $\langle p_\perp \rangle$ and $\langle p_\parallel \rangle$ by comparison of diamagnetic and asymmetry measurements if ℓ_i is known.

Because of the asymmetry of B_θ even on a magnetic surface in a toroidal machine, the measurement of $B_{\theta1}$ alone is not sufficient to determine the horizontal position and the Λ factor unless the measurement surface coincides with a magnetic surface. This is rarely guaranteed a priori, so that a further measurement is necessary. Generally, the measurement used is the $m = 1$ component of B_r. In a sense, the combination of $B_{\theta1}$ and B_{r1} is sufficient to determine the position of the outer magnetic surface and its asymmetry factor Λ. Detailed discussion of these points may be found, for example, in Mukhovatov and Shafranov (1971).

The higher Fourier components naturally give information about the shape of the plasma. For example, $m = 2$ components measure the elliptical distortion, $m = 3$ components measure the triangularity, and so on. These factors are important in specially shaped plasmas in their own right. They may also be related to the MHD equilibrium to provide additional diagnostic information. This information tends to be less universally useful than that which the $m = 0, 1$ components give and to require numerical solutions of the equilibria for its interpretation.

In strongly shaped plasmas it is usual to employ numerical solutions of the MHD equilibrium equations via major computer codes (e.g., Johnson et al. 1976) to relate plasma parameters to the observed magnetic measurements. It is inappropriate to go into detail here as to how these analysis codes operate, but it is worth noting that they and other more sophisticated analyses of magnetic equilibria generally work in terms of magnetic flux ψ.

In geometries in which there is an ignorable coordinate, such as for a toroidally symmetric plasma like a tokamak, the magnetic fields can be expressed as the gradient of scalars. In particular, the poloidal field of a toroidal plasma can be written

$$\mathbf{B}_p = \left(\hat{\boldsymbol{\phi}} \wedge \nabla \psi\right)/2\pi R, \tag{2.2.27}$$

where $\hat{\boldsymbol{\phi}}$ is the unit vector in the toroidal direction and $\mathbf{B}_p$ includes both radial and azimuthal parts of the field (B_r and B_θ of our earlier analysis). The poloidal flux ψ can be regarded as the total magnetic flux through a surface spanning the circular contour R = constant, z = constant (using cylindrical polar coordinates R, ϕ, z). Now it is possible, quite straightforwardly in many cases, to measure the poloidal flux directly using magnetic measurements, in much the same way that the toroidal flux is measured. Integrating the signal from what we have previously called a voltage loop gives directly the poloidal flux. Therefore, we could just as well have called it a (poloidal) flux loop.

Most of the measured flux through such a loop will tend to be the transformer flux used to drive the plasma current. Therefore, a single loop does not give much information about the plasma position. However, if we have a number of loops ranged around the perimeter of the plasma, then the difference in the flux measured by the different loops does give us that information. To put the point more mathematically, loops ranged around outside the plasma give the *boundary conditions* for a solution of the differential equation governing ψ in the inner region. As a simple example,

Fig. 2.14. External poloidal magnetic field measurements on a noncircular plasma typically consist of a combination of flux loops and field coils arranged around the plasma.

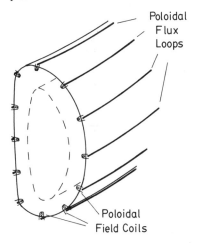

if the flux values are all the same at the loops, then the loops all lie on a single magnetic flux surface, and thereby give us the shape of the flux surface in that region of space.

In actual fact the flux alone is not sufficient to prescribe the boundary conditions fully. From an idealized viewpoint, in order to solve the equation in a closed region with bounding surface S, we require the value of ψ and also its derivative normal to the boundary $\hat{n} \cdot \nabla \psi$ everywhere on S. Thus if we regard the flux loops as giving us ψ essentially everywhere on the boundary (by interpolation provided there are enough loops), we also need $\hat{n} \cdot \nabla \psi$ at a similar number of positions. But $\hat{n} \cdot \nabla \psi$ is equal to $2\pi R$ times the tangential component of $\mathbf{B}_p$. So usually one uses magnetic coils

Fig. 2.15. Typical internal magnetic probe construction.

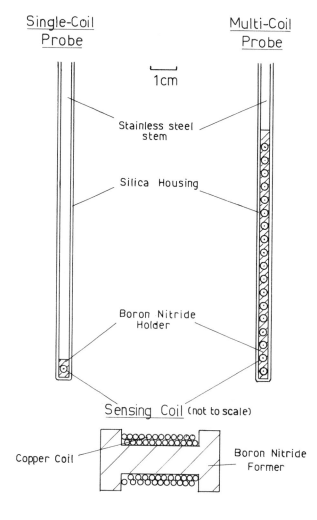

ranged around the periphery, much as we discussed for a circular plasma, to give this second condition on S. Figure 2.14 illustrates the sort of configuration used to give the information on the magnetic flux and its derivative in order to diagnose shape and position in strongly shaped plasmas.

2.3 Internal magnetic probe measurements

2.3.1 *Field measurements*

In order to measure the magnetic field inside the plasma, it is sometimes possible to use internal magnetic probes. The major restrictions upon the use of such insertable probes are that in energetic plasmas either the heat flux from the plasma may be so great as to damage the probe or else the perturbation of the plasma by the probe may be so severe as to change the whole character of the plasma being probed. Despite these difficulties, magnetic probes have proven extremely valuable in many circumstances. An example of the type of probe construction used is shown in Fig. 2.15. Individual coils, sometimes in large numbers, are usually mounted inside a vacuum-tight nonconducting jacket. These can then provide a direct measurement of the evolution of appropriate components of the magnetic field with excellent time resolution.

A typical experimental situation is pictured in Fig. 2.16, where a cylindrical or approximately cylindrical plasma is probed along a diameter, thus providing a radial profile of the relevant field components. As an example of the sort of information to be obtained, Fig. 2.17 shows the magnetic field profile evolution measured in the initial stages of a tokamak discharge. Each magnetic field profile is obtained by fitting a smooth polynomial curve to points obtained from individual probe traces at the appropriate time. For clarity, points are plotted only for five time slices; they show scatter, which gives an estimate of the measurement uncertainty. For

Fig. 2.16. Typical probe insertion geometry.

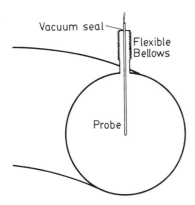

tokamaks the B_θ component illustrated is the important component in the quasicylindrical approximation, B_ϕ being almost uniform.

In other types of plasma, for example when $B_\phi \sim B_\theta$, B_ϕ may vary substantially and so be of greater importance. Figure 2.18 shows profiles obtained in a reversed field pinch (RFP), which gets its name from the reversal of the toroidal field in the outer regions, clearly seen in these measurements. The radial component B_r provides information on the shift of the plasma (assumed approximately cylindrical) perpendicular to the probe.

An important question that arises in all internal probing measurements is: How does the probe perturb the plasma? In the case of magnetic probes, provided the discharge as a whole is not qualitatively altered by inserting the probe, the degree of perturbation of the measurement due to the probe tends to be fairly small. The reason for this is that magnetic fields are

Fig. 2.17. Poloidal magnetic field evolution measured with internal magnetic probe (Hutchinson 1976b).

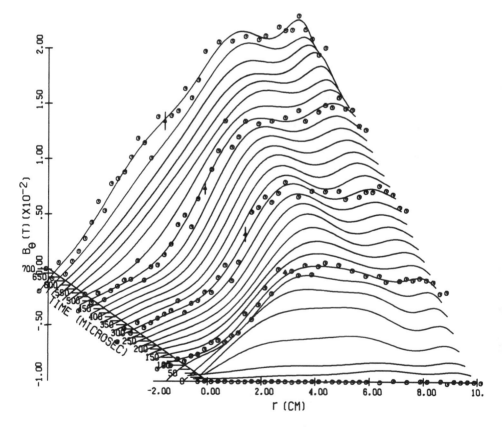

Fig. 2.18. Magnetic field profiles in a reversed field pinch. (*a*) The toroidal (ϕ) and poloidal (θ) fields are of the same order of magnitude. (*b*) The horizontal shift of the magnetic surfaces δ is deduced from the much smaller radial component [after Brotherton-Ratcliffe and Hutchinson (1984)].

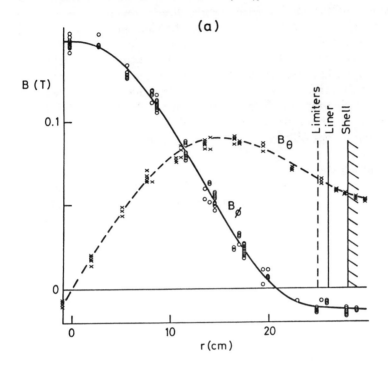

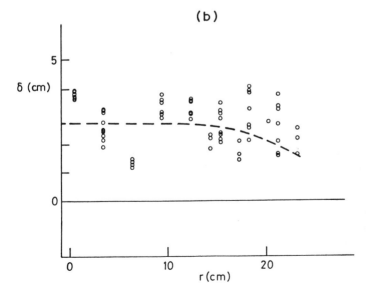

generated by currents throughout the plasma, not just locally near the probe. In other words, the field at the probe is obtained by an integral of the Biot–Savart law over all the relevant space. Therefore, even though the perturbation to the local *current density* due to the probe is large – an insulating jacket prevents all current from flowing through it, $j = 0$ – the local *magnetic field* may be scarcely perturbed at all. Of course this argument only applies to large-scale fields. Components of the field whose spatial structure has a scale size of the order of the probe size or smaller *will* be strongly perturbed, because they arise from local small-scale current structure. Thus current and field structure can only be diagnosed down to a scale on the order of the probe size. This gives a strong incentive to construct the probes with as small a diameter as possible.

2.3.2   *Current density*

From complete measurements of $\mathbf{B}(\mathbf{r})$ one can deduce the current from Ampere's law (neglecting displacement current since this is essentially a magnetostatic situation)

$$\nabla \wedge \mathbf{B} = \mu_0 \mathbf{j}. \tag{2.3.1}$$

It is normally quite impractical to attempt to obtain measurements of $\mathbf{B}(\mathbf{r})$ sufficiently complete to allow direct evaluation of $\mathbf{j}$ without additional information. Instead in many situations, symmetries or approximate symmetries exist that allow a more limited data set to suffice. In other words, we do not have to measure all the components of $\mathbf{B}$ everywhere in space. The most common case is when the plasma is cylindrically symmetric (or approximately so). Then measurements along a single radial line are sufficient and one can write the components of the current as

$$j_\phi = \frac{1}{\mu_0 r} \frac{d}{dr}(rB_\theta), \tag{2.3.2}$$

$$j_\theta = -\frac{1}{\mu_0} \frac{d}{dr} B_\phi. \tag{2.3.3}$$

By fitting a curve to measurements of B_θ, B_ϕ at a number of radial positions we obtain a function that can be differentiated to give the current. It should be noted, though, that the process of differentiation naturally tends to enhance any errors present in the measurement so that the quality of current measurements tends to be worse than that of field measurements. Nevertheless, the accuracy achievable with carefully calibrated magnetic probes is quite adequate to give current density. Figure 2.19 shows the toroidal current-density evolution from the measurements of Fig. 2.17. These are in the early stages of a tokamak discharge in which a hollow "skin" current profile is formed.

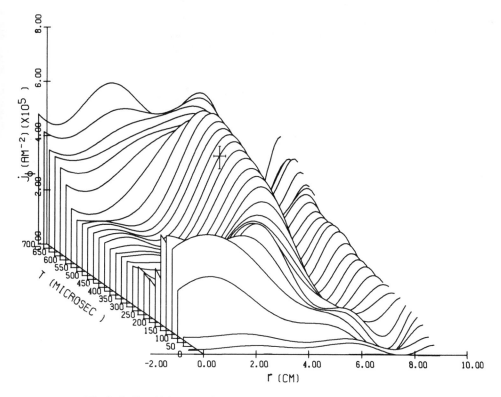

Fig. 2.19. Toroidal current density evolution derived from the measurements of Fig. 2.17.

2.3.3 Electric field

We have already mentioned the importance of inductive effects in situations where the plasma currents are not stationary. Internal magnetic probe measurements enable these effects to be measured rather directly. The basis for this is Faraday's law in the integral form

$$\oint_C \mathbf{E} \cdot \mathbf{dl} = - \int_S \dot{\mathbf{B}} \cdot \mathbf{ds}, \tag{2.3.4}$$

where S is a surface with boundary C. Applying this relation to the cylindrical situation shown in Fig. 2.20 we obtain

$$E_\phi(r) = E_\phi(a) - \int_r^a \dot{B}_\theta \, dr, \tag{2.3.5}$$

$$rE_\theta(r) = aE_\theta(a) + \int_r^a \dot{B}_\theta r \, dr = - \int_0^r \dot{B}_\phi r \, dr. \tag{2.3.6}$$

Thus we can obtain the local electric field from a combination of magnetic probe measurements and edge electric field (e.g., loop voltage). For exam-

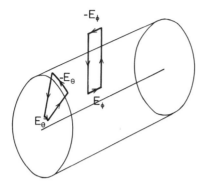

Fig. 2.20. Electric field geometry for Faraday's law.

ple, in a tokamak, for which only the first equation (2.3.5) is particularly important, the internal electric field has been measured during a disruptive instability in which rapid redistribution of current occurs. The current profile evolution is shown in Fig. 2.21 and the corresponding electric field in Fig. 2.22. The measurements indicate that during the disruption the electric field in the plasma center is extremely high, perhaps 10 times the

Fig. 2.21. Current density evolution during a disruptive instability [after Hutchinson (1976c)].

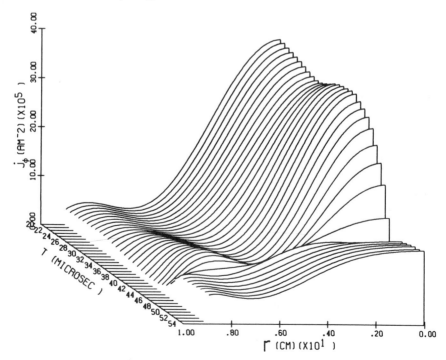

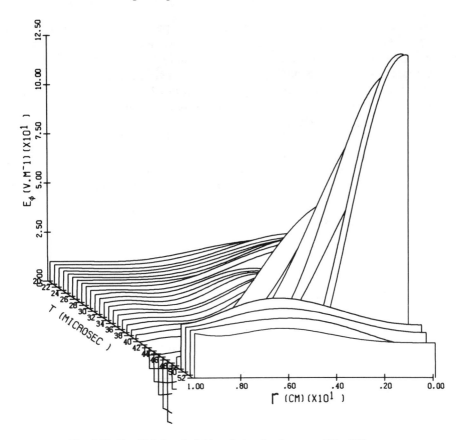

Fig. 2.22. Toroidal electric field evolution for the case of Fig. 2.21.

quiescent level; at the same time the electric field at the edge is actually reversed. These fields are important for understanding such disruptions, which are a major concern in tokamak research. The MHD instabilities that trigger disruptions are sensitive to the magnetic field distributions themselves and the field profile shapes determine, in general, whether or not the configuration is stable. Thus internal measurements of magnetic field are fundamental to the investigation of both the causes and the consequences of MHD instabilities.

It is worth noting that when we have the internal current density and the electric field, deduced from magnetic measurements, then the local conductivity of the plasma can be deduced immediately. In situations where the plasma may be expected to obey a Spitzer-type of Ohm's law, we can then deduce a local conductivity temperature and obtain its spatial profile. This avoids the averaging inherent in the use of simply the total conductance for a conductivity estimate but does not avoid all the problems associated with resistance anomalies due, for example, to impurities. In the disruptive instability measurements illustrated previously, such an approach fails because during the disruption the simple Ohm's law does *not* apply: for example, the effective conductivity at the plasma edge is negative. However, the conductivity temperature can be obtained just before and just after the disruption. It has a shape approximately proportional to the $\frac{2}{3}$ power of j, since E is approximately uniform.

2.3.4   *Pressure*

As a further illustration of the information to be gained from magnetic probe measurements, consider again the MHD equilibrium equation in the cylinder:

$$\frac{dp}{dr} + \frac{B_\phi}{\mu_0}\frac{dB_\phi}{dr} + \frac{B_\theta}{r\mu_0}\frac{d(rB_\theta)}{dr} = 0, \tag{2.3.7}$$

which may also be integrated to give

$$\left[p + \frac{\left(B_\theta^2 + B_\phi^2 \right)}{2\mu_0} \right]_{r_1}^{r_2} + \frac{1}{\mu_0}\int_{r_1}^{r_2} \frac{B_\theta^2}{r}\, dr = 0. \tag{2.3.8}$$

As is evident from this equation, knowing the profiles of B_θ and B_ϕ enables one then to deduce the internal pressure profile by taking $r_2 = a$ and assuming $p(a) = 0$. The difficulty with this procedure is that in most magnetically confined plasmas the plasma β is small ($\lesssim 10\%$). This means that, in order to deduce the plasma pressure, the field profiles must be measured with very high accuracy, requiring considerable attention to calibration and alignment (see Exercise 2.8). In some experiments sufficient accuracy has been achieved. Figure 2.23 shows an example of a pressure profile deduced from the RFP magnetic fields of Fig. 2.18, and plotted normalized to the central magnetic pressure. Notice that the cumulative uncertainty in the plasma center, indicated by the error bar, is rather large. The uncertainty varies approximately linearly to zero at the plasma edge because of the integration process.

2.3.5   *Two- and three-dimensional measurements*

Although it is essentially impossible to use internal probes to measure simultaneously the field throughout a plasma (because the number of positions required would lead to filling up the whole plasma with probes), it is possible to build up a virtually complete picture of the field, provided the plasma under study is either steady or reproducible. What is required is to make the measurements serially, moving the probe from place to place. This eventually allows one to reconstruct the entire field profile with an accuracy limited by the reproducibility of the plasma.

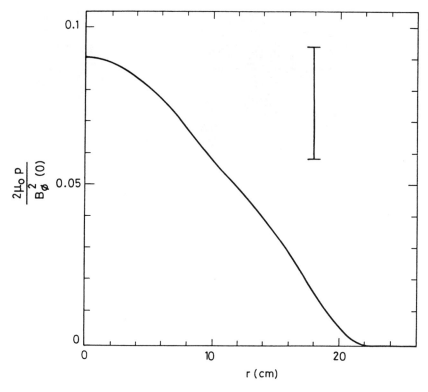

Fig. 2.23. Pressure profile deduced from the magnetic field measurements of Fig. 2.18.

Perhaps the most useful additional information that can be gained from such a two- or three-dimensional reconstruction is a plot of the magnetic surfaces and, assuming a flux function exists, the magnetic flux. Restricting our discussion to the two-dimensional case for definiteness, the transverse flux at any point can be obtained by integrating the magnetic field along a line

$$\psi(\mathbf{x}) = \psi(\mathbf{x}_0) + \int_{\mathbf{x}_0}^{\mathbf{x}} (\mathbf{B} \wedge \hat{\mathbf{z}}) \cdot \mathbf{dl}. \qquad (2.3.9)$$

Here ψ is the flux per unit length along the symmetry direction $\hat{\mathbf{z}}$. [Equation (2.3.5) is essentially the time derivative of this equation.] This quantity can be obtained for any point $\mathbf{x}$ from measurements along a single line, provided the flux is known at the reference point $\mathbf{x}_0$. It could therefore be obtained by a single multicoil probe. Then the flux along all possible different lines can be obtained on a shot-to-shot basis by moving the probe.

From complete magnetic measurements of this type, contour plots of the magnetic flux surfaces can be constructed that show the instantaneous

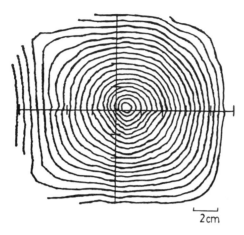

2cm

Fig. 2.24. Two-dimensional reconstruction of flux surfaces by shot-to-shot internal magnetic measurements in a shaped toroidal plasma [after Lipschultz et al. (1980)].

shape of the plasma. From such information, all the other quantities discussed earlier (j, E, possibly p, etc.) can be obtained by generalizations of the analysis presented. In Fig. 2.24 an example of flux contours reconstructed from magnetic probe measurements is shown.

2.4   Fluctuations

So far we have discussed mostly the measurements of quasistationary equilibrium values of the various parameters measurable via magnetic diagnostics. However, because magnetic measurements are continuous in time, they can also provide information on the rapidly changing or fluctuating components of such parameters. These arise, for example, from various forms of instability that can occur in magnetized plasmas.

2.4.1   *External measurements*

In a doubly periodic system, such as a torus, any field may be expressed in the form of a sum of helical Fourier modes $\exp i(m\theta + n\phi)$, m and n being the poloidal and toroidal mode numbers. The $n = 0$ components we have already discussed. They relate to radial equilibrium ($m = 0$), position ($m = 1$), and shape ($m \geq 2$). For $n \neq 0$ the modes represent helical distortions of the plasma that are generally undesired instabilities, except, of course, for machines such as stellarators in which helical fields are deliberately imposed. The poloidal mode structure of such perturbations may be determined by coils at the plasma edge in the same way as for the $n = 0$ modes. In addition, the toroidal mode structure may be determined from coils ranged around the torus in ϕ.

In general, the modes of greatest interest and importance are those whose perturbation structure lies along the direction of the magnetic field lines

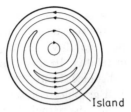

Fig. 2.25. A magnetic island structure forms in the poloidal cross section of the magnetic surfaces when a resonant field perturbation is present.

somewhere inside the plasma. They are then said to be resonant at that point, and if they possess a field component perpendicular to the equilibrium magnetic surfaces, they will cause the field topology to change there by the formation of magnetic islands as illustrated in Fig. 2.25. For tokamaks, for example, this implies $m/n \sim 1$ to 3, while for reversed field pinches $m/n \sim -0.2$ to $+0.1$. The modes that become unstable most readily in many situations and that are most easily measured are those with long wavelength and hence low m and n.

As an example of the type of phenomena observed, Fig. 2.26 shows a sequence of instabilities of decreasing m that appears in the beginning of a

Fig. 2.26. A sequence of instabilities with decreasing m number [after Granetz, Hutchinson, and Overskei (1979)].

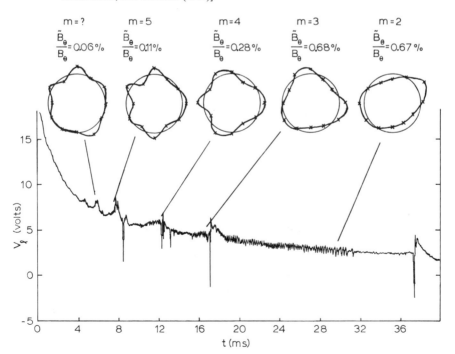

tokamak discharge. Each time a helical mode ($n = 1$) appears, perturbations of the loop voltage occur, indicating changes of the plasma inductance due to current profile modifications caused by the instability. The edge magnetic coils give the mode number m of the instability. In this example $\dot{B}_\theta$ polar plots are made in which the radial distance from the circle is proportional to $\dot{B}_\theta$ at that angle θ. This gives a graphic representation of the "shape" of the perturbation (though not really the actual shape of the plasma).

The form of the field perturbation measured by external coils is constrained by Maxwell's equations and the fact that the current density is zero outside the plasma. To illustrate these constraints let us consider a cylindrical plasma (approximately a toroidal plasma with large aspect ratio) in which the cylindrically symmetric equilibrium field **B** is perturbed by a single helical mode. Writing the field perturbation as **b**, the equations we need are

$$\nabla \cdot \mathbf{b} = \frac{\partial b_r}{\partial r} + \frac{im}{r} b_\theta + \frac{in}{R} b_\phi = 0, \tag{2.4.1}$$

$$(\nabla \wedge \mathbf{b}) \cdot \hat{\mathbf{r}} = \frac{im}{r} b_\phi - \frac{in}{R} b_\theta = 0. \tag{2.4.2}$$

From the second of these we can immediately relate the direction of the field perturbation to the mode helicity

$$\frac{b_\phi}{b_\theta} = \frac{nr}{mR}. \tag{2.4.3}$$

What this equation says is that the tangential field perturbation points parallel to the helical coordinate $m\theta + n\phi$ or, equivalently, perpendicular to the ignorable direction of the helix. For configurations such as the tokamak, in which m/n is typically of order 1 or larger, the field perturbation is almost all b_θ (because $r/R \ll 1$); b_ϕ is much smaller in magnitude. For this reason, most fluctuation measurements in tokamaks concentrate on measuring b_θ and rarely bother with b_ϕ. (Arrays of coils in a single poloidal plane measuring b_θ are often called Mirnov coils on tokamaks after an influential early user.) When m/n is small, however (as in, e.g., the reversed field pinch), measurement of b_ϕ is also important. In this case, if a single mode can be discerned, it is possible to infer the helicity (m/n) of the perturbation from measurements of b_θ and b_ϕ simultaneously at one position using Eq. (2.4.3). This ability is sometimes helpful when insufficient probes are available at different ϕ or θ positions to deduce m or n directly.

From a purely conceptual viewpoint, the use of external coils in the diagnosis of fluctuations differs little from their use in diagnosing the steady equilibria. From a practical viewpoint, however, there are several matters that are of much greater significance for fluctuations. Perhaps the most important of these is the question of magnetic field penetration into the structures of the experiment. Most modern magnetically confined plasmas are formed in vacuum vessels with metallic – and hence electrically conducting – walls. These walls have a certain time constant, say τ, for the penetration of magnetic fields through them, owing to the eddy currents induced by a time-varying magnetic field. The effect is to allow magnetic fields changing slower than τ to penetrate with minimal attenuation, but fields changing faster than τ *are* attenuated. Now it is much more convenient to implement magnetic measurements using coils that are outside the vacuum vessel than to have to overcome the various technical difficulties, associated with vacuum compatibility of the coils and their leads, involved in siting the coils inside the vacuum. Therefore, in many cases, the measurement of magnetic fluctuations is made outside the vacuum vessel, even though the high-frequency components of the fields are then subject to the attenuation just mentioned. It turns out that there is a reasonably satisfactory way of compensating for the attenuation problem, and that is to measure the unintegrated signal out of the coil rather than integrating it to get B. This works fairly well because the effect of the attenuation, when the walls are thin, is to integrate the high-frequency components.

Consider a situation in which the magnetic field is Fourier analyzed into its different frequency (ω) components. For a wall of thickness w and conductivity σ the penetration time constant for fields with spatial scale length L is approximately

$$\tau = \mu_0 \sigma L w. \tag{2.4.4}$$

Notice that different perturbation modes will have different L so that τ will generally depend on the mode under discussion. This is a cause of some awkwardness in the interpretation when more than one helical mode is of interest, but we shall ignore this problem in our brief discussion. Usually L will be of the order of the minor radius r. The penetration of transverse fields through the wall may then be expressed as a relationship between the field just inside the wall (B_i) and that just outside (B_e), in the form

$$B_e = B_i \frac{1}{1 + i\omega\tau}. \tag{2.4.5}$$

In writing this equation we are supposing that the external field is changing in response to internal perturbations; also, we are begging a number of important questions, such as the influence of other nearby conducting structures, which can really only be accounted for by a detailed electromagnetic analysis on a case-by-case basis. Nevertheless, this equation will generally represent the response reasonably well within certain frequency limits. What the equation says is that low frequencies ($\omega\tau \ll 1$) experience no attenuation, but high frequencies ($\omega\tau \gg 1$) are attenuated by the factor $\sim 1/i\omega\tau$, that is, they are integrated.

The upper frequency limit for the applicability of Eq. (2.4.5) is set by the requirement that the wall should be able to be taken as thin. In other words, there is a second penetration time constant $\tau_w \approx \mu_0\sigma w^2/2$, and for fields changing more rapidly than this (i.e., $\omega\tau_w > 1$), the finite thickness of the wall becomes important. Thus, the upper frequency limit of Eq. (2.4.5) is $\sim 1/\tau_w$, above which the external field will fall off with frequency, increasingly more rapidly than Eq. (2.4.5) indicates. Figure 2.27 illustrates the frequency response of probes outside a conducting wall.

Very often the wall structures are complex and anisotropic, so a rather careful analysis of the penetration effects is necessary to obtain accurate quantitative results. The situation most often encountered is qualitatively that the slow evolution of the equilibrium is substantially slower than τ, so that we need not worry about penetration problems in equilibrium measurements, but fluctuations due to instabilities, etcetera, are faster than τ and so are integrated. In such a situation, observing the unintegrated magnetic coil signal has the advantageous effect of suppressing the equilibrium signal but retaining the perturbation. When the very high-frequency perturbations $\omega > 1/\tau_w$ are of interest, one is generally forced to use coils inside the vacuum chamber.

So far we have been talking about tangential field perturbations b_θ and b_ϕ. What of the normal or radial perturbations b_r? These are usually of less use in fluctuation measurements, the reason being related to the electromagnetic penetration problems just discussed. The boundary conditions used in solving the field penetration problem depend on the wall conductiv-

Fig. 2.27. The frequency response of a magnetic coil outside a conducting wall.

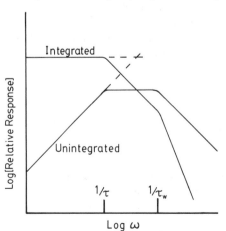

ity for the tangential components; for the *radial* component the boundary condition is just that the field should be continuous. Thus the normal field is the same just inside the wall as it is just outside. This shows, incidentally, that there is no advantage in doing radial edge-field measurements *inside* a closely fitting conducting vessel. What is more, the boundary conditions have a strong effect on the perturbations themselves. For high frequencies, $\omega \gg 1/\tau$, the effect is to enforce $b_r \approx 0$ at the conducting wall. Therefore, the radial field perturbations at high frequencies will generally tend to be rather small, for essentially the same reasons as the tangential components are smaller outside than inside: because they are integrated. What is of possibly greater interest concerning b_r is its radial derivative near the wall, $\partial b_r / \partial r$. This is where Eq. (2.4.1) comes in, since it shows that $\partial b_r / \partial r$ is directly related to b_θ and b_ϕ through the divergence equation. In other words if we measure b_θ (and b_ϕ if relevant), then for a single mode for which m and n are known, $\partial b_r / \partial r$ may be deduced straightaway.

2.4.2   *Internal fluctuation measurements*

Naturally, internal magnetic probing, particularly with multiple simultaneous measurements enables one to investigate the radial structure of magnetic fluctuations. In this respect it is complementary to edge measurements, which give the poloidal and toroidal structure. It is usually very difficult to measure the poloidal and toroidal structure internally because too many probes are required. It is rare that fluctuations are reproducible enough to allow reconstruction on a shot-to-shot basis; thus simultaneous measurements are virtually essential.

The advantage of internal probing is that, unlike external coils, it is much more sensitive to short scale-length perturbations arising as turbulence. The reason for the insensitivity of edge coils to short wavelength perturbations is easily understood. If the current in the region under discussion has negligible perturbation (of the relevant wavelength), then the field satisfies the equation

$$\nabla^2 \mathbf{b} = 0. \qquad (2.4.6)$$

This will certainly be satisfied in any vacuum region surrounding the plasma. Treating the plasma as a slab, for the purpose of this discussion, and the perturbation as having a tangential wavelength $2\pi/k$, the solution for $\mathbf{b}$ varies in the normal (radial) direction as $\exp(-kr)$. Thus if $kr \gg 1$, the field outside the plasma arising from perturbations inside will be exponentially small. Actually, in the plasma interior Eq. (2.4.6) will not generally be well satisfied, and more complicated analysis of the MHD instability structure is necessary. It turns out that the field usually falls off at least as fast as implied by the vacuum equation, which confirms that external coils are very insensitive to short wavelength perturbations.

The small-scale perturbations measured by internal probes are rarely coherent. Therefore, the usual approach to their analysis is to use simultaneous measurements at adjacent positions to calculate statistical correlation coefficients as a function of position, thus giving a measure of correlation lengths and fluctuation scale lengths. In some cases this enables the cause of the instabilities to be identified. However, this, like all turbulence topics, is one of great complexity. The methods of analysis are far from universally established and their interpretation depends on the specifics of the situation.

Further reading

An introduction to "basic measurements" in plasmas including magnetic measurements is given by:

Leonard, S. L. (1965). In *Plasma Diagnostic Techniques*. R. H. Huddlestone and S. L. Leonard, eds., p. 7. New York: Academic.

Insertable magnetic probe measurements including the important questions of plasma perturbation by the probe are covered by:

Lovberg, R. H. (1965). *Plasma Diagnostic Techniques*. R. H. Huddlestone and S. L. Leonard, eds., p. 69. New York: Academic.
Botticher, W. (1968). In *Plasma Diagnostics*. W. Lochte-Holtgreven, ed., p. 617. Amsterdam: North-Holland.

MHD equilibrium has been reviewed by various authors; for example:

Freidberg, J. P. (1982). *Rev. Mod. Phys.* 54 : 801.
Mukhovatov, V. S. and Shafranov, V. D. (1971). *Nucl. Fusion* 11 : 605.

A readable introduction to MHD equilibrium and stability is:

Bateman, G. (1978). *MHD Instabilities*. Cambridge, Mass.: MIT.

A more complete treatment of ideal MHD is by

Freidberg, J. P. (1987). *Ideal Magnetohydrodynamics*. New York: Plenum.

Exercises

2.1 Suppose we measure a magnetic field $B(t)$ using a magnetic coil with total resistance R_c whose ends are connected to measurement electronics whose resistance is R_e. If the coil has N turns of area A, what is the voltage V measured between the ends of the coil?

2.2 When a current is allowed to flow in a measurement coil it changes the magnetic field in its vicinity and thereby changes the measured voltage by induction. (This effect was implicitly ignored in Example 2.1.) Show that this leads to a full solution of the problem in Example 2.1 as

$$\frac{L}{R_e}\dot{V} + \left(1 + \frac{R_c}{R_e}\right)V = NA\dot{B},$$

where L is the self-inductance of the magnetic coil. Hence show that for sufficiently rapidly changing fields the coil is self-integrating so that

$$V = B(R_e NA/L)$$

and give the condition for "sufficiently rapidly changing."

2.3 In tokamaks, instabilities limit the maximum current density at the center of the plasma to a value

$$j_0 = \frac{2B_\phi}{\mu_0 R q_0},$$

where R is the major radius and q_0 is a number that is approximately equal to 1. Derive from this fact an expression for the central electron temperature of a $Z = 1$ plasma (using the Spitzer conductivity) in terms of toroidal loop voltage V_ϕ.

2.4 Perform the integrations to derive Eqs. (2.2.7) and (2.2.10).

2.5 Prove Eq. (2.2.11).

2.6 Prove that for a toroidally symmetric fluid satisfying the MHD equilibrium [Eq. (2.2.1)] on a magnetic surface, $p = $ constant and $B_\phi \propto 1/R$.

2.7 Show that in a plasma whose magnetic surfaces are circular in cross section, shifted from the chamber axis by a distance δ (varying with radius), measurement of the poloidal and radial components of the internal magnetic field can give the value of δ. Give an expression for δ in terms of B_θ, B_r, and radius.

2.8 The two main types of error that can arise in probe measurements (apart from interference, producing spurious signals) are (1) calibration errors, that is, incorrect values of NA or electronic amplifier gains, etcetera, and (2) angular misalignment of the coils, so that the component of field measured is in a slightly different direction from that intended. Suppose one wishes to deduce the internal pressure profile from probe measurements using Eq. (2.3.6). If the orders of magnitude of the angle of misalignment α and the fractional calibration error δ are the same, and $B_\theta < B_\phi$ show that:

(a) When all errors are *random* the most important ones are those arising from the first term in Eq. (2.3.6), giving rise to a fractional error in p:

$$\frac{\Delta p}{p} \approx \frac{2\sqrt{2}\delta}{\beta_\phi}.$$

(b) When all errors are *systematic* (the same for all positions) the most important ones come from the second term, giving rise to

$$\frac{\Delta p}{p} \sim \frac{4\alpha}{\beta_\phi}\frac{B_\theta}{B_\phi}.$$

Here $\beta_\phi = 2\mu_0 p/B_\phi^2$ and (B_θ/B_ϕ) is an order of magnitude term only.

3 Plasma particle flux

3.1 Preliminaries

Perhaps the most natural approach to diagnosing the particle distribution functions within the plasma is to propose insertion of some kind of probe that directly senses the particle fluxes. Indeed, this approach was one of the earliest in plasma diagnostics, with which the name of Irving Langmuir is most notably associated for his investigations of the operation of the electric probe often known as the Langmuir probe.

Just as with internal magnetic probes, the applicability of particle flux probes is limited to plasmas that the probe itself can survive. This means that frequently only the plasma edge is accessible, but the importance of edge effects makes the prospects bright for continued use of such probes even in fusion plasmas. In cooler plasmas, of course, the limitations are less severe and more of the plasma is accessible.

In common also, with magnetic probes, the often more important question is: What is the effect of the probe on the plasma? Because of the nonlocal nature of the source of the magnetic field (arising from possibly distant currents), in many cases the local perturbation of the plasma by a *magnetic* probe can be ignored. In contrast a particle flux measurement is essentially *local* and as a result the local perturbation of the plasma can almost never be ignored.

Thus, the difficulty with measurements of direct plasma particle flux is rarely in the measurements themselves; rather it is in establishing an understanding of just how the probe perturbs the plasma locally and how the local plasma parameters are then related to the unperturbed plasma far from the probe.

The way in which the perturbation to plasma parameters primarily occurs is through alterations of the electric potential and (hence) of particle density and energy. The nature of this perturbation depends on the potential of the probe and the electric current drawn by it. Indeed for the simplest particle flux probes, Langmuir probes, the dependence of the total current on probe potential is the main quantity measured. Most of our effort is spent in determining the total probe current as a function of probe potential. Even when more complicated particle analyzers than the simple Langmuir probe are used, the details of what they observe are usually affected by their overall electrical characteristics. Therefore, a thorough understanding of the perturbative effects of the probe on the observed particle energies and currents is important in interpreting their results.

3.1.1   *Particle flux*

In an unperturbed plasma, elementary gas–kinetic theory shows that the number of particles of a given species crossing unit area per unit time (from one side only) is

$$\Gamma = \tfrac{1}{4}n\bar{v}, \tag{3.1.1}$$

where $\bar{v}$ is the mean particle speed (in this chapter, Γ refers to particle flux density and J and I to particle and electric total currents, respectively). Suppose that a probe is present in a thermal plasma of comparable electron and ion temperatures. (We shall consider mostly only two species of plasma particles in this chapter, of equal and opposite charge, the ions, of course, being much heavier.) The mean ion speed will then be much smaller than the mean electron speed so that the total electric current from a probe of area A if the plasma were unperturbed would be dominated by the electrons:

$$I = -eA\left(\tfrac{1}{4}n_i\bar{v}_i - \tfrac{1}{4}n_e\bar{v}_e\right) \approx \tfrac{1}{4}eAn_e\bar{v}_e > 0. \tag{3.1.2}$$

The probe would thus emit a net positive current. If, for example, the probe were electrically insulated from other parts of the plasma device (a "floating" probe), then it would rapidly charge up negatively until the electrons were repelled and the net electrical current brought to zero. The potential adopted by such a floating probe is called the floating potential, which we denote V_f. Clearly it is different from the electric potential in the plasma in the absence of any probe. This later potential is called the plasma potential (or space potential) and will be denoted V_p.

Figure 3.1 shows the variation of the total electric current I (flowing out of the probe in this the usual sign convention) versus the potential of the probe V in a typical Langmuir probe experiment. Note that no zero is indicated on the voltage scale because initially we don't know what the plasma potential is, with respect to (say) the walls of the plasma chamber. Qualitatively, this probe shape arises as follows.

Roughly speaking, if the probe is at plasma potential then the perturbations to the free ion and electron currents Eq. (3.1.2) will be small. Thus, the space potential is approximately the point at which $I \approx eJ_e$, the electron current. If the voltage is increased above this level, $V > V_p$, in principle (and approximately in practice), the electron current cannot increase any further. This is because the electron current is maximized since all electrons arriving are collected. The ion current J_i decreases because of repulsion of the ions, but it is already much less than the electron current, so I is approximately constant. This region (A) is known as electron saturation and I here is equal to the electron-saturation current.

Decreasing the probe potential, $V < V_p$, the probe is now negative with respect to the surrounding plasma and an increasing fraction of impinging

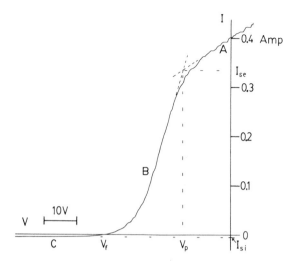

Fig. 3.1. Electric probe characteristic showing how the probe current varies with probe potential.

electrons is reflected from the negative potential (region B). Eventually the potential is sufficient to reduce J_e to a small fraction of its saturation value. The total current is zero when $J_e = J_i$ at the floating potential V_f. Decreasing the potential further, entering region C, eventually only ions are collected at approximately the constant rate given by Eq. (3.1.1) (with some important modifications to be discussed). This is the ion-saturation current $I = I_{si}$.

In order to put these qualitative remarks on sound quantitative footing we must discuss the various ways in which the presence of the probe perturbs the plasma and hence changes the currents from the simple values just used.

3.1.2   *Debye shielding*

The effects of a potential-perturbing charge in a plasma are generally much shorter-range than in a vacuum because the charges in the plasma tend to redistribute themselves so as to shield the plasma from the electric field the perturbing charge generates. The effect may be deduced readily from Poisson's equation by assuming, for example, that ions do not move but that electrons adopt a thermal equilibrium distribution in which the electron density is determined by the Boltzmann factor

$$n_e = n_\infty \exp(eV/T_e). \tag{3.1.3}$$

T_e here is the electron temperature in energy units and n_∞ is the electron density far from the perturbing charge where the potential V is taken as zero. Poisson's equation is then

$$\nabla^2 V = \frac{-\rho}{\varepsilon_0} = \frac{-e}{\varepsilon_0}(n_i - n_e) = \frac{-e}{\varepsilon_0}n_\infty\left[1 - \exp\left(\frac{eV}{T_e}\right)\right]. \quad (3.1.4)$$

If we suppose that $eV \ll T_e$, we can approximate $\exp(eV/T_e)$ by $1 + eV/T_e$ and obtain a Helmholtz type equation

$$\nabla^2 V - \frac{1}{\lambda_D^2}V = 0, \quad (3.1.5)$$

where

$$\lambda_D = \left(\frac{\varepsilon_0 T_e}{e^2 n_\infty}\right)^{1/2} \quad (3.1.6)$$

is called the (electron) Debye length. The solutions of this equation show exponential dependence upon distance far from the charge with λ_D the characteristic length. For example, in one dimension the solutions are $V \propto \exp(\pm x/\lambda_D)$. As a general rule, therefore, one expects that the perturbing effects of a charge will tend to penetrate into the plasma a distance only of the order of the Debye length, always provided the assumption of thermal equilibrium is valid for the electrons and ions. (In general, of course, for a stationary charge, the ion density is also perturbed and a similar treatment must be accorded to ions leading to a shielding length shorter by the ratio $[T_i/(T_e + T_i)]^{1/2}$. We shall use the term Debye length here to mean the electron Debye length, Eq. (3.1.6).)

Note that for laboratory plasmas, the Debye length is generally rather short. For example, a 1 eV temperature plasma of density only 10^{17} m^{-3} has $\lambda_D = 20$ μm, so the Debye length is much smaller than typical probe dimensions (say a, of order millimeters). Often the approximation $\lambda_D \ll a$ is strongly satisfied, and we shall consider mostly this limit.

3.1.3   *Collisional effects*

Probe behavior differs significantly between situations where collisions can be ignored and those where they cannot. The effect of collisions is generally to reduce the current to the probe because of the necessity for particles to diffuse up to the probe rather than arriving by free flight.

A general idea of the effect may be gained by treating the plasma around the probe as a continuum having diffusion coefficient D. Then if no perturbation to the electric potential occurs or, at least, if the effects of electric field on particle transport are ignored, and D is taken as constant, the particle current density to a perfectly absorbing spherical probe (see

Exercise 3.1) is

$$\Gamma = \frac{1}{4}n_{\infty}\bar{v}\frac{1}{1 + \bar{v}a/4D},\tag{3.1.7}$$

where n_{∞} is the density far from the probe and a is the probe radius. Note that the mean free path ℓ and the diffusion coefficient are related quite generally by

$$\ell = \frac{3D}{\bar{v}},\tag{3.1.8}$$

so that the factor by which the current drawn is less than the random current $(n_{\infty}\bar{v}/4)$ is $(1 + 3a/4\ell)^{-1}$, which for large a/ℓ is approximately $4\ell/3a$.

Thus, roughly speaking, collisions are ignorable for a spherical probe when $\ell \gg a$, while at the opposite extreme, $a \gg \ell$, the current is reduced by the ratio of mean free path to radius.

It may be noted, at this stage, that probes with "infinite" dimensions, such as cylindrical or plane probes, do not lead to a well posed problem for diffusion-limited collection unless boundary conditions (closer than infinity) or volumetric plasma source rate are included. The reason for this is that the governing equation for diffusion without sources is Laplace's equation, whose solutions [$\log(r)$ for cylinder, x for plane] are not bounded at infinity in these geometries. Thus the finite length of (say) a cylindrical probe will always be important in determining diffusive collection and it is this length that should be compared with ℓ in deciding whether a collisionless treatment is justified.

It may be shown (see Exercise 3.2) that for a highly ionized plasma, in which collisions with neutrals can be ignored, the mean free path is given roughly by

$$\ell \sim \left(n\lambda_D^3\right)\lambda_D,\tag{3.1.9}$$

so that when the usual plasma condition that there be many particles in a Debye sphere ($n\lambda_D^3 \gg 1$) is satisfied, then $\ell \gg \lambda_D$. The region of strong potential perturbation surrounding the probe is thus collisionless. In plasmas with only a very low degree of ionization it is possible for collisions with neutrals to be sufficient to cause the mean free path to be shorter than λ_D, so that the sheath is collisional. We shall not discuss such situations here. Our whole treatment can be collisionless if $\ell \gg a$. This condition is not always satisfied but in many cases it is. In the example above ($T_e = 1$ eV, $n = 10^{17}$ m^{-3}) the mean free path is $\ell \sim 5$ cm, so that it is quite easy to use a probe small enough to be collisionless in such a plasma.

We shall return later to cases in which collisions cannot be ignored, but for now we proceed on the assumption that, in the vicinity of the probe, the particles are collisionless.

3.2 Probes in collisionless plasmas without magnetic fields
3.2.1 *Sheath analysis*
When a solid probe is in contact with a plasma the potential drop between the plasma and probe is mostly confined to a region of the order of a few Debye lengths thickness surrounding the probe. This is called the sheath. In the sheath, charge neutrality is violated and the electric field is strong. Generally, the densities of electrons and ions must be determined by self-consistently solving Poisson's equation for the potential together with the equations of motion of the charged particles and hence their densities. Understandably, rather complicated equations are involved when the full orbit equations of a distribution of particle energies around a probe of given shape are to be solved. Solutions can be obtained (usually numerically) for the full problem. However, we shall here restrict our treatment to rather more tractable approximations, which nevertheless provide perfectly adequate accuracy for most diagnostic purposes and, indeed, provide the formulas almost always used in practice.

Consider, then, a thin sheath surrounding the probe; a planar approximation is then adequate. First, we consider a probe that attracts ions and

Fig. 3.2. Schematic diagram of the electric potential variation near the surface of a negatively biased probe.

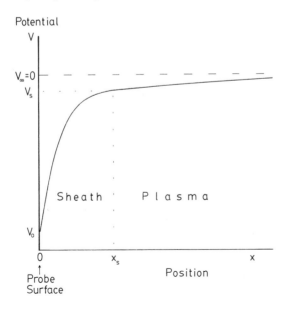

repels electrons. This is the case of greatest interest, but strictly does not cover cases where the probe is at greater than plasma potential. Figure 3.2 shows schematically the form of the potential versus position.

The electron density we determine by the observation that at any position x there is a full distribution of particles traveling toward the probe, having arrived from large distances away ($\sim \ell$) in the plasma. The distribution traveling away from the probe, however, is depleted by collection from the probe. Assuming the probe to absorb (or recombine) all particles incident on it, the result is a complete removal of all particles whose kinetic energy in the x direction exceeds the potential difference between the probe and the point x. (Motions perpendicular to x are irrelevant.) Only lower-energy electrons are reflected from the potential barrier before they reach the probe. So, as illustrated in Fig. 3.3, the distribution is cut off at the velocity $v_x = v_c = (2e[V(x) - V(0)]/m_e)^{1/2}$.

The distribution function of particles at velocities not affected by this cutoff is simply related to the distribution at large distances by the conservation of particles, that is, that the flux of particles of a certain total energy is constant:

$$f_x(v_x) v_x \, dv_x = f_\infty(v_\infty) v_\infty \, dv_\infty, \tag{3.2.1}$$

where subscripts refer to the position and f here is the distribution function in x component velocity. The relationship between corresponding energies is

$$\tfrac{1}{2}mv_x^2 - eV = \tfrac{1}{2}mv_\infty^2 \tag{3.2.2}$$

(taking $V_\infty = 0$ as our origin of potential). Hence, $v_x \, dv_x = v_\infty \, dv_\infty$ and the distribution functions are related by

$$f_x(v_x) = f_\infty(v_\infty), \tag{3.2.3}$$

Fig. 3.3. The electron distribution near a repelling probe. The cutoff above v_c is due to collection (rather than reflection) of electrons with higher energy.

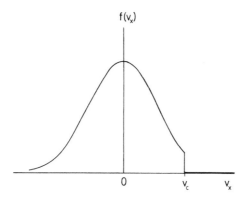

which gives, for Maxwellian f_∞, a Maxwellian distribution also at x, but corresponding to density less by the Boltzmann factor $\exp(eV/T_e)$.

The actual density obtained by integrating f over all energies is less because of the cutoff portion of the distribution function. However, if the cutoff portion is small, that is, almost all electrons are reflected before reaching the probe because $V(x) - V(0) \gg eT_e$, then this correction is small and we can make the approximation

$$n_e(x) \approx n_\infty \exp(eV(x)/T_e), \tag{3.2.4}$$

which corresponds to the thermal equilibrium distribution. (The more general formula, without making this approximation is

$$n_e(x) = n_\infty \exp\{ eV(x)/T_e \} \tfrac{1}{2} \left(1 + \text{erf}\{ [V(x) - V(0)] e/T_e \}^{1/2} \right), \tag{3.2.5}$$

where

$$\text{erf}(t) \equiv \frac{2}{\sqrt{\pi}} \int_0^t e^{-y^2} \, dy; \tag{3.2.6}$$

see Exercise 3.3.) In actual fact, the approximation $n \propto \exp(eV/T_e)$ will be valid in any geometry provided almost all electrons are reflected. This is the approximation we shall generally use.

The ion density must be calculated from the equation of motion; this calculation is straightforward only when considerable simplifying approximations are made. Let us start by supposing that the ion temperature is considerably less than the electron temperature, small enough, in fact, that we can suppose the ions to have zero energy at ∞. Then we can write the ion velocity immediately (ignoring collisions) as,

$$|v_i| = \left(\frac{-2eV}{m_i} \right)^{1/2}, \tag{3.2.7}$$

and, of course, the ion particle current density is $\Gamma_i = n_i v_i$.

The total ion current crossing a surface of area A around the probe is constant (independent of position), that is, $\nabla \cdot \Gamma_i = 0$ in equilibrium. Call this current J_i:

$$J_i = A n_i v_i = \text{constant}. \tag{3.2.8}$$

Therefore, provided the area A is known via the geometry, we have an equation relating the ion density to the known ion velocity [Eq. (3.2.7)].

In general, then, Poisson's equation becomes

$$\nabla^2 V = \frac{-e}{\varepsilon_0} [n_i - n_e] = \frac{-e}{\varepsilon_0} \left[\frac{J_i}{A} \left(\frac{m_i}{-2eV} \right)^{1/2} - n_\infty \exp\left(\frac{eV}{T_e} \right) \right]. \tag{3.2.9}$$

This equation, as implied by Fig. 3.2, splits up approximately into two distinct regions. First, at large distances from the probe there is a plasma region in which quasineutrality is satisfied:

$$n_e - n_i \ll n_e. \tag{3.2.10}$$

In this region in Eq. (3.2.9) the $\nabla^2 V$ term may be neglected and the governing equation is

$$J_i \Bigg/ \left[A \left(\frac{-2eV}{m_i} \right)^{1/2} \right] = n_\infty \exp\left(\frac{eV}{T_e} \right). \tag{3.2.11}$$

Note, though, that the solutions for V need not be $V =$ constant, so there may be some electric field in the plasma region. However, the electric field will be very small here compared to its magnitude in the second region. This second region includes a transition and a "sheath" that for present purposes we treat together and call the sheath. When a distinction between transition and sheath is required we shall denote by sheath that portion of the volume in which the electron density is small enough to be ignored in determining V.

For the sheath region (including the transition) $\nabla^2 V$ cannot be ignored, so we must use the full Poisson equation. Let us begin with this region and suppose that at the plasma–sheath boundary (x_s) the potential is V_s, different from V_∞ ($= 0$) because of the fields in the plasma region. At this boundary the solutions in the two regions must match and, because of our supposition that $\lambda_D \ll a$ and hence that x_s (which is generally several times λ_D) is also much less than a, we can approximate the geometry as planar, that is, take A constant.

Then, in the vicinity of the sheath,

$$n_i = n_{is} \left(\frac{V_s}{V} \right)^{1/2}, \tag{3.2.12}$$

where subscript s denotes values at the plasma–sheath interface. But also, considering the plasma solution,

$$n_{is} = n_{es} \quad \left(= n_\infty \exp(eV_s/T_e) \right). \tag{3.2.13}$$

Thus Poisson's equation is

$$\nabla^2 V = \frac{-e}{\varepsilon_0} n_s \left[\left(\frac{V_s}{V} \right)^{1/2} - \exp\left\{ \frac{e(V - V_s)}{T_e} \right\} \right]. \tag{3.2.14}$$

This equation is still not tractable analytically except in the region close to x_s, where we can make a Taylor expansion about $V = V_s$ and retain only the lowest order terms. This gives

$$\nabla^2 V = -\frac{en_s}{\varepsilon_0} \left[-\frac{1}{2V_s} - \frac{e}{T_e} \right] (V - V_s). \tag{3.2.15}$$

If the square bracketed term here is negative then the solutions will be monotonic (exponential in this region) and a smooth match will be possible between the plasma and sheath. However, if it is positive the solutions will be oscillatory (sinusoidal) and no sheath solution will be possible.

Thus, there is a maximum sheath edge voltage for proper sheath formation:

$$V_s \leq -T_e/2e. \qquad (3.2.16)$$

Qualitatively, this condition, first explicitly derived by Bohm (1949), can be understood thus: In the sheath region the ion density n_i must exceed the electron density since the sheath has positive charge; also the velocity v_i is given by energy conservation. The current density $n_i v_i$, therefore, must exceed some minimum value in the sheath. At the plasma edge, where n_i is of course still finite, to provide this minimum current density, a minimum velocity v_i is required. Having $|V_s| \geq T_e/2e$ provides this minimum ion velocity.

Now we examine the solution in the plasma region. To establish its behavior we differentiate the quasineutrality equation (3.2.11). The resulting equation contains a term in dV/dx (where now we need make no assumption about geometry; so that x could equally well be r in cylindrical or spherical geometry). The coefficient of this term is

$$n\left(\frac{e}{T_e} + \frac{1}{2V}\right), \qquad (3.2.17)$$

which is zero when $V = -T_e/2e$. This shows that the plasma solution has infinite derivative at this potential; therefore, the quasineutral approximation must break down, and a sheath form, at or before this potential:

$$V_s \geq -T_e/2e. \qquad (3.2.18)$$

Therefore, the only way to satisfy both sheath [Eq. (3.2.16)] and plasma region [Eq. (3.2.18)] requirements is if

$$V_s = -T_e/2e. \qquad (3.2.19)$$

This potential is where the sheath edge always forms if the probe is sufficiently negative. The question arises: What if $V(0) > -T_e/2e$, that is, the probe potential is only slightly negative? Of course, the approximation that electrons remain Maxwellian is no longer a good one in this case, but nevertheless the answer is clear: No sheath forms for $V(0) \gtrsim -T_e/2e$. Probes near the plasma potential need not be surrounded by a sheath. The plasma may be quasineutral right up to the probe surface.

Assuming that the probe is sufficiently negative for a sheath to form, we are now in a position to obtain the ion current drawn by the probe. This is simply equal to the ion current across the sheath surface:

$$J_i = A_s n_{is} v_{is} = A_s n_\infty \exp\left(\frac{eV_s}{T_e}\right)\left(-\frac{2eV_s}{m_i}\right)^{1/2}$$

$$= \exp\left(-\frac{1}{2}\right) A_s n_\infty \left(\frac{T_e}{m_i}\right)^{1/2}. \tag{3.2.20}$$

(Note $\exp(-\frac{1}{2}) = 0.61$.) Here A_s denotes the area of the sheath surface where $V = V_s = T_e/2e$. This equation is quite generally applicable, provided the sheath is thin enough to be treated locally with a planar approximation, and can be used with probes of essentially arbitrary shape so long as A_s is known.

3.2.2   *Sheath thickness*

The only remaining unknown in Eq. (3.2.20) (often called the Bohm formula) is the area of the sheath surface A_s. By presumption the sheath thickness is small compared to probe radius so that a first approximation is to take A_s equal to the surface area of the probe, A_p say. This will often provide sufficient accuracy, but in this approximation the ion current is theoretically independent of probe potential – a situation rarely exactly met in practice. When ion-saturation current is drawn, it generally shows a slow increase in magnitude with increasingly negative probe potential. In the context of the sheath analysis, the reason for this increase is that the sheath thickness increases as the probe potential is made more negative, resulting in an increase in A_s.

We can obtain an approximation to the sheath thickness by analyzing it under the assumption that electron density is negligible. (This will only work for sufficiently negative probes, but is satisfactory for most cases where the ion current is significant.) Poisson's equation (3.2.9) is then simply

$$\nabla^2 V = \frac{-e}{\varepsilon_0} \frac{J_i}{A}\left(\frac{m_i}{-2eV}\right)^{1/2}, \tag{3.2.21}$$

which in the slab approximation is

$$\frac{d^2 V}{dx^2} = \frac{-e}{\varepsilon_0}\Gamma_i\left(\frac{m_i}{-2eV}\right)^{1/2}, \tag{3.2.22}$$

where $\Gamma_i = J_i/A = \text{constant}$. This equation can now be integrated, after multiplying by dV/dx, to give

$$\left(\frac{dV}{dx}\right)^2 = \frac{2\Gamma_i}{\varepsilon_0}(2m_i e)^{1/2}\left[(-V)^{1/2} - (-V_s)^{1/2}\right]. \tag{3.2.23}$$

In performing this integration we have put dV/dx equal to zero at the sheath edge $V = V_s$. This ignores the transition region in which electron density, strictly, may not be ignored. Provided the transition region is not a large fraction of the total sheath, the error involved in this will only be modest, and since we are in any case calculating the first order correction to the ion current due to small (but finite) sheath thickness, the approximation is acceptable.

Equation (3.2.23) may be integrated again giving

$$[\sqrt{(-V)} - \sqrt{(-V_s)}]^{1/2}[\sqrt{(-V)} + 2\sqrt{(-V_s)}]$$

$$= \frac{3}{4}\left[\frac{8\Gamma_i^2 m_i e}{\varepsilon_0^2}\right]^{1/4}(x_s - x). \tag{3.2.24}$$

Here x_s is the sheath edge position. This equation is related to the Child–Langmuir law for space-charge limited current, once known to every electronic engineer, but, since the virtual demise of the vacuum diode, now of less widespread practical importance.

To determine the sheath thickness we set Γ_i equal to the Bohm value and V equal to V_0, the probe potential (again ignoring the transition region questions), after which some rearrangement allows us to express the equation in the form

$$\frac{x_s}{\lambda_D} = \frac{2}{3}\left[\frac{2}{\exp(-1)}\right]^{1/4}\left[\left(\frac{-eV_0}{T}\right)^{1/2} - \frac{1}{\sqrt{2}}\right]^{1/2}\left[\left(\frac{-eV_0}{T}\right)^{1/2} + \sqrt{2}\right].$$
$$\tag{3.2.25}$$

This equation, incidentally, confirms the previous statement that the sheath is a few Debye lengths thick. The coefficient $\frac{2}{3}[2/\exp(-1)]^{1/4}$ is equal to 1.02. If we consider a probe near the floating potential then it must be sufficiently negative to repel all but a fraction of order $\sqrt{(m_e/m_i)}$ of the electrons. In other words, the Boltzmann factor is $\sim \sqrt{(m_e/m_i)}$, which means that $eV_0/T_e \sim \frac{1}{2}\ln(m_e/m_i)$. For hydrogen this is ~ 3.75 and the sheath thickness at the floating potential is then $\sim 4\lambda_D$.

The dependence of ion-saturation current on voltage is determined by the expansion of the sheath size. For a *spherical probe*,

$$A_s \approx A_p\left(1 + \frac{x_s}{a}\right)^2, \tag{3.2.26}$$

while for a *cylindrical probe*,

$$A_s \approx A_p\left(1 + \frac{x_s}{a}\right), \tag{3.2.27}$$

where x_s is given by Eq. (3.2.25). These can be used in the Bohm formula to give the ion current.

3.2.3   *Exact solutions*

Before proceeding, we should note that it is not necessary in principle to adopt the approximations inherent in the sheath analysis. Allen, Boyd, and Reynolds (1957) proceeded to solve the Poisson equation (3.2.9) numerically to obtain the probe characteristics for a spherical probe ($A = 4\pi r^2$) with cold ions at ∞. Their results confirm that the sheath approximations are reasonably accurate.

If finite ion temperature is considered at infinity the problem is much more complicated and ion orbits in specified geometry must be analyzed. Bohm, Burhop, and Massey (1949) showed that, in spherical geometry, if the ions are monoenergetic then a solution to the orbit problem can be found. The result for ion energies 0.01 and 0.5 times T_e is to obtain coefficients 0.57 and 0.54, respectively, in Eq. (3.2.20) instead of 0.61. The dependence of ion current upon ion temperature is thus extremely weak for all $T_i < T_e$. The reason for this, from the orbit viewpoint, is that the increase in ion radial velocity over $(-2eV_s/T_e)^{1/2}$ at the sheath edge, due to ion thermal velocity, is offset by the energy taken up by angular velocity due to conservation of angular momentum, $rv_\theta = $ constant. In fact, the latter effect is slightly the greater so that ion current actually decreases initially as the ion energy increases from zero.

An essentially rigorous formulation of the orbit problem was provided in the work of Bernstein and Rabinowitz (1959). Unfortunately, even for assumed monoenergetic ions, numerical integration is necessary and no simple formulas for probe interpretation emerge. However, Lam (1965), using this formulation, has performed a complicated boundary layer analysis of the equations in order to obtain approximate solutions for the case of small λ_D/a. His results, presented in graphical form, enable a ready interpretation by hand using the so-called Lam diagram. All these analyses confirm the weak dependence of ion current on T_i for $T_i < T_e$, and therefore the Bohm formula is usually appropriate and widely used.

Exact numerical solutions with Maxwellian ion distribution far from the probe have been calculated by Laframboise (1966), providing the ion and electron currents for arbitrary T_i/T_e and λ_D/a. In Fig. 3.4 we compare the results of our approximate treatment [Eq. (3.2.26)] with some of his results at two values of λ_D/a. The agreement is quite good. Such a comparison shows that our approximate treatment will be perfectly adequate for most situations in which the sheath is reasonably thin, since probe measurements are rarely more accurate than $\sim 10\%$ anyway. When the Debye length is greater than or of the order of the probe radius, our approximations do not work. This regime is dominated by the orbital effects that have to be

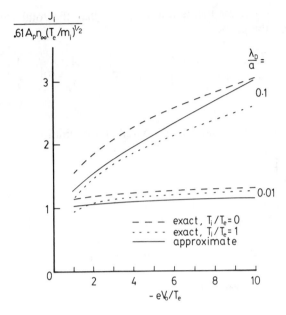

Fig. 3.4. Comparison of the approximate sheath analysis with "exact" numerical results of Laframboise (1966) for a spherical probe.

correctly taken into account. Because no clear saturation occurs and the interpretation is more difficult, this orbit-limited regime is best avoided if possible, particularly since it is rare in laboratory plasmas to achieve a probe configuration that is truly spherical (say) because of supports and leads, and so forth. In some space–plasma applications the orbital limit is unavoidable, in which case the numerical solutions must be employed.

3.2.4 Interpretation of the characteristic

Generally the more useful parts of the probe characteristic are those near the floating potential where the current is small. By operating in this region the possibly large electron-saturation currents, which may adversely affect either the probe or the plasma, are avoided. In the electron-saturation region the current is approximately equal to the random electron current across the probe area. The deviations from this value may be investigated in much the same way as for ions, but we shall not pursue this investigation here.

For our purposes, therefore, the electron current to the probe can be taken as being given by the thermal equilibrium value since most electrons are repelled. This is the random current reduced by the Boltzmann factor

$$\Gamma_e = \tfrac{1}{4} n_\infty \bar{v}_e \exp(eV_0/T_e). \qquad (3.2.28)$$

The ion current is given by the Bohm value so that the total electrical current drawn from the probe is

$$I = n_\infty e A_p \left(\frac{T_e}{m_i} \right)^{1/2} \left[\frac{1}{2} \left(\frac{2m_i}{\pi m_e} \right)^{1/2} \exp\left(\frac{eV_0}{T_e} \right) - \frac{A_s}{A_p} \exp\left(-\frac{1}{2} \right) \right],$$

(3.2.29)

where we have used the fact that for a Maxwellian distribution

$$\bar{v} = 2 \left(\frac{2T}{\pi m} \right)^{1/2}.$$

(3.2.30)

$A_s/A_p \sim 1$, given by Eq. (3.2.26) or (3.2.27).

This equation for the probe characteristic, which is valid for $T_i < T_e$ and $\lambda_D \ll a$, shows negligible dependence upon ion temperature apart from slight changes in the term $\exp(-\frac{1}{2})$, so that the characteristic cannot be used to determine T_i.

The floating potential can immediately be obtained by setting $I = 0$ so that

$$\frac{eV_f}{T_e} = \frac{1}{2} \left[\ln\left(2\pi \frac{m_e}{m_i} \right) - 1 \right]$$

(3.2.31)

(the difference between A_s and A_p is irrelevant here because it appears only inside the logarithm). It might seem easy, therefore, to deduce T_e from the floating potential. Unfortunately, the problem is that we do not, in general, have a good measurement of the plasma potential V_p, which is the reference point of voltage (i.e., V_∞) in our treatment. V_p may be estimated as the voltage at which electron saturation is reached. This then gives an estimate of T_e from $V_f - V_p$, but usually not a very accurate one.

A more satisfactory approach is to use the slope of the characteristic

$$\frac{dI}{dV_0} = \frac{e}{T_e} (I - I_{si}) + \frac{dI_{si}}{dV_0},$$

(3.2.32)

where $I_{si} = -eJ_i$ is the ion-saturation current. (We put subscript 0 on V in this equation to remind us that it is the probe potential, but from now on we drop this, leaving it understood.) Taking a point where dI_{si}/dV (which arises from dA_s/dV) may be ignored compared to dI/dV, one then has

$$T_e = e(I - I_{si}) \bigg/ \frac{dI}{dV}.$$

(3.2.33)

These quantities can be obtained numerically from the characteristic, for example by fitting a line to $\ln|I - I_{si}|$ versus V, or read graphically as illustrated in Fig. 3.5.

Once T_e has been determined in this way, the density may be determined from the ion-saturation current using the primitive form of Bohm's formula (taking $A_s = A_p$). If the Debye length λ_D, based on T_e and n thus deduced, is significant compared to a, then the effective area may be corrected using Eq. (3.2.26) or (3.2.27) with this value of λ_D. This provides a corrected value of n (and hence λ_D). The process may be iterated if desired, but the accuracy of our approximations, or indeed probe measurements generally, rarely warrants more than this first correction.

The density deduced n_∞ is the electron (and ion if singly charged) density at large distances. If the ions are not singly charged but have charge Ze, the sheath potential V_s is unchanged; the ion electric current is increased by $Z^{1/2}$ over the Bohm value obtained using the electron density. The density deduced by using the Bohm formula will then be equal to the electron density provided m_i is replaced by m_i/Z (see Exercise 3.4). This will hold too for a mixture of ions provided m_i/Z is constant; otherwise no simple formula is available.

Finally, returning to the question of the plasma potential, because of the uncertainties involved in determining V_p from the electron saturation "knee" in the characteristic, the most satisfactory way to obtain it is to use the floating potential and the electron temperature determined from the slope [Eq. (3.2.33)]. Then the difference between V_p and V_f is given by Eq. (3.2.31), which we regard not as a means of determining T_e but as a means

Fig. 3.5. Graphical analysis of a probe characteristic.

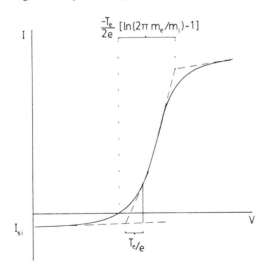

of determining V_p – the reference of potential in our analysis.

3.3   The effects of a magnetic field

3.3.1   *General effects*

All the analysis thus far has tacitly assumed that no magnetic field is present, so that the particle dynamics are determined only by the electric field. Many situations arise where probes are to be used in a magnetized plasma, so we need to know how this magnetic field will affect the results.

The main effect of the field is to cause the electrons and ions to move no longer in straight lines but to orbit around the magnetic field lines in circular orbits of radius $\rho = mv/eB$. This radius is called the Larmor or gyro radius. The particle motion across the magnetic field is thus greatly restricted although the motion along the field is essentially as before.

The importance of the magnetic field effects is obviously determined by the ratio of ρ to the typical dimension a of the probe. If $\rho \gg a$, then the previous treatment should apply.

Clearly the electron Larmor radius is smaller than the ion radius (for comparable T_e and T_i) by the factor $\sqrt{(m_e/m_i)}$. As a result, the electrons are more strongly affected than the ions. The first thing that happens to the probe characteristic when a magnetic field is present is that the electron-saturation current is decreased since the electron flow is impeded. This will be most immediately evident as a reduction in the ratio of electron- to ion-saturation currents.

In many cases, even though the electron current is impeded, because $\rho_e < a$, the ion Larmor radius remains larger than the probe. In such a situation, if the probe is significantly negative so that most electrons are reflected, then the electron density will be governed, as before, by the thermal Boltzmann factor

$$n_e = n_\infty \exp(eV/T_e). \tag{3.3.1}$$

The ions, being relatively unaffected (for $\rho_i \gg a$) by the magnetic field, satisfy the same equations as before. So the ion current, which depends only on n_e and ion dynamics, is just as before. The electron current will also maintain its exponential dependence on V_0 so that analysis of the current slope will again provide T_e. In summary, if $\rho_i \gg a$ and most of the electrons are repelled, the previously discussed interpretation of the probe characteristic should provide accurate results.

When the magnetic field is sufficiently strong that the ion Larmor radius is smaller than the probe size, $\rho_i < a$, considerable modifications to the ion collection can occur and a fully satisfactory theory has not yet been established. The reason for this is that it is no longer possible to formulate a completely collisionless theory.

To understand this we note that for $\rho_i \ll a$ the particle flows, in the absence of collisions, are effectively one dimensional along the magnetic field. Now consider the quasineutral equation that we supposed to hold outside the sheath:

$$\frac{J_i}{A}\left(\frac{m_i}{-2eV}\right)^{1/2} = n_\infty \exp\left(\frac{eV}{T_e}\right). \tag{3.3.2}$$

In a three- or two-dimensional situation (e.g., a sphere or cylinder with no magnetic field), A is a function of position, so that the solution of this equation for V gives a potential varying with position, and in fact tending to zero within a distance of the order of the probe radius (see Exercise 3.6). However, when A is a plane (one-dimensional flow) it is independent of position and $J_i/A = \text{constant}$. The solution of the equation is then $V = \text{constant}$ and no well behaved solution satisfying $V = 0$ at large distances and $V = -T_e/2e$ at the sheath edge is possible. The result is that the quasi neutral "presheath" region expands until plasma-source or collisional terms become significant, allowing a solution to be obtained; hence the comment that no collisionless theory is possible in the one-dimensional strong magnetic field case.

As a practical matter, despite the theoretical difficulties, the usual approach to probe interpretation in a strong magnetic field is to take the electrons to be governed by the Boltzmann factor as before, when V is negative, so that the temperature can be deduced from the slope of the characteristic. The ion-saturation current is estimated by noting that the Bohm formula for the zero field case corresponds to ions flowing to the probe at approximately the ion sound speed $(T_e/m_i)^{1/2}$. In a strong field this sonic flow can occur only along the field not across it. Therefore, it is reasonable to suppose that the Bohm formula would still apply except that the effective collection area is not the total probe surface but the *projection* of the surface in the direction of the magnetic field. So an estimate of the plasma density may be obtained from Eq. (3.2.20) using this projected area in place of A_s. A justification for this procedure and a discussion of some of the complications is given in the following sections.

3.3.2 *Quasicollisionless ion collection in a strong magnetic field*

The general characteristics of the situation of strong magnetic field are illustrated in Fig. 3.6. The sheath is still thin and, hence, located close to the probe. The presheath, however, is a long flux tube extending in the direction of B. When the probe is attracting ions, the ion flux into the sheath is made up almost entirely of ions that have diffused into the collection tube across the field rather than entering from the end, since, for

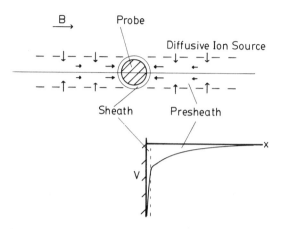

Fig. 3.6. Schematic representation of sheath and presheath in a strong magnetic field.

small T_i, the inflow of ions at the end is very small. It may be, in some situations, that sources of ions due to ionization within the presheath are also important, but these may be treated in much the same way as ions entering the flux tube via diffusion.

If ions, once having entered the presheath, have a high probability of reaching the probe without making a collision, then a quasicollisionless treatment of the collection can be used. ("Quasi" here reminds us that some form of collisions is essential to provide the ion source within the presheath via diffusion or ionization.) Such a treatment requires the mean free path along the field to be longer than the presheath, a criterion that we shall discuss later; let us assume, for now, that the criterion is satisfied.

The model we adopt, therefore, is that the presheath can be taken as one dimensional, ignoring variation perpendicular to the field, and has a source of ions within it, due to cross-field diffusion or ionization. To simplify the treatment we take the ions to be born with zero energy so the source rate is described by a single function $S(x)$, particles born per unit volume per unit time.

Ions born at x_1 are accelerated toward the probe and at a position x ($< x_1$) have acquired a velocity

$$v = -\left[\frac{2e\{V(x_1) - V(x)\}}{m_i}\right]^{1/2}. \tag{3.3.3}$$

Assuming the potential distribution in the sheath to be monotonic so that there is a one-to-one relationship between velocity (at x) and birth position x_1, the ion distribution function $f(v, x)$ is given by equating the number of ions passing x per second to their source rate:

$$f v \, dv = S(x_1) \, dx_1,$$
(3.3.4)

where x_1 and v are related via Eq. (3.3.3).

The ion density is then

$$n_i(x) = \int f \, dv = \int \frac{S_1 \, dx_1}{v} = \int_v^0 \left(\frac{m_i}{2e} \right)^{1/2} S_1 \frac{dx_1}{dV_1} \frac{dV_1}{(V_1 - V)^{1/2}}.$$
(3.3.5)

The electrons, being the repelled species for the case of most interest, are governed by a simple Boltzmann factor as before:

$$n_e = n_\infty \exp(eV/T_e).$$
(3.3.6)

These are the two particle densities to be substituted into Poisson's equation.

As before, we treat the plasma region (the presheath) as quasineutral $n_i = n_e$, with the result that we obtain an integral equation that may be written

$$F(\eta) = \int_0^\eta \Phi(\eta_1) \frac{d\eta_1}{(\eta - \eta_1)^{1/2}},$$
(3.3.7)

where

$$\eta \equiv -eV/T_e,$$
(3.3.8)

$$F(\eta) = \exp(-\eta),$$
(3.3.9)

$$\Phi(\eta) \equiv -\left(\frac{m_i}{2T_e} \right)^{1/2} \frac{S}{n_\infty} \frac{dx}{d\eta},$$
(3.3.10)

and we regard the potential, in its dimensionless form η, as the independent variable and x, implicitly in Φ, as the dependent variable.

Equation (3.3.7) is a form of Abel's integral equation, which we shall encounter again in a different context (Section 4.4). Its solution is

$$\Phi(\eta) = \frac{1}{\pi} \left[\frac{F(0)}{\eta^{1/2}} + \int_0^\eta F'(\eta_1) \frac{d\eta_1}{(\eta - \eta_1)^{1/2}} \right],$$
(3.3.11)

where the prime denotes differentiation with respect to argument. The reader may care to verify this solution, as Exercise 4.7 indicates. Substituting for F we obtain

$$\Phi(\eta) = \frac{1}{\pi} \left[\frac{1}{\eta^{1/2}} - 2\exp(-\eta) \int_0^{\sqrt{\eta}} \exp(t^2) \, dt \right].$$
(3.3.12)

Now we shall need the ion current density, which is

$$\Gamma_i = \int f v \, dv = \int S_1 \, dx_1 = \int_\eta^0 S_1 \frac{dx_1}{d\eta_1} \, d\eta_1$$

$$= \int_0^\eta n_\infty \left(\frac{2T_e}{m_i} \right)^{1/2} \Phi(\eta_1) \, d\eta_1. \tag{3.3.13}$$

The value of Φ is substituted from Eq. (3.3.12) and integration by parts then gives

$$\Gamma_i = n_\infty \left(\frac{2T_e}{m_i} \right)^{1/2} \frac{2}{\pi} \exp(-\eta) \int_0^{\sqrt{\eta}} \exp(t^2) \, dt. \tag{3.3.14}$$

We must now determine the position (or rather the potential) of the sheath edge. We proceed, as before, to identify this as the place where the plasma solution breaks down. This occurs at or before the point where $d\eta/dx$ becomes infinite, that is, where $\Phi = 0$. So, on this basis, we should take the sheath potential to be the solution of

$$\frac{1}{\eta_s^{1/2}} - 2 \exp(-\eta_s) \int_s^{\sqrt{\eta_s}} \exp(t^2) \, dt = 0. \tag{3.3.15}$$

This proves to be $\eta_s = 0.854$ ($V_s = -0.854 T_e/e$), a value originally obtained by Tonks and Langmuir (1929) for a specific choice of S. Notice, though, that this value is in fact completely independent of any assumptions about the spatial variation of the source S, a fact first demonstrated by Harrison and Thomson (1959) (whose treatment we are loosely following). Indeed, not only is the sheath potential independent of S but so also is the ion current at the sheath:

$$\Gamma_i(\eta_s) = n_\infty \left(\frac{2T_e}{m_i} \right)^{1/2} \frac{1}{\pi \eta_s^{1/2}} = 0.49 n_\infty \left(\frac{T_e}{m_i} \right)^{1/2}. \tag{3.3.16}$$

This expression is just like what we had for collisionless ion collection without a magnetic field, the Bohm current [Eq. (3.2.20)], except that the coefficient is 0.49 instead of 0.61.

It should be noted that we have not strictly demonstrated that the sheath forms exactly at $\Phi = 0$. It might form somewhat before that, although not by much, as an analysis of the sheath criterion along our earlier lines [Eqs. (3.4.12)–(3.4.16)] would show (see Harrison and Thomson 1959). However, because $\Gamma_i \propto \int \Phi \, d\eta$, the derivative $d\Gamma_i/d\eta$ is equal to zero when $\Phi = 0$, that is, at the potential $\eta = \eta_s$. Thus, the difference in ion current due to some small correction in the sheath potential will be zero to lowest order.

Because the sheath itself is thin, we may ignore the source of ions within it, so the ion current to the probe is equal to the ion current across the sheath: $0.49n_\infty(T_e/m_i)^{1/2}$. This gives us all the information we require in order to determine the probe characteristic because the electron current is given, as before, by a simple Boltzmann factor. Therefore, we obtain a characteristic given by just the same expression as in the field-free case with two small modifications. First, the area to be used in determining the ion current is the *projection* of the probe surface in the direction of the magnetic field, for example, $2\pi a^2$ instead of $4\pi a^2$ for a sphere. Second, the coefficient in the Bohm current is modified. This modification is so small, though, and the approximations involved so substantial, that the difference is hardly significant. Very often the coefficient is just taken as $\frac{1}{2}$, regardless of magnetic field strength, recognizing that the analysis is only good to perhaps 10–20%.

3.3.3 *Collisions in a magnetic field*

We must now consider in somewhat more detail the question of whether the quasicollisionless treatment is valid. This involves the relative magnitude of the ion mean free path along the magnetic field ℓ and the length of the collection (presheath) region L, say. If $\ell > L$, then the treatment is justified; otherwise not.

It should be noted here that the distinction between ion–ion and ion–electron collisions needs to be borne in mind. For the purposes of the present discussion, only the ion–electron collisions are really important. Ion–ion collisions, though they lead to modification of the ion distribution, do not change the total ion momentum. They therefore give rise neither to cross-field diffusion nor parallel momentum loss, which are the crucial factors determining the ion collection. Therefore, all the collision lengths and rates in this section should be regarded as referring to ion–electron momentum transfer, which takes place more slowly by a factor $\sim \sqrt{(m_e/m_i)}$ than ion–ion transfer.

The length of the collection region must be just great enough to allow sufficient ion sources within it to make up the ion flow to the probe. If we were to assume some form for $S(x)$ we could, in principle, integrate Eq. (3.3.10) to determine L. However, the mathematical labor would not be justified since we mostly want an approximate result for present purposes. Therefore, we proceed on a very rough basis as follows.

In our case the source of interest is perpendicular diffusion of ions across the field. Suppose this is governed by a diffusion coefficient $D_\perp$; then the cross-field particle flux is $D_\perp \nabla_\perp n$, which we approximate by taking the perpendicular scale length as a, the probe radius, and the difference in density between the presheath and the surrounding plasma as $\sim n_\infty/4$ on average. Then the cross-field flux along the whole tube length for a circular tube is

$$J_i \sim L2\pi a D_\perp n_\infty / 4a. \tag{3.3.17}$$

We put this equal to the Bohm value of the ion flux to the probe,

$$J_i \approx \tfrac{1}{2} n_\infty (T_e/m_i)^{1/2} \pi a^2, \tag{3.3.18}$$

and deduce

$$L \approx \frac{a^2}{D_\perp} \left(\frac{T_e}{m_i} \right)^{1/2}. \tag{3.3.19}$$

On the other hand, if ions have a parallel diffusion coefficient $D_\parallel$, then the mean free path along the field lines is

$$\ell \approx \frac{D_\parallel}{v_{ti}}. \tag{3.3.20}$$

Hence, the ratio of ℓ to L is

$$\frac{\ell}{L} \approx \frac{D_\perp D_\parallel}{a^2 c_s v_{ti}}, \tag{3.3.21}$$

where we have written $c_s \equiv (T_e/m_i)^{1/2}$, the ion sound speed. This shows that, provided diffusion is strong enough, a quasicollisionless approach is valid.

However, one can readily show that classical collisional diffusion is *never* strong enough. Classical diffusion gives

$$D_\perp \approx \rho_i^2 \nu_c, \tag{3.3.22}$$

$$D_\parallel \approx v_{ti}^2 / \nu_c, \tag{3.3.23}$$

where ν_c is the collision frequency and ρ_i is the ion Larmor radius. Thus classically

$$\frac{\ell}{L} \approx \left(\frac{\rho_i}{a} \right)^2 \frac{v_{ti}}{c_s}. \tag{3.3.24}$$

But the ion thermal speed is always less than the sound speed (remember that c_s really includes $T_e + \gamma T_i$), and by presumption we are discussing the strong magnetic field case $\rho_i < a$. Therefore, $\ell/L < 1$ and collisions *are* important.

It might seem, therefore, that all our treatment of the quasicollisionless case of the previous section is in vain. Fortunately, this is not so because in most plasmas the cross-field particle diffusion is substantially enhanced above the classical value by a variety of collective effects. This so-called anomaly in the transport is frequently sufficient to make ℓ/L large and thus the quasicollisionless treatment appropriate, in which case our analysis using the Bohm value of the ion current may be expected to give reliable results.

Even so, there are undoubtedly cases where the diffusion is sufficiently slow as to have an appreciable effect upon the current and it is to these cases that we now turn. We shall consider the plasma to be sufficiently well described by a continuum approach as in Section 3.1.3, but now we take the diffusion coefficient to be anisotropic. We note also that particle fluxes will be driven by potential gradients as well as density gradients, so we write

$$\Gamma = -\mathbf{D} \cdot \nabla n - n\boldsymbol{\mu} \cdot \nabla V. \tag{3.3.25}$$

The tensor $\boldsymbol{\mu}$ is the mobility and is anisotropic like the diffusivity $\mathbf{D}$.

In a simple isotropic collisional treatment, provided the drift velocity is much less than the thermal velocity, the mobility and diffusivity are related by (Chapman and Cowling 1970)

$$\mu \approx \frac{q}{T} D, \tag{3.3.26}$$

where q is the species charge. It is far from obvious that this expression remains true in an anisotropic case, but we shall assume that it does, so that

$$\Gamma = -\mathbf{D} \cdot \left(\nabla n + \frac{nq}{T} \nabla V \right). \tag{3.3.27}$$

Primarily, we want to calculate the ion current, so we apply this diffusive approach to the ions and assume (as usual) that the electrons are governed by a simple Boltzmann factor. This is reasonable when they are the repelled species, even if collisions are important, since the electron diffusion tends to be much quicker than that of the ions. Also, we treat the outer plasma region, which is quasineutral; therefore

$$\nabla n_i = \nabla n_e = n_\infty \nabla \exp\left(\frac{eV}{T_e} \right) = \frac{e}{T_e} n_i \nabla V. \tag{3.3.28}$$

Substituting and taking $q = e$, we get

$$\Gamma = -\mathbf{D} \cdot \left(1 + \frac{T_e}{T_i} \right) \nabla n_i. \tag{3.3.29}$$

In order to obtain a tractable problem let us take $\mathbf{D} = $ constant. Then the equation of continuity becomes

$$\nabla \cdot \Gamma = 0 = -\nabla \cdot (\mathbf{D} \cdot \nabla n_i)\left(1 + \frac{T_e}{T_i} \right)$$

$$= \left(1 + \frac{T_e}{T_i} \right)\left\{ D_\parallel \frac{d^2 n_i}{dx^2} + D_\perp \left(\frac{d^2 n_i}{dy^2} + \frac{d^2 n_i}{dz^2} \right) \right\}, \tag{3.3.30}$$

where x is the parallel direction. A simple scaling of the parallel coordinate putting

$$\xi = \left(D_\perp / D_\parallel \right)^{1/2} x \qquad (3.3.31)$$

leads to Laplace's equation in the (ξ, y, z) coordinate system. It may be solved for specified geometry to give a relationship between n_i and the total current J_i (Bohm, et al. 1949). In view of the rather gross assumption involved in taking $\mathbf{D} = \text{constant}$, it is sufficient to obtain this relationship by a more physically transparent approximation.

We recognize that the perpendicular scale length of the solution is of order the transverse dimension of the probe (a), therefore, we replace $\nabla_\perp^2 n_i$ by $(n_\infty - n_i)/a^2$, to get

$$\frac{d^2 n_i}{dx^2} + \frac{D_\perp}{D_\parallel} \frac{(n_\infty - n_i)}{a^2} = 0. \qquad (3.3.32)$$

This gives a solution

$$n_i - n_\infty = (n_0 - n_\infty) \exp\left[-\left(\frac{D_\perp}{a^2 D_\parallel} \right)^{1/2} x \right], \qquad (3.3.33)$$

where n_0 is the density at the inner surface $(x = 0)$ of the diffusion region. The total ion current across this inner surface is then

$$J_i \approx \pi a^2 D_\parallel \left(1 + \frac{T_e}{T_i} \right) \frac{dn_i}{dx}\bigg|_0 = \pi a \left(D_\parallel D_\perp \right)^{1/2} \left(1 + \frac{T_e}{T_i} \right) (n_\infty - n_0). \qquad (3.3.34)$$

We must now equate this ion current to the current obtained by a local analysis of the sheath region, based on the local density n_0. If we write this as

$$J_i = n_0 v_0 \pi a^2, \qquad (3.3.35)$$

so that v_0 represents an effective flow velocity, then by combining Eqs. (3.3.34) and (3.3.35) to eliminate n_0, we find

$$J_i = n_\infty v_0 \pi a^2 \left(\frac{R}{1 + R} \right), \qquad (3.3.36)$$

where the factor

$$R = \left(\frac{D_\parallel D_\perp}{a^2 v_0^2} \right)^{1/2} \left(1 + \frac{T_e}{T_i} \right) \qquad (3.3.37)$$

is a reduction factor determining the magnitude of ion current reduction from its expected value if diffusion were ignored and n_∞ used in Eq.

(3.3.35). Normally for ion collection v_0 will be given by the Bohm value $\frac{1}{2}c_s$. Notice then that apart from the correction factor due to mobility $(1 + T_e/T_i)$ and the distinction betwen c_s and v_{ti}, the factor R is equal to $2(\ell/L)^{1/2}$ from Eq. (3.3.21).

Alternatively, we can regard R as representing, as in Section 3.1.3, approximately the ratio of a mean free path ℓ' to probe size. In the strong field case ℓ' represents the geometric mean of the transverse and longitudinal diffusion lengths, that is,

$$ R \approx \frac{4\ell'}{3a}, \qquad \ell' \equiv \frac{3\left(D_{\parallel}D_{\perp}\right)^{1/2}}{c_s}. \tag{3.3.38} $$

(In this final form we have taken notice of the fact that the mobility enhancement factor $1 + T_e/T_i$ will not be appropriate if $T_e \gg T_i$ because the drift velocity will be too large. Therefore, we have just put $T_e = T_i$ in this factor.)

3.3.4 *Discussion of probe theories in magnetic fields*

It should be obvious from the previous sections that probe theory in magnetic fields is extremely complicated even when rather gross assumptions are adopted in the analysis. There are a number of additional complications that we shall not treat mathematically but just mention as a caution to the reader. First, one may criticize our quasicollisionless approach as being an inadequate model in allowing ions to be born only at zero energy. Unfortunately, this does not model even a low ion-temperature result very well because it does not allow for the possible loss of ions out of the presheath (by diffusion) once they have been accelerated in the collection region but before reaching the probe. Such a loss would correspond to a negative source at nonzero velocity, excluded in the model. A more elaborate analysis of one-dimensional presheaths by Emmert et al. (1980), allowing for a Maxwellian ion birth velocity distribution, has been used [by Stangeby (1982)] to extend the approach we have taken to finite ion temperature, but suffers from the same criticism when applied to probe analysis. What is more, for the finite ion-temperature case, clearly unphysical results are obtained if $T_i \gg T_e$. One may anticipate a priori then that $\Gamma_i = \frac{1}{4}n\bar{v}_i$, since the ions will be unperturbed by potentials $\sim T_e/e$. However, the result actually derived for Maxwellian birth ions is $\Gamma_i = \frac{1}{2}n\bar{v}_i$. Thus the analysis of Emmert et al. is not directly applicable to modelling probe current in a magnetic field; it gives an unphysical result for $T_i \gg T_e$. Recent analysis (Hutchinson 1987), using fluid approximations, indicates that a consistent treatment of diffusive exchange of ions between the presheath and the outer plasma gives essentially the Bohm result [Eq. (3.3.16)] for $T_i < T_e$.

Another phenomenon that appears (theoretically) when the reduction factor R is important for electrons (i.e., when electron collection is diffusive) is that a potential "hill" forms in front of the probe [see, e.g., Cohen (1978)]. This hill may substantially affect the ion collection since ions are repelled by the hill. This problem is not fully solved.

In view of these and other difficulties that arise because of the inherently nonlinear mathematical nature of probe analysis and that make probe theory a still active area of research, one may ask whether one can have much confidence in the accuracy of the results we have obtained. In other words: How robust are the results? A result that *is* very robust, reemerging in various cases with or without magnetic field, is the Bohm formula for ion current when $T_i \le T_e$. There is a good physical reason for this. In essence, the Bohm result is that the sheath edge forms when the ion drift velocity equals the sound speed. From the viewpoint of nonlinear plasma theory, one may regard this drift velocity as that necessary for a "shock" to form. In this sense the sheath *is* a shock, and the wide applicability of the Bohm criterion is then seen to depend upon the underlying physics of this shock formation.

A fluid derivation gives the ion sound speed as generally $[(T_e + \gamma T_i)/m_i]^{1/2}$, where γ is the ratio of specific heats for the ions. A common approach to including finite ion temperature is to replace T_e by $T_e + T_i$. However, even on a simple fluid picture, it is not clear that $\gamma = 1$ is the correct value to use. Besides, ion sound waves are normally strongly damped unless $T_i \ll T_e$. So it seems that a fluid approach leaves somewhat open the exact coefficient of T_i to use. As a reasonable resolution of this problem, recognizing the inherently approximate nature of the approach, it seems appropriate to use a coefficient that will correctly give the gas-kinetic value $\frac{1}{4}n\bar{v}_i$ at high ion temperatures $T_i \gg T_e$. In the absence of a magnetic field this requires us to use a coefficient $e/2\pi = 0.43$ for T_i in the Bohm formula. Presumably a comparable value $\sim \frac{1}{2}$ is appropriate also for the strong magnetic field case $\rho_i \ll a$. Therefore, a reasonable generalization of the Bohm formula to finite ion temperature is to replace T_e by $T_e + \frac{1}{2}T_i$. Notice, though, that a Langmuir probe does not provide a measurement of T_i, so other information must be used to provide it.

3.4   Applications
3.4.1   *Some practical considerations*

A typical Langmuir probe construction is shown in Fig. 3.7. The tip of the probe is usually subject to the greatest energy flux and is usually made of some appropriate refractory metal (platinum, molybdenum, etc.). The end of the insulating jacket also needs to be robust. Alumina and fused silica are materials that have been widely used.

Various complicating factors influence the design and operation of the probe. These have been reviewed by Chen (1965) and include: surface

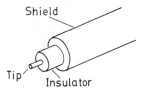

Fig. 3.7. A typical Langmuir probe.

layers either insulating on the probe tip or conducting on the insulator, in either case changing the effective probe area; secondary emission of electrons from the probe, possibly leading to arcing; photoemission of electrons; negative ions.

A major factor in many cases is to minimize the heat flux to the probe in order to avoid damage. Because the electron-saturation current is so much larger than the ion, a probe biased to draw electrons receives a much higher heat load. Thus, it is often advantageous to avoid the electron saturation part of the characteristic, especially since it provides little extra information beyond what can be deduced from the characteristic in the ion-saturation and floating potential region.

Of course, the electron current can be minimized by simply choosing appropriate probe voltage biasing. However, another method that ensures that one never collects more than ion-saturation current is to use a floating double probe. Figure 3.8 shows schematically the operation of such a probe. The two probe tips, although their potential *difference* is set by the circuit, float in *mean* potential and rapidly adopt a potential such that their total

Fig. 3.8. A circuit for operating a floating double probe.

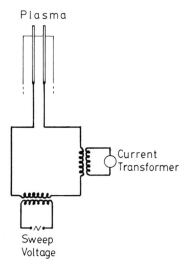

current is zero. Since their difference current flows out of one and into the other, its magnitude can never exceed the ion-saturation current.

The characteristic of the difference current versus the difference voltage may be calculated by using our previous results for the current to each probe separately [Eq. (3.2.29)] and eliminating the floating mean potential by using the knowledge that the total current is zero. The result for the usual case of identical probes is

$$I = I_{si}\tanh(eV/2T_e) \tag{3.4.1}$$

(see Exercise 3.5), where the ion-saturation current I_{si} is just as before. Analysis of the characteristic similar to that for the single-probe case provides the measurement of T_e and n_∞.

The use of Langmuir probes from spacecraft naturally enforces a double-probe configuration upon the experiment because there is no independent "Earth"! Normally, though, the spacecraft itself will be one probe and the other will be much smaller. Provided the ion-saturation current to the spacecraft is greater than the electron-saturation current to the probe, the result will be an effective single-probe measurement. In a deeper sense even a lab experiment is double, the second electrode being, for example, the plasma chamber. One normally does not need to take account of this because the chamber surface is so much bigger than the probe.

Often, Langmuir probes are used to measure the *fluctuations* in the local plasma parameters, not merely their average value. If the fluctuations are sufficiently rapid, then the technique of sweeping the probe voltage, normally used for obtaining the mean characteristic, cannot be used. Instead the probe is either left floating and measurements made of floating potential fluctuations or else the probe is biased at constant voltage (often such that saturation ion current is drawn) and current fluctuations measured. In the case of ion-saturation current, the signal is a function of both density and electron temperature although the density dependence ($I_{si} \propto n$) is stronger than that of the temperature ($I_{si} \propto T_e^{1/2}$). When floating potential fluctuations are measured, the signal depends strongly on plasma potential. Unfortunately the dependence is not direct. However, if temperature fluctuations are relatively small, there is a constant difference between V_p and V_f, so the floating potential fluctuations are equal to the plasma potential fluctuations.

The difficulties mentioned in Section 3.2.4 in determining the plasma potential from the Langmuir probe characteristic may be overcome in large measure by the use of an *emissive probe*. Such a probe is continually emitting electrons, usually thermionically from a heated surface. The idea behind the use of the emissive probe is this: If the probe potential is positive with respect to the plasma, then electrons emitted with low energy cannot escape from the probe but are simply attracted back to it. The probe current is therefore unchanged by the electron emission. On the other hand, if the probe is negative with respect to the plasma, the electrons can escape and so the probe current is decreased compared to what it would have been without the electron emission. Thus, if we can obtain probe characteristics with the same probe hot (i.e., emitting electrons) and cold (not emitting), the characteristics will differ for $V < V_p$ but not for $V > V_p$. This should enable us to identify the plasma potential, V_p.

Since obtaining hot and cold characteristics is rather troublesome and requires a reproducible plasma, it would be helpful if a more direct way of obtaining the plasma potential were available. One idea is to use a floating emissive probe. The notion is that if the electron emission exceeds the electron-saturation current and also takes place with an effective electron temperature much less than the plasma electron temperature, then the probe will tend to float at a potential close to the plasma potential, just negative enough (by a fraction of the emission temperature) to reduce the emission current to that of electron saturation.

Unfortunately, this simple picture does not describe what actually occurs. In fact, a potential minimum usually forms between the probe surface and the plasma. This tends to reflect a fraction of the emitted electrons back to the probe, even when the probe is still negative with respect to the plasma potential. This double layer thus prevents the emission probe from floating at the plasma potential except in very special circumstances over which the experimenter, in general, has no control. Therefore, as discussed briefly by Chen (1965), the floating emissive probe does not usually float at the plasma potential.

The amount of electron emission required to use an emissive probe, even in the hot/cold mode, to give a reasonable indication of V_p is obviously of the order of magnitude of the electron-saturation current; otherwise the difference between hot and cold would be undetectable. This tends to limit the use of emissive probes to plasmas of low density and temperature since there are limits to the emission current density possible. These limitations and the cumbersome procedure required for the hot/cold comparison make the use of emissive probes unattractive except in rather special circumstances.

3.4.2 *More sophisticated analyzers*

Because the current flow to any probe is intimately connected with its potential relative to the plasma, our Langmuir probe analysis is important even if the electric current is not the primary quantity measured. There are, however, various techniques that rely on other quantities related to the plasma particle flux for diagnostic purposes.

As we have seen, for most cases when $T_i \lesssim T_e$, the Langmuir probe current characteristic is insensitive to ion temperature. Here then, is an area where a more sophisticated analyzer can be useful. The reason why we cannot easily obtain information on the ion distribution function from a

Langmuir probe is that when the probe is at positive potential, repelling ions, it is drawing electron-saturation current. This is usually large enough to swamp completely any variations in the ion current that might have told us about T_i. A solution to this problem is to use a "gridded energy analyzer," in which a system of grids at different potentials is used. In Fig. 3.9 we illustrate schematically the type of approach to be used.

The plasma particles are allowed to approach the current collector only after passing first through a grid. This grid is biased strongly negative so as to repel essentially all electrons. Therefore, only the ions are able to penetrate past to the collector. Now, when the collector voltage is varied, only ions with energy greater than eV_c will be collected, so that the logarithmic slope of the collection current for $V_c > 0$ should give the ion temperature.

The main shortcoming of the single-grid analyzer illustrated in Fig. 3.9 arises because of secondary electron emission caused by ions or electrons striking the surfaces of the electrodes. Although the electron repeller may prevent all the plasma electrons from passing through to the collector, the ions all pass the grid (some of them more than once!). In doing so they may collide with it and liberate secondary electrons. The secondary electron coefficient (i.e., the number of electrons liberated per colliding ion) may be

Fig. 3.9. Simplified illustration of the operation of a gridded energy analyzer.

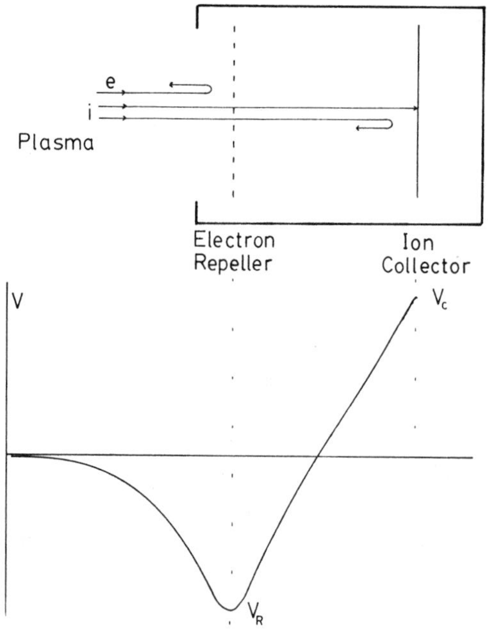

typically a few percent. Therefore, when the collector is repelling all but a few percent of ions, the electron current due to secondaries from the grid will become dominant and obscure the desired signal.

This problem may be overcome by using a more elaborate series of grids such as is illustrated in Fig. 3.10. In this case the first grid repels all the plasma electrons; the second is the ion repeller whose potential is varied in order to obtain the characteristic; the third grid is the electron suppressor, which prevents secondaries from either of the first two grids from reaching the collector. The collector is negative (but not as much as the suppressor) ensuring good ion collection. Other configurations are also possible; for example, sometimes the ion repeller is made the first grid, but the configuration shown overcomes most of the difficulties with secondary electrons. It also has the important merit that the electron and ion repellers are separate from the collector, so that any spurious currents carried by them do not obscure the required collection current.

Another difficulty with gridded analyzers, particularly in higher-density plasmas, arises because of space-charge limitations. In Fig. 3.10 we plotted the potential due just to the grids. However, when plasma flows into the analyzer and the positive or negative charge species are selectively removed from the flow, additional charge density exists between the grids, which can change the potential. This change will affect the operation of the analyzer if the potential between two grids is altered so much that it is larger than the

Fig. 3.10. Potential plot of a practical gridded energy analyzer including secondary electron suppression.

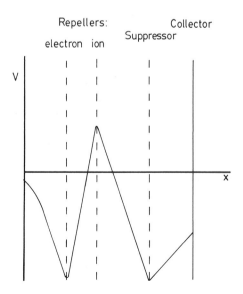

greater or smaller than the lesser of the two grid potentials. If this happens, then either ions or electrons will encounter a repulsive potential hill higher than that given by the grids. The result will be an unexpected lowering of the current of that species.

The space-charge effects may be calculated using just the same approach as in Section 3.2.2, where we discussed the sheath thickness. The worst case occurs between the electron and ion repellers when the bulk of the ions have just enough energy to reach the ion repeller. The marginal case, when the potential has zero slope at the ion repeller so that higher density would lead to space-charge limitation, then corresponds approximately to the sheath solution of Section 3.2.2 with $V_s = 0$. In other words, the relationship between the grid spacing x and the potential difference between the grids V is [see Eq. (3.2.25)]

$$\frac{x}{\lambda_D} = \frac{2}{3}\left[\frac{2}{\exp(-1)}\right]^{1/4}\left(\frac{eV}{T_e}\right)^{3/4} = 1.02\left(\frac{eV}{T_e}\right)^{3/4}. \qquad (3.4.2)$$

The Debye length λ_D to be used here is that corresponding to the plasma density outside the electron repeller grid, but reduced by the attenuation factor of that grid for ions (due to collisions with the mesh). As before, in order to repel all electrons, V is generally a few times T_e and $x \sim 4\lambda_D$.

What we have shown, therefore, is that to avoid space-charge limitation the distance between the electron and ion repellers must be less than $\sim 4\lambda_D$ in this one-dimensional case. For high-density plasmas in which the Debye length is short, this condition becomes too difficult to satisfy. (For example, grids closer than perhaps 0.5 mm would be very hard to maintain.) Two main approaches to solving this problem are available. The first is to attenuate the plasma flux before it encounters the repellers using, for example, meshes with low transparency. This has the effect of increasing the relevant λ_D (because the density is decreased). The second is to restrict the transverse dimensions of the plasma flow using an entrance slit. This has the effect of transforming the Poisson equation into a two- or three-dimensional problem, rather than the one dimension we implicitly assumed. Oversimplifying slightly, one may say that space-charge limitations will then be avoided if the *transverse* dimension (rather than the longitudinal dimension x) of the flow is less than a few Debye lengths.

The problem with both these solutions is that they tend to introduce greater uncertainties in the total effective collection area. The result is that the absolute plasma density tends to be more uncertain. Nevertheless, the relative energy distribution is usually not affected much.

Finally, let us mention a difficulty that may arise if the plasma presheath region is not collisionless, for example, in a strong magnetic field. Then the starting potential of the ions may not be exactly zero, but may be some fraction of the sheath potential ($\sim T_e/2e$). This will tend to cause additional broadening of the observed ion distribution and obscure the ion temperature if $T_i \lesssim T_e$.

Implicit in the discussion of Fig. 3.10 is the assumption that the particles are flowing mostly in the direction along the axis of the analyzer; otherwise, of course, they would tend to hit the sides of the analyzer before being collected or repelled. In a strong magnetic field this collimation is provided by aligning the axis along the field. The analysis is then of the parallel component of the particle velocity and using a gridded analyzer the parallel distribution function may, in principle, be deduced either for the ions or the electrons. Note that a Langmuir probe characteristic may also be used for the same purpose for the electron distribution (but not for the ions) by analysis of the shape in the intermediate region. This will give information on the shape of the distribution function if it is not Maxwellian. For probe or gridded analyzer the analysis is straightforward, as follows.

The current collected (for a single species) per unit area is

$$\Gamma(V_0) = \int_{-\infty}^{0} f(v)v\,dv, \tag{3.4.3}$$

where quantities are those at the probe (or repeller) potential V_0 and f is the parallel distribution function. Using Eqs. (3.2.1) and (3.2.2) this may be written

$$\Gamma(V_0) = \int_{-\infty}^{-\sqrt{|2eV_0/m|}} f_\infty(v_\infty)v_\infty\,dv_\infty. \tag{3.4.4}$$

Differentiating this we get

$$\frac{d\Gamma}{dV_0} = \left(\frac{2e}{m}\right)^{1/2}\frac{1}{2|V_0|^{1/2}}f_\infty(v_\infty)v_\infty = \frac{e}{m}f_\infty(v_\infty), \tag{3.4.5}$$

where $v_\infty = -\sqrt{|2eV_0/m|}$. Thus the distribution function is proportional to the derivative of the collection current for a single species.

When the magnetic field is absent or small so that $\rho \gg a$, it does not provide the collimating effect we require. In this case, collimation must be achieved for a gridded analyzer by some kind of material apertures. A particularly effective method is to use a microchannel plate consisting of a honeycomb of small channels all pointing in the same direction. This method has been used to great effect by Stenzel et al. (1983) in obtaining the distribution functions of electrons and ions in essentially arbitrary directions in a large low-field plasma with non-Maxwellian distributions.

In an unmagnetized plasma whose distribution function is known to be isotropic, the characteristic of a spherical Langmuir probe (without collimation) may be analyzed to give the electron distribution function. The method, due originally to Druyvesteyn (1930), involves taking the second derivative of the characteristic and so tends to be rather sensitive to errors. It is described by Schott (1968).

These and related techniques are widely used in space plasmas to provide extremely detailed measurements of the distribution functions from spacecraft. In lab plasmas it is much more unusual to have such detail because plasmas are rarely large enough, cold enough, or unmagnetized enough for nonperturbing gridded analyzer experiments to be possible.

Another approach to obtaining additional information from a probe is to use a *bolometric probe*. This measures the total power deposition on the probe, usually by sensing the probe temperature rise. Usually the electric current is simultaneously monitored in the manner of a Langmuir probe. The additional information obtained on the particle energy may (in rare circumstances) be sufficient to give an estimate of ion temperature, although the ion energy gained in the sheath is usually substantially greater than the thermal energy unless $T_i \gg T_e$. (Remember that near the floating potential the probe has a negative potential of order a few T_e/e.) More often such probes are used as a direct measure of heat flux to edge structures by replacing a representative part of the structure with a bolometric probe.

Finally, mention should be made of *trapping* probes. They are used to analyze the species impinging on the probe by trapping them in some otherwise virgin probe surface by burial in the material. Subsequent analysis of the probe surface can tell, for example, the different impurity species in the plasma and something about their energy (and hence possibly charge state because of acceleration in a known sheath potential) from measurements of how deeply they are buried in the surface.

Many and varied are the methods used to glean extra information from the measurement of particle flux inside the plasma using sophisticated analyzers. For the most part, the techniques employed are variations on the themes we have discussed here and so we leave the interested reader to pursue further studies in the extensive literature.

Further reading

The theory and practice of electric probe diagnostics is reviewed by

Chen, F. F. (1965). In *Plasma Diagnostic Techniques*. R. H. Huddlestone and S. L. Leonard, eds. New York: Academic.
Schott, L. and Boyd, R. L. F. (1968). In *Plasma Diagnostics*. W. Lochte-Holtgreven, ed. Amsterdam: North-Holland.

Probe techniques also have devoted to them two monographs:

Swift, J. D. and Schwar, M. J. R. (1971). *Electric Probes for Plasma Diagnostics*. London: Iliffe.

Chung, P. M., Talbot, L., and Touryan, K. J. (1975). *Electric Probes in Stationary and Flowing Plasmas*. New York: Springer.

The latter monograph has considerable discussion of the orbit-limited case whose treatment we have omitted.

Examples of modern discussions in the context of magnetic fusion are:

McCracken, G. M. (1982). In *Diagnostics for Fusion Reactor Conditions*, Proc. Int. School of Plasma Physics. P. E. Stott et al., eds. Vol. II, p. 419. Brussels: Commission of E.E.C.
Stangeby, P. C. (1982). *J. Phys. D: Appl. Phys.* 15:1007.

Exercises

3.1 Prove Eq. (3.1.7) for the current density drawn to a perfectly absorbing spherical probe from the equation $\nabla^2 n = 0$ in the diffusion region surrounding the probe and the boundary condition

$$\Gamma = D|\nabla n| = \tfrac{1}{4} n \bar{v}$$

at the surface of the probe.

3.2 Take the electron collision cross section in a fully ionized plasma to be

$$\sigma = \left(\frac{e^2}{4\pi\varepsilon_0} \right)^2 \frac{4\pi \ln \Lambda}{m_e^2 v_e^4},$$

where $\ln \Lambda \sim 15$ is the Coulomb logarithm. Hence, show by substituting thermal values that the mean free path is

$$\ell \approx \frac{4\pi}{\ln \Lambda} n_e \lambda_D^4.$$

3.3 (a) Show that the electron density at position x where the potential is V_x (relative to the potential at infinity) near a plane probe of potential V_0 (< 0) that absorbs all particles incident on it is given by Eq. (3.2.5).

(b) Hence, calculate the minimum potential at which the approximate formula,

$$n_e(x) = n_\infty \exp\left(\frac{eV_x}{T_e} \right),$$

is accurate to 1%.

3.4 In a plasma with ions of charge Ze show:
(a) The sheath potential is still $-T_e/2e$.
(b) The ion electric current is

$$I_i = Z^{1/2} e \exp\left(-\tfrac{1}{2} \right) A_S n_{e\infty} (T_e/m_i)^{1/2}.$$

Hence, show that for a mixture of ions of different Z but constant Z/m

the electron density may be deduced from the Bohm formula by replacing m_i with m_i/Z.

3.5 Prove Eq. (3.4.1).

3.6 Solve the quasineutral equation governing the potential in the plasma region surrounding a probe for radius as a function of potential, when the probe is (a) spherical and (b) cylindrical. Hence, calculate at what multiple of the probe radius the value of $|e(V - V_\infty)/T_e|$ is equal to 0.01 for these two cases.

3.7 Discuss the relative merits of operating a trapping probe, designed to diagnose impurities by trapping them on its surface, under the following conditions:

(a) Electrically floating (i.e., isolated from electrical conduction to other parts of the machine).

(b) Electrically grounded (connected by a low resistance path to, say, the vacuum vessel).

What would you expect to be the differences in heat load and trapping efficiency?

4 Refractive-index measurements

In many plasmas it is unsatisfactory to use material probes to determine internal plasma parameters, so we require nonperturbing methods for diagnosis. Some of the most successful and accurate of these use electromagnetic waves as a probe into the plasma. Provided their intensity is not too great, such waves cause negligible perturbation to the plasma, but can give information about the internal plasma properties with quite good spatial resolution. In this chapter we are concerned with the uses of the refractive index of the plasma, that is, the modifications to free space propagation of the electromagnetic waves due to the electrical properties of the plasma.

The way in which waves propagate in magnetized plasmas is rather more complicated than in most other media because the magnetic field causes the electrical properties to be highly anisotropic. This is due to the difference in the electron dynamics between motions parallel and perpendicular to the magnetic field. Therefore, we begin with a brief review of the general problem of wave propagation in anisotropic media before specializing to the particular properties of plasmas.

Interferometry is the primary experimental technique for measuring the plasma's refractive properties and we shall discuss the principles of its use as well as some of the practical details that dominate plasma diagnostic applications.

4.1 Electromagnetic waves in plasma
4.1.1 *Waves in uniform media*

We must first consider the nature and properties of electromagnetic waves in a plasma. We treat the plasma as a continuous medium in which current can flow, but that is otherwise governed by Maxwell's equations in a vacuum. The important equations are then

$$\nabla \wedge \mathbf{E} = -\frac{\partial \mathbf{B}}{\partial t}, \qquad \nabla \wedge \mathbf{B} = \mu_0 \mathbf{j} + \varepsilon_0 \mu_0 \frac{\partial \mathbf{E}}{\partial t}, \qquad (4.1.1)$$

where all the electromagnetic properties of the plasma appear explicitly in the current $\mathbf{j}$. There may be equilibrium values of fields $\mathbf{B}_0$, $\mathbf{j}_0$, $\mathbf{E}_0$ that are present in the absence of the waves we are considering. If so, then Eq. (4.1.1) applies to the equilibrium and wave parts separately and we shall consider now that unsubscripted variables refer to the wave quantities only.

We can eliminate **B** by taking $\nabla \wedge$ (the first) and $\partial/\partial t$ (the second equation) to get

$$\nabla \wedge (\nabla \wedge \mathbf{E}) + \frac{\partial}{\partial t}\left(\mu_0 \mathbf{j} + \varepsilon_0 \mu_0 \frac{\partial \mathbf{E}}{\partial t}\right) = 0. \tag{4.1.2}$$

Now we suppose, first, that the wave fields are small enough that the current is a linear functional of the electric field (not at all a severe restriction in general). This means that if a certain spatial and temporal variation of **E**, say $\mathbf{E}_1$, gives rise to current $\mathbf{j}_1$ and similarly $\mathbf{E}_2$ to $\mathbf{j}_2$, then $\mathbf{E}_1 + \mathbf{E}_2$ gives rise to $\mathbf{j}_1 + \mathbf{j}_2$. Second, we take the plasma to be homogeneous in space and time. These properties allow us to Fourier analyze the fields and currents so that, for example,

$$\mathbf{E}(\mathbf{x}, t) = \int \mathbf{E}(\mathbf{k}, \omega) e^{i(\mathbf{k}\cdot\mathbf{x} - \omega t)} d^3 k \, d\omega, \tag{4.1.3}$$

and treat each Fourier mode $\mathbf{E}(\mathbf{k}, \omega)$ separately since each separately satisfies Eq. (4.1.2). Then, for each Fourier mode, the assumption of linearity allows us to write the relationship between current and electric field, usually called Ohm's law, as

$$\mathbf{j}(\mathbf{k}, \omega) = \boldsymbol{\sigma}(\mathbf{k}, \omega) \cdot \mathbf{E}(\mathbf{k}, \omega), \tag{4.1.4}$$

where $\boldsymbol{\sigma}$ is the conductivity of the plasma. In general, a plasma may be an anisotropic medium so that $\boldsymbol{\sigma}$ is a tensor conductivity; we shall explore its form later.

Writing Eq. (4.1.2) for a single Fourier mode we get

$$\mathbf{k} \wedge (\mathbf{k} \wedge \mathbf{E}) + i\omega(\mu_0 \boldsymbol{\sigma} \cdot \mathbf{E} - \varepsilon_0 \mu_0 i\omega \mathbf{E}) = 0, \tag{4.1.5}$$

which may be written (noting $\varepsilon_0 \mu_0 = 1/c^2$) as

$$\left(\mathbf{k}\mathbf{k} - k^2 \mathbf{1} + \frac{\omega^2}{c^2} \boldsymbol{\varepsilon}\right) \cdot \mathbf{E} = 0, \tag{4.1.6}$$

where **1** is the unit dyadic and $\boldsymbol{\varepsilon}$ is the dielectric tensor

$$\boldsymbol{\varepsilon} = \left(\mathbf{1} + \frac{i}{\omega \varepsilon_0} \boldsymbol{\sigma}\right). \tag{4.1.7}$$

In tensor subscript notation this is

$$\left(k_i k_j - k^2 \delta_{ij} + \frac{\omega^2}{c^2} \varepsilon_{ij}\right) E_j = 0, \tag{4.1.8}$$

$$\varepsilon_{ij} = \left(\delta_{ij} + \frac{i}{\omega \varepsilon_0} \sigma_{ij}\right). \tag{4.1.9}$$

The properties of the plasma may be equally well specified through the permittivity $\varepsilon_0 \boldsymbol{\varepsilon}$ as through the conductivity $\boldsymbol{\sigma}$. These contain equivalent information since they are related by Eq. (4.1.7); it is often more convenient to think of the plasma as a dielectric medium whose permittivity is specified, rather than its conductivity.

Now Eq. (4.1.6) [or equivalently Eq. (4.1.8)] represents three homogeneous simultaneous equations for the three components of **E**. In order for these to have a nonzero solution, the determinant of the matrix of coefficients must be zero:

$$\det\left(\mathbf{kk} - k^2 \mathbf{1} + \frac{\omega^2}{c^2}\boldsymbol{\varepsilon} \right) = 0. \tag{4.1.10}$$

This equation relates the ks and ωs for different waves and is the "dispersion relation" for these waves. For any given wave vector **k** it determines the corresponding frequency ω or vice versa. Mathematically, one may regard the propagation equation (4.1.6) as a matrix eigenvalue problem. The eigenvalue, making the determinant zero, provides the dispersion relation, while the eigenvector, which is the solution for **E** corresponding to a particular eigenvalue, determines the characteristic polarization of the wave with that **k** and ω.

The simplest case to consider is when the medium is isotropic, that is, when

$$\boldsymbol{\sigma} = \sigma\mathbf{1}, \qquad \boldsymbol{\varepsilon} = \varepsilon\mathbf{1}. \tag{4.1.11}$$

In such a case the possible waves separate into two types, one in which the electric field polarization is transverse ($\mathbf{k} \cdot \mathbf{E} = 0$) and one in which it is longitudinal ($\mathbf{k} \wedge \mathbf{E} = 0$) to the propagation direction. Taking **k** to be along the z axis, we can write out explicitly the matrix

$$\mathbf{kk} - k^2\mathbf{1} + \frac{\omega^2}{c^2}\boldsymbol{\varepsilon} = \begin{bmatrix} -k^2 + \dfrac{\omega^2}{c^2}\varepsilon & 0 & 0 \\[2ex] 0 & -k^2 + \dfrac{\omega^2}{c^2}\varepsilon & 0 \\[2ex] 0 & 0 & \dfrac{\omega^2}{c^2}\varepsilon \end{bmatrix},$$

$$\tag{4.1.12}$$

whose determinant is zero if

$$-k^2 + \frac{\omega^2}{c^2}\varepsilon = 0, \qquad \text{**E** transverse,} \tag{4.1.13}$$

or

$$\frac{\omega^2}{c^2}\varepsilon = 0, \qquad \text{E longitudinal.} \tag{4.1.14}$$

The transverse wave dispersion relation is just the familiar expression of simple optics

$$N \equiv \frac{kc}{\omega} = \varepsilon^{1/2}, \tag{4.1.15}$$

where N is the refractive index. The longitudinal wave dispersion relation is simply $\varepsilon = 0$, which in the plasma case can represent a nontrivial solution, but which we shall not explore further here.

When $\boldsymbol{\varepsilon}$ is not isotropic there is no such simple division into transverse and longitudinal waves. The electric field is in general partly transverse and partly longitudinal, and naturally the refractive index depends on the direction of propagation as well as the frequency. With the choice of axes such that $\mathbf{k}$ is along z, the matrix of Eq. (4.1.12) will have nonzero off-diagonal terms arising from those of $\boldsymbol{\varepsilon}$. However, if these terms contain no explicit dependence on k, the determinant will be a quadratic in k^2. There will then in general be two solutions for k^2, given ω. These correspond to the transverse waves of the isotropic case (in the isotropic case the two solutions are degenerate); these two waves will be the focus of our interest.

So far everything we have said applies quite generally to any uniform linear medium in which electromagnetic waves may propagate. Our treatment shows that all that is needed in order to calculate the wave propagation properties in this medium is a knowledge of the (ac) conductivity $\boldsymbol{\sigma}(k, \omega)$ or, equivalently, the permittivity $\boldsymbol{\varepsilon}(k, \omega)$. The particular properties of plasmas as media enter the treatment through their particular form for these quantities.

4.1.2   *Plasma conductivity*

Our task now is to calculate the plasma conductivity and hence, permittivity on the basis of an understanding of the nature of the plasma. This can be done at various levels of sophistication, from a kinetic-theory treatment (using the Boltzmann equation) down to a fluid treatment (ignoring plasma temperature). These correspond, in part, to the various levels of detail to which we have noted that the plasma parameters can be measured. The kinetic treatment incorporates details of the distribution function, whereas the simplest fluid treatment ignores all but the zeroth moment, the density. For present purposes it turns out that the simplest of treatments is perfectly adequate. The reason for this is that we shall be concerned with waves traveling at phase velocities close to the speed of light in plasmas whose thermal electron speed is $v_t \ll c$. We are, therefore, able to ignore thermal particle motions and adopt what is called the *cold plasma* approximation.

The electrons are taken to be at rest, except for motions induced by the wave fields, and we must calculate the electron current caused by a specified electric field. We could obtain this from the Ohm's law for an electron fluid including electron inertia [e.g., Boyd and Sanderson (1969) follow this approach]. However, it is probably more transparent to proceed from first principles as follows. The equation of motion of a single electron is

$$m_e \frac{\partial \mathbf{v}}{\partial t} = -e(\mathbf{E} + \mathbf{v} \wedge \mathbf{B}_0), \qquad (4.1.16)$$

where we include, in general, a static magnetic field $\mathbf{B}_0$ but ignore collisions. In view of our cold plasma assumption, for a single Fourier wave mode, $\mathbf{v}$ is purely harmonic, $\propto \exp(-i\omega t)$. In order to simplify the algebra we take the z axis in the direction $\mathbf{B}_0$. Then the three components of Eq. (4.1.16) are

$$-m_e i\omega v_x = -eE_x - eB_0 v_y,$$
$$-m_e i\omega v_y = -eE_y + eB_0 v_x, \qquad (4.1.17)$$
$$-m_e i\omega v_z = -eE_z.$$

We may readily solve for v in terms of E:

$$v_x = \frac{-ie}{\omega m_e} \frac{1}{1 - \Omega^2/\omega^2} \left(E_x - i\frac{\Omega}{\omega} E_y \right),$$
$$v_y = \frac{-ie}{\omega m_e} \frac{1}{1 - \Omega^2/\omega^2} \left(i\frac{\Omega}{\omega} E_x + E_y \right), \qquad (4.1.18)$$
$$v_z = \frac{-ie}{\omega m_e} E_z,$$

where $\Omega \equiv eB_0/m_e$ is the electron cyclotron frequency. In this cold plasma approximation all electrons move alike and so the current density is simply

$$\mathbf{j} = -en_e \mathbf{v} = \boldsymbol{\sigma} \cdot \mathbf{E}, \qquad (4.1.19)$$

where

$$\boldsymbol{\sigma} = \frac{in_e e^2}{m_e \omega} \cdot \frac{1}{1 - \Omega^2/\omega^2} \begin{bmatrix} 1 & -i\Omega/\omega & 0 \\ i\Omega/\omega & 1 & 0 \\ 0 & 0 & 1 - \Omega^2/\omega^2 \end{bmatrix}. \qquad (4.1.20)$$

This is the conductivity tensor we require.

Strictly speaking, this is only the electron current conductivity. The ions may be treated in exactly the same way and an identical equation obtained with ion parameters (mass, charge, density) substituted for electron. The total conductivity is then the sum of the electron and ion parts. Because $m_i \gg m_e$, the ion contribution is usually small provided the frequency is high enough; we shall consider here only cases where the ion motion may be ignored.

The dielectric tensor may be written down immediately from our knowledge of $\boldsymbol{\sigma}$ using Eq. (4.1.7):

$$\boldsymbol{\varepsilon} = \begin{bmatrix} 1 - \dfrac{\omega_p^2}{\omega^2 - \Omega^2} & \dfrac{i\omega_p^2 \Omega}{\omega(\omega^2 - \Omega^2)} & 0 \\[3ex] \dfrac{-i\omega_p^2 \Omega}{\omega(\omega^2 - \Omega^2)} & 1 - \dfrac{\omega_p^2}{\omega^2 - \Omega^2} & 0 \\[3ex] 0 & 0 & 1 - \dfrac{\omega_p^2}{\omega^2} \end{bmatrix}, \qquad (4.1.21)$$

where $\omega_p = (n_e e^2 / \varepsilon_0 m_e)^{1/2}$ denotes the electron plasma frequency. To simplify the notation we adopt the common practice of using the nondimensional quantities

$$X = \omega_p^2 / \omega^2, \qquad Y = \Omega/\omega, \qquad N = kc/\omega. \qquad (4.1.22)$$

Then we substitute $\boldsymbol{\varepsilon}$ from Eq. (4.1.21) into Eq. (4.1.10). In doing so we can choose axes such that $k_x = 0$, that is,

$$\mathbf{k} = k \cdot (0, \sin\theta, \cos\theta),$$

where θ is the angle between $\mathbf{k}$ and $\mathbf{B}_0$. The determinantal equation then becomes

$$\begin{vmatrix} -N^2 + 1 - \dfrac{X}{1 - Y^2} & \dfrac{iXY}{1 - Y^2} & 0 \\[3ex] -\dfrac{iXY}{1 - Y^2} & -N^2\cos^2\theta + 1 - \dfrac{X}{1 - Y^2} & N^2\sin\theta\cos\theta \\[3ex] 0 & N^2\sin\theta\cos\theta & -N^2\sin^2\theta + 1 - X \end{vmatrix} = 0.$$

$$(4.1.23)$$

In the cold plasma approximation $\boldsymbol{\varepsilon}$ is independent of $\mathbf{k}$; thus, as we have previously noted, this dispersion relation represents a quadratic equation for N^2. It may, of course, be solved, though the algebra is heavy. The solutions are usually written in the form

$$N^2 = 1 - \frac{X(1-X)}{1 - X - \frac{1}{2}Y^2\sin^2\theta \pm \left[\left(\frac{1}{2}Y^2\sin^2\theta\right)^2 + (1-X)^2Y^2\cos^2\theta\right]^{1/2}}.$$

(4.1.24)

This expression is called the Appleton–Hartree formula for the refractive index.

Two cases provide much simplified results that may be obtained from Eq. (4.1.24) or directly from Eq. (4.1.23).

Parallel propagation $(\theta = 0)$: When waves propagate parallel to the magnetic field, the solutions are

$$N^2 = 1 - \frac{X}{1-Y^2} \pm \frac{XY}{1-Y^2} = 1 - \frac{X}{1 \pm Y},$$

(4.1.25)

for which the characteristic polarization of the wave electric field is

$$\frac{E_x}{E_y} = \pm i, \qquad E_z = 0,$$

(4.1.26)

that is, circularly polarized waves with left and right handed E rotation, respectively.

Perpendicular propagation $(\theta = \pi/2)$: When waves propagate perpendicular to the magnetic field, the solutions are

$$N^2 = 1 - X \quad \text{or} \quad N^2 = 1 - \frac{X(1-X)}{1 - X - Y^2},$$

(4.1.27)

for which the characteristic polarizations of the electric field are

$$E_x = E_y = 0$$

(4.1.28)

and

$$\frac{E_x}{E_y} = -i \cdot \frac{1 - X - Y^2}{XY}, \qquad E_z = 0,$$

(4.1.29)

respectively.

In addition to dependence upon the wave frequency, the parameter X $(= \omega_p^2/\omega^2)$ depends only on electron density n_e and Y $(= \Omega/\omega)$ only on magnetic field B_0. It should be no surprise that the values of the refractive index obtained depend only on density and magnetic field, since our cold plasma treatment has explicitly excluded other effects, such as finite temperature. However, the fact that this treatment represents a typical plasma very well allows the refractive index to be used to measure these two parameters with excellent confidence and accuracy, as indicated in Table 1.2.

4.1.3   *Nonuniform media: The* WKBJ *approximation and full-wave treatments*

Naturally, no practical plasma or any other medium satisfies the condition of being uniform throughout all space. It is important to consider, then, what happens when there are spatial gradients in the electromagnetic properties. Mathematically, the results are that fields of the form $\exp i(\mathbf{k} \cdot \mathbf{x} - \omega t)$ no longer separately satisfy Maxwell's equations (4.1.2). One can still express any solution as a sum of such Fourier modes, but these will, in general, be coupled together by the nonuniformities of the medium. For example, a wave propagating in one direction that encounters a gradient in the refractive index will be partially reflected (coupled to the oppositely directed wave).

If the properties of the plasma vary sufficiently slowly, then locally the wave can be thought of as propagating in an approximately uniform medium and, hence, behaving as if all the previous treatment applied. Thus, for any frequency and propagation direction, there is locally a well defined k and refractive index N corresponding to the local values of the plasma parameters.

The names of Wentzel, Kramers, and Brillouin (and sometimes Jeffreys, hence WKBJ) have become associated with a very widespread technique for solving such wave-type equations in slowly varying media. The approach is also often called the eikonal or more simply geometric optics approximation. In this approximation the propagation of the wave amplitude for a given frequency is expressed in the form

$$E \approx \exp i \left\{ \int \mathbf{k} \cdot \mathbf{dl} - \omega t \right\}, \qquad (4.1.30)$$

where l is the distance along the ray path and $\mathbf{k}$ is the solution of the homogeneous plasma dispersion relation for the given ω, based on the local plasma parameters. This will be a good approximate solution provided that the fractional variation of $\mathbf{k}$ in one wavelength of the wave is small. That is,

$$|\nabla k|/k^2 \ll 1. \qquad (4.1.31)$$

In this case the coupling to other waves is small and may generally be ignored as far as the single transmitted wave is concerned. For our present purposes we need only note that the phase of the emerging wave is given by $\int \mathbf{k} \cdot \mathbf{dl}$, which may be written $\int \frac{\omega}{c} N \, dl$ provided the ray direction and $\mathbf{k}$ approximately coincide (for example if the plasma is approximately isotropic). (The usual $k^{-1/2}$ amplitude variation, which provides an even better approximation, is not important for the present discussion.)

When the condition Eq. (4.1.31) is not satisfied, the WKBJ approach breaks down and it is necessary to return to the original wave equation (4.1.2), prior to Fourier transformation, to describe the propagation. One therefore has to deal with partial differential equations rather than algebraic equations and the solutions are naturally much more difficult. When the spatial nonuniformity can be regarded as being in a single coordinate, for example, with a cylindrical plasma (variation only in the r direction) or in a slab, then it is possible to Fourier analyze the problem in the remaining coordinates and arrive at an ordinary differential equation governing the wave. The application of boundary conditions again leads to an eigenvalue problem (under some conditions to a second order Sturm–Liouville problem) whose eigenvalues relate ω and the k vector in the direction of uniformity and whose eigenfunctions give the wave field solution in the direction of nonuniformity.

The sort of situation in which this approach is essential is for a plasma column of low electron density, such that the wavelength of radiation in the vicinity of the plasma frequency is greater than the plasma radius. For example, for a column of dimension 10 cm, say, the frequency of radiation having this wavelength is about 3 GHz, which is the plasma frequency of a plasma of density $\sim 10^{17}$ m^{-3}. Therefore, refractive-index diagnosis of such a plasma would require a full-wave treatment. This density tends to be low enough to allow the use of internal probes for diagnostics, which reduces the need for refractive-index methods. What is more, the full-wave solutions and their interpretation in terms of plasma parameters are highly dependent upon the exact geometry, so a general description is difficult. For these reasons no detailed discussion of situations requiring a full-wave solution will be given here. The interested reader may refer to treatments given elsewhere [e.g., Heald and Wharton (1965)].

4.2 Measurement of electron density

If the magnetic field is negligible ($Y \to 0$) then the refractive index is

$$N^2 = 1 - X = 1 - \omega_p^2/\omega^2 \tag{4.2.1}$$

for all modes, independent of the direction of propagation because the plasma is then isotropic. There is also, even when $Y \neq 0$, one polarization for perpendicular propagation that has this same refractive index. This mode is called the *ordinary wave*. It corresponds to the positive sign in Eq. (4.1.24). By extension, at other angles of propagation, $\theta \neq \pi/2$, the solution having the positive sign in Eq. (4.1.24) is also referred to as the ordinary wave, although its refractive index is then more complicated. The other solution is called the *extraordinary wave*.

4.2.1 *Interferometry*

Measurements of the refractive index of any medium are most often made by some form of interferometry. An interferometer is any device in which two or more waves are allowed to interfere by coherent addition of electric fields. The intensity then observed is modulated according to whether the fields interfere constructively or destructively, that is, in phase or out of phase.

Consider a simple two-beam interferometer in which monochromatic fields $E_1 \exp i\omega t$ and $E_2 \exp i(\omega t + \phi)$ are added together, with some phase difference ϕ between them. Then the total field is given by

$$E_t = (E_1 + E_2 \exp i\phi) \exp i\omega t. \qquad (4.2.2)$$

The power detected, for example, by a square-law detector is proportional to $|E_t|^2$, which may readily be shown to be

$$|E_t|^2 = [E_1^2 + E_2^2]\left[1 + \frac{2E_1 E_2}{E_1^2 + E_2^2}\cos\phi\right]. \qquad (4.2.3)$$

The output intensity (power) thus has a constant component plus a component varying like $\cos\phi$, as shown schematically in Fig. 4.1.

There are many different types of interferometers used for a variety of different purposes. The configurations in widest use are as follows:

1. *The Michelson interferometer* is a two-beam interferometer, as illustrated schematically in Fig. 4.2, with one beamsplitter, two arms in which the beams travel in both directions, and two outputs, one of which is along the input. In this, as in other configurations, the arms may be free space straight optical paths or, for example,

Fig. 4.1. Variation in the (power) output signal of a two-beam interferometer with relative phase.

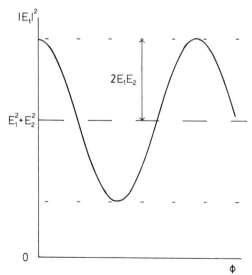

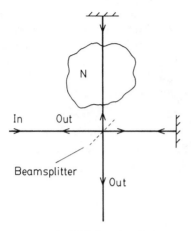

Fig. 4.2. The Michelson interferometer configuration.

microwave wave guides, and the beamsplitter may be some form of partial reflector of optical character or, for example, some type of microwave coupler. The principles are the same in all cases. Phase differences between the two components of one of the output beams arise, in the case of refractive-index measurements, by changes in the refractive index in one of the arms of the interferometer.

2. *The Mach–Zehnder configuration* is also a two-beam interferometer but differs from the Michelson in having two arms in which the beams travel in only one direction. Both outputs are separate from the input (see Fig. 4.3). Again phase changes are caused by variations of the refractive index in one arm.

Fig. 4.3. The Mach–Zehnder interferometer configuration.

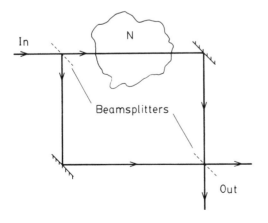

3. *The Fabry–Perot interferometer* is a multiple-beam interferometer in which there are two beamsplitters and two composite output beams. Because it has multiple beams (see Fig. 4.4), the output is not a simple cosine as in two-beam interferometers. This makes phase shift interpretation more difficult, so it is less often used for plasma refractive-index measurements.

Consider, then, a situation in which we wish to measure the refractive index of a plasma that we arrange to be in one arm of a two-beam interferometer such as the Mach–Zehnder. The total phase lag in the plasma arm, assuming we can apply a geometrical optics (WKBJ) type of solution, will be

$$\phi = \int k\,dl = \int N\frac{\omega}{c}\,dl. \tag{4.2.4}$$

Of course, a significant proportion of the length of this arm along which this integration is taken will be outside the plasma. Also the reference arm has a length and corresponding phase lag that we may not know with great precision. These effects are all removed by comparing the phase difference between the two arms [deduced from Eq. (4.2.3)] with plasma present to that without plasma. The difference in these phases is then simply that introduced into the plasma arm by the plasma, namely

$$\Delta\phi = \int \left(k_{\text{plasma}} - k_0\right) dl = \int (N-1)\frac{\omega}{c}\,dl, \tag{4.2.5}$$

where we assume that in the absence of plasma $k_0 = \omega/c$, that is, the wave propagates effectively in vacuo. The integral may now be considered to be limited to that part of the path that lies in the plasma.

The measurement of the interferometer phase shift $\Delta\phi$ thus provides us with a measure of the mean refractive index along the line of the interferometer beam through the plasma.

Fig. 4.4. The Fabry–Perot interferometer configuration.

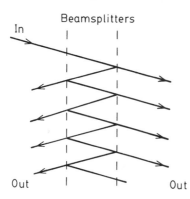

When the plasma refractive index is given by Eq. (4.2.1), we can write this as

$$N^2 = 1 - \omega_p^2/\omega^2 = 1 - n_e/n_c,$$ (4.2.6)

where n_c is called the cutoff density,

$$n_c \equiv \omega^2 m \varepsilon_0/e^2.$$ (4.2.7)

Plotting N^2 versus n_e gives a linear relationship as shown in Fig. 4.5. For $n_e < n_c$ the interferometer then gives a measure of electron density from

$$\Delta\phi = \frac{\omega}{c} \int \left[\left(1 - \frac{n_e}{n_c} \right)^{1/2} - 1 \right] dl,$$ (4.2.8)

which is a form of average along the beam path. If the plasma density is sufficiently small, $n_e \ll n_c$, then an approximate expansion is

$$N \approx 1 - \tfrac{1}{2}(n_e/n_c)$$ (4.2.9)

and the phase shift simplifies to

$$\Delta\phi = \frac{\omega}{2cn_c} \int n_e \, dl,$$ (4.2.10)

that is, the simple chord averaged density.

If the density exceeds the cutoff value, $n_e > n_c$, then N^2 becomes negative and N pure imaginary. This means that the wave is no longer propagating but evanescent, falling off exponentially with distance. The

Fig. 4.5. The variation of the square of the refractive index N^2 of the ordinary mode with plasma density.

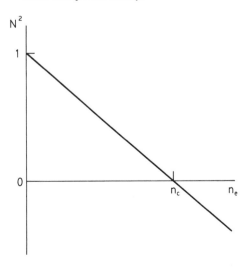

result is that, normally, very little power is transmitted through the plasma and the interferometer ceases to function.

4.2.2   *Determining the phase shift*

To determine the mean density requires us to interpret the output *power* of the interferometer – the quantity directly measured by a detector – in terms of *phase shift*, using Eq. (4.2.3). Various problems arise in this interpretation. There are two main sources of ambiguity and error.

1. Amplitude variations (of E_1 or E_2) due to absorption or refraction of the beam. These must be distinguished from the phase variations in which we are interested.
2. Ambiguity of the phase change direction. This arises primarily where $\phi = 0$, π, 2π, and so forth, because at these points $d|E_t|^2/d\phi = 0$, so that the interferometer has a null in its phase sensitivity. Consider a time-dependent phase, passing through such point, say $\phi = 0$; it is impossible to tell a priori whether or not there may have been a change in the sign of $d\phi/dt$ at the instant $\phi = 0$. For example, variations of phase illustrated in Fig. 4.6 would give identical outputs from the interferometer. This figure also illustrates the ambiguity arising from knowledge of ϕ only modulo 2π. Sometimes other knowledge about the plasma allows one to determine whether ϕ is increasing or decreasing; however, many situations arise when the ambiguity remains.

Fig. 4.6. Possible evolution paths of phase that would give identical outputs from a simple interferometer.

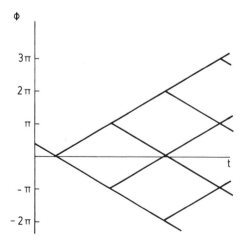

The problem of amplitude variations can be alleviated somewhat by monitoring both outputs of the interferometer since the total powers in both outputs (for a lossless beamsplitter) is equal to the sum of the power in the interferometer arms. The second problem is not solved by this means, however, and generally requires an additional output whose power is proportional to $\sin \phi$ rather than $\cos \phi$. That is, it requires a second output in quadrature with the first. It is possible to provide this by operating essentially a second interferometer, with phase different by $\pi/2$, along the same path as the first, although this solution is rarely adopted.

The most satisfactory way to resolve both problems is to modulate the phase of the interferometer. One can think of this as a method of causing the interferometer to read alternatively sine and cosine functions of the phase. If it does so more rapidly than variations in ϕ occur, then ϕ is unambiguously determined.

The frequency of a wave is simply the time rate of change of its phase. Therefore, variations of phase may be regarded as variations in frequency. It is instructive, then, to consider the phase determination problem and its resolution by phase modulation in terms of frequency. We shall then see that the problem is essentially one of FM detection and is common in radio reception and other familiar situations.

Consider the final beamsplitter in our interferometer, illustrated in Fig. 4.7. Two waves are mixed there and the power detected in the output. The wave that has passed through the plasma has been phase modulated or equivalently frequency modulated by the changing refractive index of the plasma. The reference arm serves as a local oscillator in the detection of the received wave from the plasma arm. The output then contains sum and difference frequency components, that is, if the frequency in the reference

Fig. 4.7. The final beamsplitter of an interferometer can be regarded as a mixer in a heterodyne receiver.

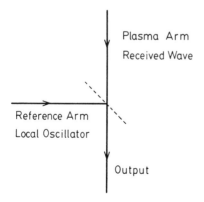

arm is ω_1 and that in the plasma arm is ω_2, the output contains frequencies $\omega_2 + \omega_1$ and $\omega_2 - \omega_1$. Only the low frequency $\Delta\omega = \omega_2 - \omega_1$ will be of any interest here.

Now, in the simple interferometer configurations we have considered so far, if the rate of change of the plasma arm phase is zero (denoted by superscript zero) (e.g., if density is constant), then $\omega_2^0 = \omega_1^0$ and $\Delta\omega^0 = 0$; the output of the interferometer is constant. When phase changes occur, $\Delta\omega = d\phi/dt$, and the output frequency is nonzero. However, it is generally not possible to distinguish between positive and negative $\Delta\omega$; both give an output with frequency $|\Delta\omega|$. This is the cause of the ambiguity of phase change direction.

Suppose, then, that we introduce an extra phase modulation, not present in the simple interferometer, in addition to the plasma effects. Then even when the plasma phase shift is constant, the frequencies ω_1^0 and ω_2^0 are no longer equal and the output contains a signal at frequency $\Delta\omega^0 = \omega_2^0 - \omega_1^0$, which for simplicity we take as constant. The final mixer is thus acting as a heterodyne receiver with intermediate frequency (IF) $\Delta\omega^0$ (rather than as

Fig. 4.8. Illustration of the frequency relationships in homodyne and heterodyne reception.

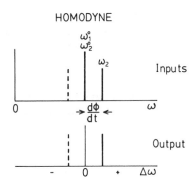

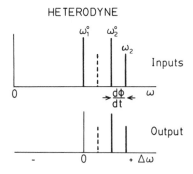

before, a homodyne receiver with zero IF frequency). In this case, when additional frequency modulation occurs, due to plasma phase changes, the output signal is shifted in frequency to

$$\Delta\omega = \Delta\omega^0 + d\phi/dt. \tag{4.2.11}$$

The output frequency thus increases or decreases according to the direction of phase change, and the ambiguity in direction is no longer present. These relationships are illustrated in Fig. 4.8, which shows the locations of the input and output frequencies for positive and negative $d\phi/dt$ that can be distinguished in a heterodyne configuration but not in a homodyne one (because opposite signs of $\Delta\omega$ are indistinguishable).

There are additional advantages to the heterodyne interferometer arising from the shift of $\Delta\omega^0$ from zero. First, there is now no need to sense the dc signal level and all the detecting electronics can be ac coupled. Second, the amplitude of the output modulation is not crucial in determining phase; only its frequency is required. These factors ensure that the interferometer can be made very insensitive to changes of amplitude in either plasma or reference beam or in the phase contrast (see Section 4.2.4).

4.2.3 *Modulation and detection methods*

Various methods for producing the required frequency shift can be employed, depending upon the radiation frequency of the interferometer. At its simplest level some kind of mechanical vibration of one of the mirrors in an interferometer arm may be employed.

A more satisfactory technique (Veron 1974) involving mechanical modulation, is illustrated in Fig. 4.9. Inserted into one interferometer arm is a reflection from a rotating wheel, usually having an appropriate diffraction grating cut into its rim to optimize the reflection. Because of the Doppler effect the radiation frequency emerges shifted to $\omega' = \omega(1 + v_i/c)/(1 - v_r/c)$, where v_i and v_r are the components of wheel velocity along the incident and reflected directions, respectively, thus providing the heterodyne frequency required. This technique is most useful for far infrared and submillimeter wavelengths.

Fig. 4.9. Rotating wheel Doppler modulator for producing a frequency shift.

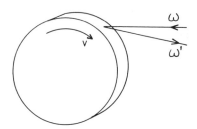

A functionally equivalent result [e.g., Jacobson (1978)] is available for shorter wavelengths (~ 10 μm to visible) from commercial acousto-optic modulators, in which the radiation partially reflects (scatters) from a traveling sound wave in a refractive medium. The wave acts like a moving diffraction grating and the scattered wave is shifted in frequency by the sound wave frequency.

A common technique used with microwave interferometers (Heald and Wharton 1965) is to sweep the frequency of the source. Then if one arm of the interferometer is much longer than the other, the frequencies of the waves, when they interfere, are different by

$$\Delta\omega^0 = \frac{d\omega}{dt}\frac{L}{v_p},$$
(4.2.12)

where $d\omega/dt$ is the frequency sweep rate, L the difference in arm lengths, and v_p the radiation phase velocity in the interferometer ($v_p = c$ in free space).

It is appropriate, in some schemes, to obtain the heterodyne detection by simply employing two different sources of radiation at slightly different frequency. This poses quite serious constraints upon the frequency stability because $\Delta\omega/\omega$ is usually extremely small. However, it can be a very satisfactory solution. As an example, Fig. 4.10 shows schematically an interferometer used for density measurements on the Alcator tokamak (Wolfe et al. 1976). It employs two lasers at 119 μm wavelength, detuned so that their frequencies are different by ~ 1 MHz, thus providing heterodyne phase detection.

The detection of the phase shift in a heterodyne interferometer is usually performed automatically. What is required is, in essence, to count the number of periods (and fractional periods) of the IF (beat frequency) and

Fig. 4.10. Schematic of a dual-laser heterodyne interferometer [after Wolfe et al. (1976)].

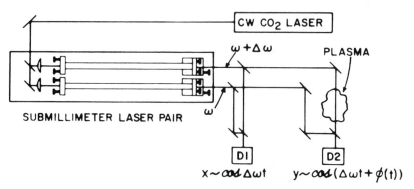

subtract from it the number $(2\pi\Delta\omega^0 t)$ that would have been observed in the same time duration if no change of refractive index had occurred. This difference measures the number of wavelengths or "fringes" by which the phase has changed ($\Delta\phi/2\pi$). Usually some kind of fast digital circuitry is used to perform this measurement. Meddens and Taylor (1974) describe one widely used scheme; Veron (1979) describes another. Sometimes analog techniques prove more satisfactory for measuring very small phase shifts $\Delta\phi \ll 1$ [e.g., Jacobson (1982)]. Naturally, these methods strongly resemble standard techniques of FM reception.

4.2.4 *Coherence, diffraction and refraction*

As mentioned in Section 4.2.1, in order to propagate through the plasma, the radiation used for interferometry must have a frequency greater than the plasma frequency. The choice of frequency to be used thus depends on the plasma to be probed. Plasmas of fusion research interest, and many other laboratory plasmas, may have density from, say, 10^{17} to 10^{21} m^{-3} and beyond. The corresponding plasma frequencies range from ~ 3 to 300 GHz, which lie in the microwave and millimeter wave spectral range, that is, wavelength $\lambda = 10$ cm to 1 mm.

The radiation used must also generally have a very narrow bandwidth. The reason for this is that the contributions to the output power from different frequencies across the source bandwidth must all experience the same phase shift, otherwise the degree of modulation of the output power due to phase changes – called the phase contrast of the interferometer – will be decreased by the different contributions adding up with random phase. The exact bandwidth limitation depends upon the precise configuration of the interferometer, but the requirements are met almost always by using a coherent source of radiation such as a microwave generator (klystron, Gunn diode oscillator, etc.) or else some form of laser. Such a source will generally guarantee adequate temporal coherence – narrow bandwidth.

Now the propagation of coherent beams of radiation in free space is governed by the equations of Gaussian optics, so-called because spatial eigenmodes of beam propagation are products of Laguerre polynomials with Gaussian curves, the lowest eigenmode being simply a Gaussian beam profile. The feature that distinguishes such beams from incoherent pencils of rays is that they are diffraction limited, meaning that the angular divergence and beam size are uniquely related by virtue of the principles of diffraction. Although multimoded lasers can be used for interferometry, it is more usual in plasma interferometry to employ single-mode generators and lasers, which tend naturally to give diffraction-limited beams for their propagation through the plasma. We shall not employ here the detailed mathematics of Gaussian optics but rather adopt an elementary approximate treatment of the diffraction from first principles. This is sufficient to

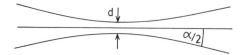

Fig. 4.11. Diffraction-limited beam focus.

understand and illustrate the important diffraction and coherence questions that arise.

Suppose we have a diffraction-limited beam brought to a focus as illustrated in Fig. 4.11. Then the angular half width of the beam far from the focus (Fraunhofer limit) is given by the condition for the difference in path length across the wavefront to be no more than about half a wavelength, as illustrated in Fig. 4.12, so that the Huygens "wavelets" add up in phase. Thus the angular width of the beam is

$$\alpha = \frac{\lambda}{d}. \tag{4.2.13}$$

This will be the basis of our simplified treatment.

At the final beamsplitter of an interferometer two beams are added. They are then generally focused onto some form of detector. We desire to optimize the interferometer by maximizing the interference signal of these two beams. To do this it is not sufficient merely to ensure that all the power in both beams falls on the detector. It is also required that the signal be coherent in space, that is, that all the power in the beams' spatial extent should add simultaneously either in phase or out of phase. This condition maximizes the phase contrast. There are essentially two requirements for the alignment of the beams that will ensure this optimization. One is that the beams should coincide (in position and width) and the other is that the wavefronts should be parallel. Two forms of misalignment are illustrated in Fig. 4.13. In (*a*) the beams coincide on the detector but phase contrast is decreased because their wavefronts are not parallel and, instead of a single

Fig. 4.12. Diffraction angle determined by a Huygens construction.

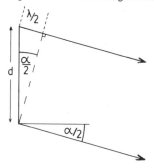

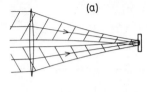

Lens   Detector

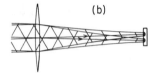

Fig. 4.13. Possible types of interferometer misalignment.

maximum or minimum power detected, a modulated interference pattern exists across the detector at all times. Phase shifts in the interferometer cause the pattern to move across the detector, but the total power modulation is small. In (*b*) the wavefronts are parallel, but because they do not overlap significantly, little interference, and hence power modulation, occurs.

If the detector is at the focus of a lens, then far away from the focus, for example, at the lens or perhaps at the beamsplitter, the types of misalignment are reversed; that is, the situation of nonparallel wavefronts at the focus (detector) corresponds to poor beam coincidence at the lens and poor beam coincidence at the detector arises from nonparallel wavefronts at the lens.

It is easy to see that these types of misalignments are entirely equivalent; phase contrast will be lost if the relative displacement of the wavefronts is more than $\sim \lambda/2$ across the beam. This corresponds to an angular deviation of the beam directions of $\alpha/2 = \lambda/2d$, just as with the calculation of beam diffraction size. So the condition for wavefronts to be parallel is just the condition for the far-field beams to overlap.

Suppose, now, that we have an interferometer that is adequately aligned in the absence of plasma. If the plasma were uniform, then its presence would introduce a phase shift but no significant change in alignment. However, no laboratory plasmas are uniform, of course, because they have boundaries. So the plasma may act so as to cause deterioration in the alignment of the interferometer, by the processes of refraction. Figure 4.14 illustrates the type of process we have in mind. In this case a cylindrical

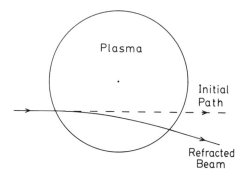

Fig. 4.14. Refraction of an interferometer beam by a nonuniform plasma.

plasma of nonuniform density acts like a lens and refracts the originally well aligned beam away from its original path.

We can rapidly arrive at a criterion governing the importance of these effects as follows. Consider a beam that traverses a refractive slab in which the total phase difference along the beam path $[\phi = \int(\omega/c)N\,dl]$ varies uniformly across the beam, as illustrated in Fig. 4.15. Then the wavefront emerging from the slab will have an angle θ to the incident wavefront:

$$\theta = \frac{d\phi}{dy}\frac{\lambda}{2\pi} = \frac{d}{dy}\left\{\int N\,dl\right\} \tag{4.2.14}$$

(assuming $\theta \ll 1$). This will also be the angular deviation of the ray. If the transverse beam dimension is d, then clearly the coherence across the wavefront will be maintained if

$$\pi > \Delta\phi = d\frac{d\phi}{dy} = \frac{2\pi\theta d}{\lambda}, \tag{4.2.15}$$

Fig. 4.15. Slab approach to calculating the refractive ray deviation.

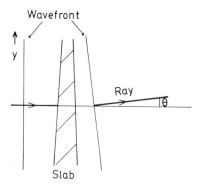

that is,

$$\theta < \frac{\lambda}{2d}. \qquad (4.2.16)$$

Again, naturally, for a diffraction-limited beam this is just the condition for the far-field beam patterns to overlap. Thus the refractive deviation can equivalently be thought of as a problem of whether the far-field beams remain coincident or whether the wavefronts near the plasma remain coherent.

The latter approach is perhaps more useful since it may immediately be generalized to a situation in which ϕ varies arbitrarily across the beam. The criterion for maintaining phase contrast remains that $\Delta\phi \leq \pi$ over the width of the beam.

It is generally the case that this criterion for maintaining phase contrast is much more demanding than that of simply avoiding cutoff of the radiation. One can show (see Example 4.2) that for a cylindrical plasma having a parabolic density profile, the angular deviation has a maximum value

$$\theta = n_0/n_c, \qquad (4.2.17)$$

where n_0 is the peak (central) density. Now the acceptance angle of an interferometer diagnosing such a plasma will usually be limited by practical matters such as the available port size and so on. But the largest acceptance angle possible, in principle, if we wish to sample the density along M different chords is $\alpha \approx 1/M$. Therefore, we certainly need

$$n_0/n_c < 1/M, \qquad (4.2.18)$$

and usually a more stringent condition is required.

The requirements of maintaining phase contrast in the presence of refraction thus dictate that not only should $\omega > \omega_p$ to avoid cutoff, but also actually for off-axis chords

$$\omega \gtrsim \omega_p \alpha^{-1/2}, \qquad (4.2.19)$$

where α is the interferometer's acceptance angle. This usually requires the use of frequencies considerably higher than the plasma frequency.

4.2.5   *Choice of frequency, vibration*

As we have just seen, it is generally advantageous to use a frequency significantly above the maximum expected plasma frequency to avoid cutoff and refraction difficulties. The constraints upon the *maximum* desirable frequency tend to involve primarily issues other than those of the plasma itself.

The use of high frequencies, such as optical, has the advantage that well-developed technology exists. However, it places extremely severe demands on the mechanical stability of the interferometer. The angular alignment must be set and maintained within angles proportional to the wavelength; so shorter wavelengths are more demanding. By far the most important issue, however, is the stability of the path length.

Any interferometer, particularly one in the usually noisy environment of a plasma experiment, is subject to spurious changes in path length due to vibration of its optical components. If a total vibrational path disturbance l is present, then the change of the interferometer phase is $2\pi l/\lambda$; it is larger for shorter wavelengths. Moreover, the phase shift introduced by the plasma is $\Delta\phi \propto \lambda$, so the ratio of spurious vibrational phase error to plasma phase change is proportional to λ^{-2}. The consequence is that vibration becomes a very serious problem for short wavelengths.

It is usually desirable, therefore, when available technology permits, to employ a frequency large enough to satisfy the demands of refraction, and so forth, but not much larger than this; otherwise vibration becomes more important. This is the reason why many interferometers in fusion research operate at millimeter to submillimeter wavelengths ($\lambda \sim 100$–2000 μm), which tend to be optimum for typical densities.

Another alternative when vibration is not or cannot be avoided is to compensate for it by measuring simultaneously the phase shift at two widely different wavelengths. An early example of this is Gibson and Reid (1964). Veron (1979) discusses more recent applications. The shorter wavelength (usually visible) measures mostly the mechanical vibration, while the longer wavelength has a stronger dependence upon plasma refractive index. The effects of vibration can be removed by subtracting from the long wavelength phase shift an appropriate proportion of the phase shift of the shorter wavelength (see Exercise 4.9).

Another way to implement vibration compensation using such "two color" interferometry is to use the output of the short wavelength detector to stabilize the path length in the interferometer. This is accomplished by feedback control of one of the mirrors, using, for example, a piezoelectric transducer to maintain constant phase difference at the short wavelength. This has the advantage of avoiding the necessity for fringe counting at the short wavelength, but it requires a transducer that has sufficient speed and displacement to follow the vibrations (see Exercise 4.10).

4.2.6   *Interferometric imaging*

So far, we have implicitly considered a case in which all the power in the interferometer beam is to be focused onto a single detector. Any spatial information is then to be obtained by using multiple beams and multiple interferometers. In this case, phase variations across the beam are a cause of decreased phase contrast and hence decreased signal-to-noise ratio, an effect one usually wishes to avoid.

If, however, sufficiently intense sources or sufficiently sensitive detectors are available, it is of value to consider deliberately sampling only a small proportion of the beam wavefront. If one does this with a sufficiently large array of detectors or, for example, a photographic plate, then a one- or two-dimensional interferometric image of the plasma can be formed.

In this case the spatial coherence requirement is that $\Delta\phi \leq \pi$ not over the whole beam but over that portion of the beam that is imaged onto each spatially resolved element of the detecting array (or photographic plate).

In this application it is usually convenient to introduce some deliberate misalignment into the interferometer so that a pattern of linear interference fringes appears across the detecting plane in the absence of plasma as illustrated in Fig. 4.16. The phase shift introduced by the plasma then shifts the position of the fringes to produce a pattern in the image plane indicating the phase shift. The deliberate misalignment has the merit that it allows unambiguous interpretation of the pattern produced without any need for heterodyne techniques. This is basically because the initial phase of the interferometer is then continuously varying across the beam so that quadrature information is available directly on the plate. One can think of this as *spatial* phase modulation as opposed to *temporal* phase modulation of the heterodyne systems.

An example of the type of result that may be obtained is shown in Fig. 4.17. The technique has been used most successfully in the visible wavelength range using pulsed lasers and photographic techniques. Unfortunately, these wavelengths require dense plasmas to produce significant phase shifts so that many plasmas are not suitable for diagnosis using visible interferometry.

The requirements on flatness of optical components to maintain straight initial fringes of an optical interferometer are quite severe; however, it is possible to circumvent these restrictions by double-exposure holography, a

Fig. 4.16. Deliberate misalignment is used to provide a pattern of reference fringes even in the absence of plasma.

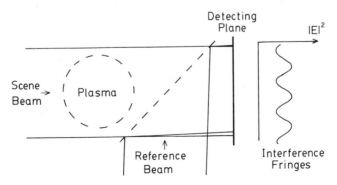

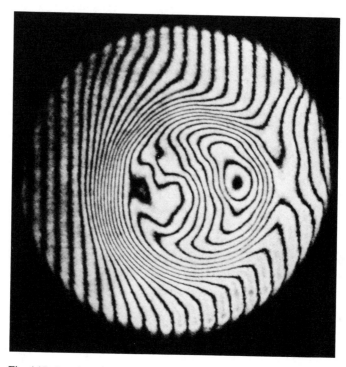

Fig. 4.17. Interferometric image of a θ pinch [courtesy F. Jahoda [(1985)] showing the circular plasma compressed away from the outer low-density regions in which the fringes are straight.

more complicated interferometric technique, described in the plasma context by Jahoda and Sawyer (1971). Some authors use the expression "holographic interferometry" to refer to the simple interferometric imaging just described. However, the spatial resolution of the phase information in such an image is generally much lower than in a normal hologram [see, e.g., Jones and Wykes (1983)] and the image reconstruction does not require coherent illumination. Therefore, the expression is not fully appropriate.

4.2.7   *Schlieren and shadowgraph imaging*

There are ways other than interferometry to obtain information on the spatial variation of the refractive index across a probing beam. These rely directly upon the deviation of the different parts of the beam due to refraction. In these techniques no separate reference beam is used; rather, the intensity variations arise by virtue of local intensification of the probing beam due to refraction.

The distinction between schlieren and shadowgraph methods is generally that schlieren techniques are sensitive to the first (spatial) derivative of the refractive index while shadowgraphy depends on the second derivative.

The basic principles of the schlieren approach are illustrated in Fig. 4.18. An approximately plane parallel beam illuminates the refractive plasma slab, whose thickness must be much less than the distance to the imaging lens. The ray bundle at any point is deviated by an angle

$$\theta = \frac{d}{dy} \int N \, dl, \qquad (4.2.20)$$

as shown earlier. A knife edge at the focal point of the lens partially obstructs the image formed there of the undeviated beam. The ray deviation causes this obstruction either to decrease or to increase according to the sign of θ. The image of the plasma itself is not shifted, but its intensity is altered by the variation in the obstructing effect of the knife edge. The result is that, for small enough deviations, the change in image intensity is proportional to the local value of θ and hence of $d(\int N \, dl)/dy$.

Shadowgraphy is even simpler in experimental layout, as illustrated in Fig. 4.19. No imaging optics need be present between the refractive plasma

Fig. 4.18. Principle of the schlieren approach to refractive imaging.

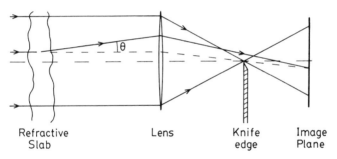

Refractive Lens Knife Image
Slab   edge Plane

Fig. 4.19. Shadowgraph imaging.

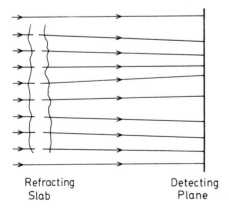

Refracting Detecting
Slab   Plane

slab and the detecting plane a distance L away. Variations in intensity occur because rays are deviated by angle θ, which varies across the slab. The effect is to cause the electromagnetic energy that would have fallen upon the point y to be moved to the point $y' = y + L.\theta(y)$, where again

$$\theta = \frac{d}{dy} \int N \, dl. \tag{4.2.21}$$

The same effect occurs in the orthogonal direction within the plasma denoted by coordinate x. Thus, the ray incident at position (x, y) is moved to the position on the detecting plane

$$(x', y') = \left(x + L \frac{d}{dx} \left[\int N \, dl \right], \; y + L \frac{d}{dy} \left[\int N \, dl \right] \right). \tag{4.2.22}$$

If the incident beam is of uniform intensity I_i then the detected intensity I_d will be given by

$$I_d \, dx' \, dy' = I_i \, dx \, dy, \tag{4.2.23}$$

so that

$$\frac{I_i}{I_d} = 1 + L \left[\frac{d^2}{dx^2} + \frac{d^2}{dy^2} \right] \left(\int N \, dl \right). \tag{4.2.24}$$

This gives, for small fractional intensity variations,

$$\frac{\Delta I_d}{I} = L \left[\frac{d^2}{dx^2} + \frac{d^2}{dy^2} \right] \int N \, dl, \tag{4.2.25}$$

as stated earlier, a signal proportional to the second derivative of the refractive index.

The limitations on these two methods may be seen as follows. For a schlieren technique, if $d\phi/dy$ is constant, the maximum angular deviation detectable is equal to the angular beam divergence of the illuminating beam. Larger deviations will leave the intensity at either the maximum brightness or fully dark because the focal spot will be obscured by the knife edge either not at all or completely. The minimum detectable deviation will depend upon the dynamic range of the detector; that is, how small an intensity variation is detectable. Note that this might mean that to use a diffraction-limited illumination beam is not optimum when large deviations are expected. If a diffraction-limited beam is used, however, the maximum detectable angle corresponds to a refractive-index gradient that, continued constant across the whole beam, corresponds to a total phase difference across the beam of $\Delta\phi \approx \pi$. If $d\phi/dy$ is not constant, that is, the scale length of the perturbation is smaller than the beam size, the effective diffraction angle corresponds to the perturbation size rather than the beam size and is proportionately larger than for constant $d\phi/dy$. The result is that $\Delta\phi \approx \pi$ is the maximum detectable phase shift for all scale lengths. The shadowgraph's allowed maximum phase shift before rays cross and hence cause ambiguities is $\Delta\phi \approx \pi d^2/\lambda L$, where d is the scale size of the refractive-index variation. Since $d^2/\lambda L > 1$ is the condition for the intensity enhancements not to be "washed out" by diffraction before the screen is reached, it is thus clear that the maximum $\Delta\phi$ is greater than π and, depending upon d and L, may be much greater. Thus, shadowgraphy is generally less sensitive than schlieren but can handle larger phase shifts.

4.2.8 *Scattering from refractive-index perturbations*

There is one further large topic that can be thought of as being a refractive-index based measurement, namely the scattering of electromagnetic waves from density (and hence refractive index) nonuniformities. Here we are thinking of perturbations whose scale length is generally much smaller than the plasma or beam size.

Of course, the shadowgraph and schlieren techniques are sensitive to such perturbations and so we have in part already ventured into this area. However, the problem of electromagnetic wave scattering is not most easily approached by the geometrical optics approximation that we have used (with due note of diffraction limitations) so far. For this reason, consideration of scattering will be deferred to a later chapter, even though it can be shown to be mathematically equivalent in some cases to some of the methods we have discussed here.

4.3 **Magnetic field measurement**
4.3.1 *Effect of magnetic field*

As we have seen, the refractive index of the plasma is primarily determined by the electron density and also the magnetic field, through the two quantities $X \equiv \omega_p^2/\omega^2$ and $Y \equiv \Omega/\omega$, respectively. In our consideration of density measurement by plasma interferometry we assumed that the magnetic field effects were negligible either because $Y \ll 1$ or because only the ordinary mode at perpendicular propagation was used, so that the refractive index was given by $N^2 = 1 - X$. Let us begin our discussion of magnetic field effects by considering the case of small but finite Y and examine the accuracy of our previous assumptions about N^2.

The Appleton–Hartree dispersion relation, retaining only first order terms in the assumed small parameter Y, can be written

$$N^2 \approx 1 - X \pm XY\cos\theta, \qquad (4.3.1)$$

where $+$ refers to the ordinary wave. Thus the lowest order correction to the refractive index is the additional term in N^2: $\pm XY\cos\theta$. (This term tends to zero as $\theta \to \pi/2$, so for perpendicular propagation one must go to the next order and the correction to N^2 for the extraordinary wave is $\sim XY^2/(1 - X)$, the ordinary wave having $N^2 = 1 - X$ exactly, of course.)

From the viewpoint of density measurement, then, the presence of a magnetic field introduces an extra term leading to a fractional error of magnitude $Y\cos\theta$ (for $\theta \neq \pi/2$) in the density deduced by the previous analysis. This may occasionally be significant but usually can be ignored provided $\omega \gg \Omega$.

On the other hand, the finite magnetic field introduces anisotropy and hence birefringence into the wave propagation. That is, the characteristic waves now have different refractive indices. The difference is perhaps only small, but if it is measured it can give information on the magnetic field inside the plasma and hence offer important diagnostic possibilities.

4.3.2   *Faraday rotation*

Consider a wave propagating through a medium in which the polarizations of the two characteristic modes are circular, that is, in a coordinate system with **k** along z the polarization is

$$\frac{E_x}{E_y} = \pm i. \tag{4.3.2}$$

Suppose also that these characteristic waves have different refractive indices N_+, N_-. Then the progress of a wave of arbitrary polarization $\mathbf{E}(z)$ is determined by resolving **E** into two circular components, corresponding to a superposition of characteristic waves, and then allowing these two waves to propagate with their known refractive indices. The wave amplitude at any other position is then determined by the superposition of the waves there. Their phase will, in general, be different because of the different Ns, so that the polarization will vary with position.

In particular, suppose that at $z = 0$ the wave is linearly polarized such that $E_y = 0$, $E_x = E$. Then this must be written as

$$\mathbf{E}(0) = \frac{E}{2}\left[(1, -i) + (1, +i)\right]. \tag{4.3.3}$$

At $z \neq 0$ this decomposition will then become

$$\mathbf{E}(z) = \frac{E}{2}\left[(1, -i)\exp\left(iN_+\frac{\omega}{c}z\right) + (1, +i)\exp\left(iN_-\frac{\omega}{c}z\right)\right]$$

$$= E\exp\left[i\left(\frac{N_+ + N_-}{2}\right)\frac{\omega}{c}z\right]\left(\cos\frac{\Delta\phi}{2}, \sin\frac{\Delta\phi}{2}\right), \tag{4.3.4}$$

where

$$\Delta\phi \equiv (N_+ - N_-)\frac{\omega}{c}z \tag{4.3.5}$$

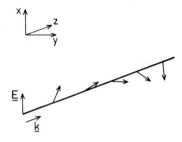

Fig. 4.20. Faraday rotation causes a linearly polarized wave's **E** vector to rotate as the wave propagates.

is the phase difference between the characteristic waves arising because of the difference in refractive index. Thus, the polarization of the wave after propagating this distance z is linear still, but rotated by an angle $\Delta\phi/2$ with respect to the initial polarization. This is the effect known as Faraday rotation, illustrated in Fig. 4.20.

Now consider the case of magnetized plasma propagation. We have already obtained the refractive index for propagation at a general angle θ to the magnetic field. For the purposes of evaluating Faraday rotation we also require the characteristic polarization. This requires solving Eq. (4.1.6) for **E**. In our present coordinate system, with k along the z axis and taking the x axis perpendicular to B [not the same system as used for Eq. (4.2.3)], as illustrated in Fig. 4.21, the transverse components of **E** are related by

$$\frac{E_x}{E_y} = -\frac{iY\sin^2\theta}{2(1-X)\cos\theta} \pm i\left[1 + \frac{Y^2\sin^4\theta}{4(1-X)\cos^2\theta}\right]^{1/2} \tag{4.3.6}$$

(see Exercise 4.5). Now provided $Y\sec\theta \ll 1$ (and $1-X$ is not small), this may be approximated to lowest order as

$$\frac{E_x}{E_y} \approx \pm i, \tag{4.3.7}$$

Fig. 4.21. Coordinate system for expressing the wave polarization.

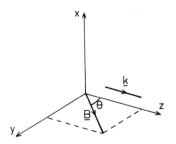

which could also have been obtained using the approximate form of the refractive index Eq. (4.3.1). Thus, we find the perhaps surprising result that for weak magnetic fields, at all angles not too close to perpendicular ($Y \sec \theta \ll 1$), the characteristic polarizations are indeed circular.

We can therefore immediately apply our previous analysis and find that the Faraday rotation angle is

$$\alpha = \frac{\Delta\phi}{2} = \frac{1}{2}(N_+ - N_-)\frac{\omega}{c}z \approx \frac{1}{2}\left[\frac{XY\cos\theta}{(1-X)^{1/2}}\right]\frac{\omega}{c}z. \qquad (4.3.8)$$

If, in addition, $X \ll 1$, this approximates to

$$\alpha = \frac{\Delta\phi}{2} \approx \frac{\omega_p^2 \Omega \cos\theta}{2\omega^2 c}z. \qquad (4.3.9)$$

Whichever of these forms is appropriate, the Faraday rotation is proportional to $\Omega \cos\theta$, which is $(e/m)\mathbf{B}\cdot\mathbf{k}/k$. Thus, Faraday rotation is proportional to the *parallel* component of the magnetic field. In the approximation $X \ll 1$ it is also proportional to ω_p^2, that is, to electron density.

For nonuniform plasmas, as with the treatment of interferometry, we suppose that a WKBJ approach can be adopted. Then the total Faraday rotation along the beam is given by

$$\alpha = \frac{\Delta\phi}{2} = \frac{1}{2}\int \frac{\omega_p^2 \Omega \cos\theta}{\omega^2\left(1 - \omega_p^2/\omega^2\right)^{1/2}}\frac{dl}{c}, \qquad (4.3.10)$$

which may be written

$$\alpha = \frac{\Delta\phi}{2} = \frac{e}{2m_e c}\int \frac{n_e \mathbf{B}\cdot\mathbf{dl}}{n_c(1 - n_e/n_c)^{1/2}}$$

$$\approx \frac{e}{2m_e c}\int \frac{1}{n_c}\cdot n_e \mathbf{B}\cdot\mathbf{dl}, \qquad \text{for } \frac{n_e}{n_c} \ll 1. \qquad (4.3.11)$$

4.3.3   *Propagation close to perpendicular*

When the direction of propagation is sufficiently close to perpendicular that $Y \sec\theta \gtrsim 1$, effects other than simply polarization rotation become important, because the characteristic waves are no longer approximately circularly polarized. A linearly polarized incident wave then generally acquires some degree of ellipticity in its polarization: This is known as the Cotton–Mouton effect. As a practical matter it is usually more convenient when possible to choose a frequency and propagation angle such that $Y \sec\theta \ll 1$, so that our previous approximations do hold. When this is impossible a more elaborate analysis is necessary.

Elegant, but more abstract, mathematical apparatus exists to deal with such cases, using a representation of the polarization known as the Poincaré sphere [see, e.g., DeMarco and Segre (1972)]. It would take us too far from our main theme to master these methods here, but the important results may be obtained by a generalization of our more elementary treatment. Even so, the algebra is somewhat tedious and is omitted in places.

We use the general expression for the polarization ratio Eq. (4.3.6), which we write as

$$E_x/E_y = iq = i\left\{-1 \pm [F^2 + 1]^{1/2}\right\}/F, \qquad (4.3.12)$$

where

$$F \equiv 2(1 - X)\cos\theta/Y\sin^2\theta. \qquad (4.3.13)$$

We shall require two identities, which follow by elementary manipulation:

$$\frac{2q}{q^2 + 1} = \pm\frac{F}{(F^2 + 1)^{1/2}}; \qquad \frac{q^2 + 1}{q^2 - 1} = \mp(F^2 + 1)^{1/2}. \qquad (4.3.14)$$

We shall also need a more general form of the difference between the refractive indices of the two characteristic waves:

$$N_+ - N_- \approx \frac{1}{2}(N_+^2 - N_-^2) = \frac{XY\cos\theta}{2(1 - Y^2)}\frac{(F^2 + 1)^{1/2}}{F}, \qquad (4.3.15)$$

for $X \ll 1$ (after considerable algebra). Now we express the wave electric field in terms of its decomposition into the two characteristic wave polarizations,

$$\mathbf{E}_+ = (iq, 1)/(q^2 + 1)^{1/2}, \qquad \mathbf{E}_- = (1, iq)/(q^2 + 1)^{1/2}, \qquad (4.3.16)$$

in the coordinate system of Fig. 4.21, where we now choose to make q correspond explicitly to the upper sign in Eq. (4.3.6) (in other words, $E_x/E_y = iq$ for $\mathbf{E}_+$; $E_x/E_y = -i/q$ for $\mathbf{E}_-$). So we write at $z = 0$,

$$\mathbf{E}(0) = a_+\mathbf{E}_+ + a_-\mathbf{E}_-. \qquad (4.3.17)$$

The characteristic waves propagate separately for a small distance z, acquiring different phases $\exp(iN_\pm\omega z/c)$, so that then

$$\mathbf{E}(z) = \exp\left[i\left(\frac{N_+ + N_-}{2}\right)\frac{\omega}{c}z\right]$$

$$\times\left[a_+\mathbf{E}_+\exp\left(+i\frac{\Delta\phi}{2}\right) + a_-\mathbf{E}_-\exp\left(-i\frac{\Delta\phi}{2}\right)\right], \qquad (4.3.18)$$

where, as before, $\Delta\phi \equiv (N_+ - N_-)(\omega/c)z$. We take the input wave at $z = 0$

to be linearly polarized (with unit magnitude) at an angle β to the x axis, $\mathbf{E}(0) = (\cos\beta, \sin\beta)$. It is most convenient to work in a coordinate system in which $\mathbf{E}(0)$ is aligned along the (new) 1 axis. So we transform all our vectors to this new coordinate system by multiplying them by the rotation matrix

$$\begin{bmatrix} \cos\beta & \sin\beta \\ -\sin\beta & \cos\beta \end{bmatrix}. \tag{4.3.19}$$

In the new coordinate system

$$\begin{aligned} \mathbf{E}_+ &= \left(iq\cos\beta + \sin\beta, \ -iq\sin\beta + \cos\beta\right)/\left(q^2+1\right)^{1/2}, \\ \mathbf{E}_- &= \left(\cos\beta + iq\sin\beta, \ -\sin\beta + iq\cos\beta\right)/\left(q^2+1\right)^{1/2}, \end{aligned} \tag{4.3.20}$$

and in either coordinate system

$$\begin{aligned} a_+ &= \left(-iq\cos\beta + \sin\beta\right)/\left(q^2+1\right)^{1/2}, \\ a_- &= \left(\cos\beta - iq\sin\beta\right)/\left(q^2+1\right)^{1/2}. \end{aligned} \tag{4.3.21}$$

We substitute these values in our expression for $\mathbf{E}(z)$ and retain only the lowest order terms in $\Delta\phi$, to get

$$\begin{aligned} \mathbf{E}(z) &\approx \exp\left[i\left(\frac{N_+ + N_-}{2}\right)\frac{\omega}{c}z\right] \\ &\times \left(1, -i\frac{\Delta\phi}{2}\frac{q^2-1}{q^2+1}\sin 2\beta + \frac{\Delta\phi}{2}\frac{2q}{q^2+1}\right). \end{aligned} \tag{4.3.22}$$

The electric field thus acquires two components perpendicular to its original direction, one real and one imaginary, which become, on substitution for $\Delta\phi$, etcetera,

$$E_2(z) = \left[\frac{XY\cos\theta}{2(1-Y^2)} + i\frac{XY^2\sin^2\theta}{2(1-Y^2)}\sin 2\beta\right]\frac{\omega}{c}z. \tag{4.3.23}$$

The imaginary part corresponds to an ellipticity gained by the polarization, whilst the real part corresponds to a rotation of the polarization (major axis), as illustrated in Fig. 4.22.

The Faraday rotation angle is essentially as in the $Y\sec\theta \ll 1$ case,

$$\alpha = \frac{XY\cos\theta}{2(1-Y^2)}\frac{\omega}{c}z \approx \frac{1}{2}XY\cos\theta\frac{\omega}{c}z, \tag{4.3.24}$$

when both X and Y are small. Thus Faraday rotation, for small $\Delta\phi$, is unchanged by propagation close to perpendicular. The ellipticity is such that the ratio of the major axes is

$$\frac{b}{a} = \left|\frac{XY^2\sin^2\theta}{2(1-Y^2)}\sin 2\beta\frac{\omega}{c}z\right|. \tag{4.3.25}$$

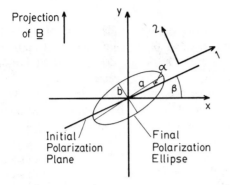

Fig. 4.22. The polarization ellipse and the polarization coordinate systems for the case when propagation is close to perpendicular.

This is not necessarily small compared with α unless $Y \sec\theta \sin 2\beta \ll 1$. This criterion is thus what is required for the ellipticity, acquired by linear birefringence, to be negligible. It shows that the optimum input polarization for minimizing the ellipticity is perpendicular or parallel to the magnetic field (i.e., $\beta = 0$ or $\pi/2$). Note that the criterion we used in the previous section for ignoring linear birefringence ab initio, $Y \sec\theta \ll 1$, guarantees that the present less restrictive condition be satisfied.

All of this analysis requires the phase difference $\Delta\phi$ between the characteristic waves to be small. This is the other requirement for Faraday rotation to remain dominant even at perpendicular propagation. Provided $\Delta\phi \ll 1$ then a spatially inhomogeneous plasma gives rise to the immediate generalizations

$$\alpha \approx \int \frac{1}{2} XY \cos\theta \frac{\omega}{c}\, dz,$$

$$\frac{b}{a} \approx \left| \int \frac{1}{2} XY^2 \sin^2\theta \sin 2\beta \frac{\omega}{c}\, dz \right|.$$

$$(4.3.26)$$

If $\Delta\phi$ is not small, then our elementary treatment fails and numerical integration of the full polarization equations is generally required. Moreover, the information about the internal magnetic field is eventually lost when $\Delta\phi \gg 1$ since the polarization observed becomes dependent only on the magnetic field at the plasma edge. This is because the wave anisotropy is so great that the two characteristic waves are decoupled and propagate through the plasma retaining their separate identities. The fraction of power that enters the plasma in the ordinary (or extraordinary) mode thus exits in that same mode, regardless of the intervening internal plasma characteristics. The phase difference between the modes rapidly becomes virtually random (because it is large).

4.3.4   *Measurement of the polarization*

The ratio of the Faraday rotation angle to the phase shift of an interferometer at the same frequency in a given plasma is approximately $2(\Omega/\omega)\cos\theta$, as may be verified by comparing Eqs. (4.2.10) and (4.3.11). As we have noted, Faraday rotation is generally easiest to interpret when the condition $Y\sec\theta \ll 1$ is satisfied. This implies that $\Omega/\omega = Y$ is small, and so the rotation angle will be much less than the interferometer phase shift, even more so in cases when the propagation angle is substantially different from zero.

Even with optimum choice of frequency, the Faraday rotation angle may thus be quite small ($\ll \pi$) and so sensitive techniques need to be employed to measure it. One method that is particularly appropriate involves modulation of the polarization. For example, suppose that the linear input polarization is arranged to be at an angle that rotates at frequency $\Delta\omega/2$ ($\ll \omega$). The reason for writing the rotation frequency thus is that this situation is equivalent to having two collinear incident beams, circularly polarized with opposite E rotation direction, but now, instead of them having identical frequency (which would give linear polarization at constant angle), they have frequencies different by $\Delta\omega$, so that their relative input phase changes as $\Delta\omega \cdot t$ and the linear superposition rotates at frequency $\Delta\omega/2$.

This approach is essentially identical to the heterodyne interferometer discussed earlier, except that in this case both beams pass through the plasma and the phase difference being measured is the difference between two polarizations, rather than the difference between the two spatially separated plasma and reference beams.

Because the Faraday rotation measures approximately the integral of $n_e B$, in order to deduce B we must know n_e as well. The most satisfactory

Fig. 4.23. Principle of a heterodyne scheme for simultaneous Faraday rotation and interferometry.

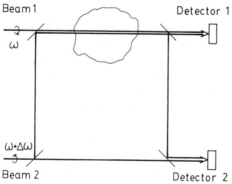

way to get this is generally to measure n_e by interferometry simultaneously along the same path. This may be done using the same radiation sources, by providing an additional reference beam path, which is then appropriately mixed with a proportion of the transmitted beam to give an ordinary interferometer. This amounts to measuring simultaneously the refractive indices of both characteristic waves N_+, N_- (or equivalently measuring N_+ and ΔN). Schematically, such a measurement is illustrated in Fig. 4.23 (but see examples) in approximately the form proposed by Dodel and Kunze (1978).

Despite the importance of internal magnetic field measurements, the use of Faraday rotation is only at a fairly preliminary stage in plasma diagnostics. Nevertheless, recent experiments, illustrated in Fig. 4.24, have shown that measurements with useful accuracy can be made even of the relatively small poloidal field in a tokamak. It seems probable that the future will see its increasing development and implementation.

Fig. 4.24. Simultaneous measurements of interferometer phase shift and Faraday rotation angle as a function of chord major radius in the Textor tokamak [after H. Soltwisch (1983)]. The rotation reverses with the direction of the vertical component of the (poloidal) magnetic field. From such data information about the toroidal current density can be obtained.

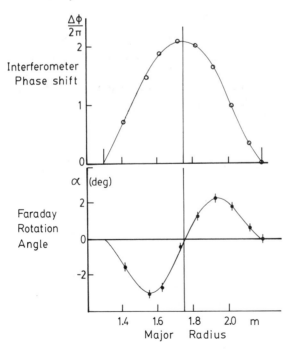

4.4   Abel inversion

Interferometry and polarimetry share with many other diagnostic techniques the property that they measure the average value of some quantity along a chord through the plasma. A recurrent problem is then to deduce local values of the quantity under consideration from the available chordal measurements. Naturally this problem is of much wider interest than just plasma diagnostics. However, very many plasmas have the property that they are cylindrically symmetric, that is, they are independent of θ (and z) in a cylindrical coordinate system (r, θ, z). This fact enables one to address the problem of deducing the radial distribution from the chordal measurements using the known mathematical properties of the Abel transform. It is to this generic problem that we now turn.

Consider, then a cylindrically symmetric quantity (such as refractive index or radiative emissivity) $f(r)$ of which the accessible measurements are chord integrals,

$$F(y) = \int_{-\sqrt{(a^2 - y^2)}}^{+\sqrt{(a^2 - y^2)}} f(r)\, dx, \qquad (4.4.1)$$

as illustrated in Fig. 4.25. We may change the x integral into an r integral to get

$$F(y) = 2\int_{y}^{a} f(r)\, \frac{r\, dr}{(r^2 - y^2)^{1/2}}. \qquad (4.4.2)$$

This relationship between F and f is an integral equation for F (of the Volterra type) first studied by the mathematician Abel in the nineteenth century. F is sometimes said to be the Abel transform of f. The inverse transform relates the quantity we seek $[f(r)]$ to an integral of F, as follows:

$$f(r) = -\frac{1}{\pi}\int_{r}^{a} \frac{dF}{dy}\, \frac{dy}{(y^2 - r^2)^{1/2}}, \qquad (4.4.3)$$

Fig. 4.25. Chordal measurement in a cylinder.

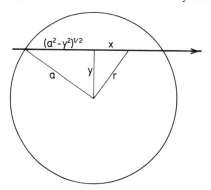

provided that $f(a) = 0$ (see Exercise 4.7). We therefore have a simple formula for obtaining the radial profile of f from measurements of chord integrals F. This process is often called Abel inversion.

Naturally, in practical situations, we have measurements of F at only a finite number of y values, so that to perform the required integral requires some kind of interpolation scheme to be adopted. Moreover, it is clear that a small number of chord measurements will give only limited information about $f(r)$. Thus, if an effectively continuous $f(r)$ is deduced using interpolation, it must be remembered that much of the details of f deduced will depend upon the *assumptions* inherent in the interpolation scheme.

Finally, note that in the inverse transform, it is the spatial derivative of F that appears. This tends to make the Abel inversion $[f(r)]$ rather sensitive to any errors in $F(y)$. One can see this easily by realizing that deduction of dF/dy will require essentially taking the difference (ΔF) of adjacent F measurements. If these are such that ΔF is considerably smaller than F, as will usually be the case if F is measured at a reasonable number of y values, then the fractional error in ΔF is much greater than that in F. Fortunately, this effect is partially compensated by the integration occurring in Eq. (4.4.3), which "smooths out" some of the errors generated by differentiation. Nevertheless, enhanced error sensitivity remains to errors in dF/dy near $r = y$.

We shall discuss in Chapter 5 some more general inversion problems when cylindrical symmetry cannot be assumed.

4.5   Reflectometry

When a wave of a certain frequency propagates through a plasma with density increasing in the direction of propagation, it may arrive at a point where the electron density equals the critical density n_c. As we have seen, the wave is evanescent at higher density; therefore, what will happen is that the wave will be *reflected* from the cutoff point and propagate back down the density gradient and out of the plasma the way it came in. If the reflected wave is detected, it is possible to use it to diagnose the plasma density. This is then called reflectometry.

Naturally reflectometry depends upon the presence of a cutoff within the plasma; therefore, it requires that somewhere there is a point at which

$$\omega_p > \omega. \tag{4.5.1}$$

This is obviously the opposite inequality for the effective operation of a transmission interferometer, so reflectometry generally uses lower frequency than interferometry (for the same plasma).

The mere observation of substantial reflected power indicates that reflection is occurring and hence gives information about the electron density, namely that somewhere $n_e > n_c$. However, the objective of practical re-

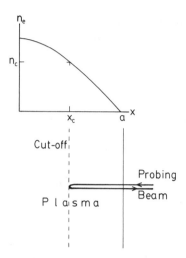

Fig. 4.26. The principle of reflectometry.

flectometry is to say much more than this in order to be really useful. To fulfill this objective it is necessary to measure the *phase* of the reflected wave not just its power. Measurement of phase generally involves using coherent interference effects, so that reflectometry is in reality a type of interferometry.

The form of experiment we are considering is illustrated in Fig. 4.26. A wave is launched into the plasma, is reflected from the cutoff layer, and detected again near its launch point. The similarities with radar are obviously very strong (although in radar the group delay of a pulse is generally used for ranging rather than the phase delay of a continuous wave as is more usual in reflectometry). If we measure the relative phase of the reflected and incident waves this gives us information about where in the plasma the reflection occurs, and hence about density as a function of position.

It might be thought, by analogy with radar, that the phase delay provides a single number that is directly proportional to the distance of the cutoff layer from the launching point, independent of the density profile elsewhere in the plasma. It is important to emphasize that this is *not* the case. The analogy with radar breaks down because the plasma between the launching point and the reflection point acts, as it does in interferometry, to alter the phase delay. In other words, we must take into account the refractive index of the plasma all along the wave path, unlike radar, in which the refractive index is essentially unity everywhere.

4.5.1   *Calculation of the phase delay*
We shall discuss a one-dimensional plasma slab model in which the density n_e varies with position (x) giving rise to a plasma frequency that is

a function of position $\omega_p(x)$. We consider only the ordinary wave, for which the refractive index is given by Eq. (4.2.1). The WKBJ form of the phase difference between points A and B is then

$$\phi(B) - \phi(A) = \frac{1}{c}\int_A^B \left(\omega^2 - \omega_p^2\right)^{1/2} dx. \qquad (4.5.2)$$

Near the reflection point $\omega = \omega_p$ the usual conditions for the applicability of the WKBJ approximation break down because $k \to 0$. However, it can be shown that, provided the density gradient can be taken as uniform in the region of reflection (a condition not always satisfied in practice), the phase difference at position a between forward and reflected waves can be written

$$\phi = \frac{2}{c}\int_{x_c}^a \left(\omega^2 - \omega_p^2\right)^{1/2} dx - \frac{\pi}{2} \qquad (4.5.3)$$

(Ginzburg 1961), where x_c is the position where $\omega_p = \omega$, that is, the cutoff position. This equation states that the phase is just what would be obtained from simple-minded application of the WKBJ approach, regarding the cutoff layer as a mirror, except that an additional $\pi/2$ phase change at reflection must be included.

Identifying a as the edge of the plasma, where the wave is launched, and recalling that $\omega_p^2 \propto n_e$ it is clear that ϕ is determined, via Eq. (4.5.3), by an integral of $(1 - n_e/n_c)^{1/2}$ from the plasma edge to the cutoff point. This is one of the unattractive features of reflectometry, that measurement of a single phase delay cannot be related to some simple average value of (say) the density in the way that interferometry can.

If, however, we are more ambitious and wish to deduce a complete density profile rather than just an average density the comparison between interferometry and reflectometry is not so unfavorable. In fact, there are close mathematical parallels between the deduction of radial density profiles from interferometric measurements along *different chords* of a cylindrical plasma and the deduction of density profiles along a reflectometer's wave path from phase measurements at *different frequencies*. Both involve Abel transforms, as we can show readily as follows.

Suppose that we have (ideally) measurements of the phase delay ϕ at lots of different frequencies so that we can construct the function $\phi(\omega)$ (by interpolation if necessary) for all relevant frequencies. Differentiation of Eq. (4.5.3) with respect to ω and substitution of the (vacuum) wavelength $\lambda = 2\pi c/\omega$ as the integration variable leads straightforwardly to

$$\frac{d\phi}{d\omega} = 2\int_\lambda^\infty \left(\frac{dx_c}{c\,d\lambda_p}\right)\frac{\lambda_p\,d\lambda_p}{\left(\lambda_p^2 - \lambda^2\right)^{1/2}}. \qquad (4.5.4)$$

This is exactly the same integral equation (though with infinite upper limit, which is unimportant) as occurs in the chordal Abel inversion problem Eq. (4.4.2), provided we make the identifications

$$\frac{d\phi}{d\omega} \to F(y), \qquad \frac{dx_c}{c\,d\lambda_p} \to f(r), \qquad \lambda \to y, \qquad \lambda_p \to r. \qquad (4.5.5)$$

Perhaps the most convenient form of the inverse transform for the reflectometry case is

$$x_c(\omega) = a + \frac{c}{\pi} \int_0^\omega \frac{d\phi}{d\omega'} \frac{d\omega'}{(\omega^2 - \omega'^2)^{1/2}}, \qquad (4.5.6)$$

which may be verified directly from Eq. (4.5.3) (Exercise 4.8).

One can, therefore, deduce the position of the cutoff $x_c(\omega)$ provided one knows the phase delay $\phi(\omega)$ for all frequencies less than ω. In reality it is not necessary to have information for frequencies below which $d\phi/d\omega$ is negligible. For low enough frequencies the wave is essentially reflected immediately at the plasma edge so that $d\phi/d\omega \approx 0$ and no contribution is made to the inverse integral Eq. (4.5.6).

4.5.2   *Relative merits of reflectometry and interferometry*

In order to obtain a density profile, interferometry requires measurements over many chords while reflectometry requires measurements over many frequencies. Therefore, the relative ease of application of these techniques depends a great deal on the details of the plasma experiment under consideration. In practice the awkwardness of having to make measurements over a wide range of frequencies for reflectometry is usually sufficient to make interferometry more attractive. Moreover, reflectometry frequencies must be specifically chosen to match a particular density plasma whereas the operation of an interferometer is usually satisfactory over a much wider density range.

The fact that reflectometry cannot deal with density profiles that are not monotonic (since it cannot "see over the horizon" of a local density maximum) is a significant handicap. Also the resolution of the phase ambiguity (modulo 2π) and directional ambiguities, which we discussed in detail for the interferometer, tends to be more difficult for reflectometry. All these factors contribute to the conclusion that, in most situations, interferometry (when feasible) is more satisfactory than reflectometry in the laboratory, except possibly for obtaining the details of density at the plasma edge.

On the other hand, in ionospheric work, for example, where transmission interferometry is rarely possible, reflectometry naturally dominates the picture. In this case, very often, the group delay of a wave pulse is used, as in radar, so a formulation of the problem different from that presented here must be used.

Further reading
Wave propagation in plasmas is treated in most introductory plasma physics textbooks, and the principles of interferometry in most introductions to optics.

A monograph devoted primarily to refractive-index based plasma diagnostics, which is still valuable, even though some of the applications are dated, is:

Heald, M. A. and Wharton, C. B. (1965). *Plasma Diagnostics with Microwaves*. New York: Wiley.

Another excellent introduction is

Jahoda, F. C. and Sawyer, G. A. (1971). Optical refractivity of plasmas. In *Methods of Experimental Physics*. R. H. Lovberg and H. R. Griem, eds., Vol. 9B. New York: Academic.

A detailed discussion of interferometers and associated technology in the far-infrared for plasma diagnostics is given by:

Veron, D. (1979). In *Infrared and Millimeter Waves*. K. J. Button, ed., Vol. 2. New York: Academic.

A wider ranging discussion is given by:

Luhmann, N. C. (1979). In *Infrared and Millimeter Waves*. K. J. Button, ed., Vol. 2. New York: Academic.

Exercises
4.1 Solve Eq. (4.1.23) to obtain Eq. (4.1.24).

4.2 Consider a beam propagating along a chord of a refractive cylinder at a distance y from the axis. Suppose the cylinder has a refractive index $N(r)$, where r is the radius. Obtain a general equation for the angular deviation of the beam θ due to refraction when $\theta r \ll y$ so that the chord can be approximated as straight. In the case where $\omega \gg \omega_p$ and $n_e = n_0(1 - r^2/a^2)$, calculate the value of y at which θ is greatest and prove that this maximum θ is n_0/n_c [Eq. (4.2.17)].

4.3 In some plasmas with high magnetic shear, one cannot guarantee doing interferometry with the ordinary mode. It is of interest to calculate the error occurring in the deduced density if we use the expression $N^2 = 1 - \omega_p^2/\omega^2$ when really the mode is the extraordinary mode. Consider perpendicular propagation and calculate an approximation for the difference in refractive index between the ordinary and extraordinary wave for $\omega \gg \omega_p$. Hence calculate the fractional error in using the above expression to determine density if the extraordinary mode is used in a plasma with $n_e = 10^{20}$ m^{-3}, $B = 6$ T, and $\omega/2\pi = 10^{12}$ Hz.

4.4 Starting with the general equations for propagation with the k vector in the z (3) direction and arbitrary dielectric tensor

$$\left(\sum_j\right)\left(k_i k_j - k^2 \delta_{ij} + \frac{\omega^2}{c^2} \varepsilon_{ij}\right) E_j = 0,$$

eliminate E_3 and k and hence obtain the equation for the polarization, $\rho \equiv E_1/E_2$, in the form

$$-\left(\varepsilon_{21}\varepsilon_{33} - \varepsilon_{31}\varepsilon_{23}\right)\rho^2 + \left(\varepsilon_{11}\varepsilon_{33} - \varepsilon_{22}\varepsilon_{33} + \varepsilon_{32}\varepsilon_{23} - \varepsilon_{31}\varepsilon_{13}\right)\rho$$
$$+ \left(\varepsilon_{12}\varepsilon_{33} - \varepsilon_{32}\varepsilon_{13}\right) = 0.$$

4.5 If B is in the 2/3 plane at an angle θ to z, then

$$\varepsilon_{ij} = \begin{bmatrix} 1 - \dfrac{X}{1 - Y^2} & \dfrac{iXY\cos\theta}{1 - Y^2} & -\dfrac{iXY\sin\theta}{1 - Y^2} \\[2mm] -\dfrac{iXY\cos\theta}{1 - Y^2} & 1 - \dfrac{X(1 - Y^2\sin^2\theta)}{1 - Y^2} & \dfrac{XY^2\cos\theta\sin\theta}{1 - Y^2} \\[2mm] \dfrac{iXY\sin\theta}{1 - Y^2} & \dfrac{XY^2\cos\theta\sin\theta}{1 - Y^2} & 1 - \dfrac{X(1 - Y^2\cos^2\theta)}{1 - Y^2} \end{bmatrix}.$$

Using these values obtain the equation for ρ,

$$\rho^2 + \frac{iY\sin^2\theta}{\cos\theta(1 - X)}\rho + 1 = 0,$$

and hence prove Eq. (4.3.6).

4.6 One may generalize the cold plasma treatment to include collisions by writing the momentum equation for $B_0 = 0$ as

$$m_e \frac{\partial \mathbf{v}}{\partial t} = -e\mathbf{E} - \nu_{ei}m_e\mathbf{v}.$$

Show that this leads to spatial damping of the wave and calculate the damping exponent.

4.7 Substitute Eq. (4.4.3) into the right hand side of Eq. (4.4.2) and reverse the order of integration. Hence verify that Eq. (4.4.3) is the correct solution.

4.8 Prove Eq. (4.5.6) by substituting from Eq. (4.5.3).

4.9 Consider an interferometer that operates at a radiation wavelength $\lambda_C = 10.6\ \mu\text{m}$ (CO_2 laser) in the presence of spurious vibrations of the optical components. To compensate for these vibrations, interferometry is performed simultaneously using the same optical components at $\lambda_H = 0.633$ μm (HeNe laser). The HeNe interferometer is affected much less than the CO_2 by the plasma phase shift, but still somewhat. If $\omega \gg \omega_p$ for both wavelengths:

(a) Derive an expression for the plasma density $\int n_e \, dl$ in terms of the phase shifts ϕ_C and ϕ_H, of the CO_2 and HeNe interferometers.

(b) If ϕ_H can be measured to an accuracy of $\pm\pi$, what uncertainty does this introduce into the plasma density measurement?

(c) Thus evaluate the fractional error in measuring a 1 m thick plasma of density 10^{20} m^{-3}, assuming ϕ_C is measured exactly.

4.10 Use the formulas developed in Exercise 4.9 to calculate the residual error in the density measurement in an interferometer with feedback stabilization that maintains the HeNe phase shift constant. The density will be deduced straight from the CO_2 phase shift assuming that the physical path length is constant. However, the path length will not in fact be constant because of the small phase shift of the HeNe interferometer due to the plasma. Express the density error as a fraction of the true density for general values of the two wavelengths λ_C and λ_H when $\omega \gg \omega_p$. Notice that this result makes it possible to compensate very easily for this error effect in the interpretation.

5 Electromagnetic emission by free electrons

5.1 Radiation from an accelerated charge

There are several ways in which free electrons can emit radiation as well as simply affecting its passage through the plasma via the refractive index. This radiation proves to be of considerable value in diagnosing the plasma, especially since the emission depends strongly upon the electron energy distribution, for example the temperature of a Maxwellian plasma, as well as the electron density. Before moving on to discuss the specific emission processes of interest, we briefly review the fundamentals of single particle radiation. Since the derivations are fairly standard, readers desiring a more detailed treatment should consult a textbook on electromagnetism, such as one of those mentioned in the further reading suggestions.

5.1.1 The radiation fields

Our starting point is Maxwell's equations, now with the fields expressed in terms of scalar potential ϕ and vector potential $\mathbf{A}$ in the Lorentz gauge:

$$\nabla \cdot \mathbf{A} + \frac{\partial \phi}{\partial t} = 0, \tag{5.1.1}$$

$$\mathbf{B} = \nabla \wedge \mathbf{A}, \tag{5.1.2}$$

$$\mathbf{E} = -\frac{\partial \mathbf{A}}{\partial t} - \nabla \phi. \tag{5.1.3}$$

then the potentials satisfy inhomogeneous wave equations

$$\left(\nabla^2 - \frac{1}{c^2} \frac{\partial^2}{\partial t^2} \right) \phi = -\frac{\rho}{\varepsilon_0},$$

$$\left(\nabla^2 - \frac{1}{c^2} \frac{\partial^2}{\partial t^2} \right) \mathbf{A} = -\mu_0 \mathbf{j}. \tag{5.1.4}$$

These can be solved using the Green's function for the wave operator, namely

$$g(\mathbf{x}, t | \mathbf{x}', t') = \frac{\delta(t' + |\mathbf{x} - \mathbf{x}'|/c - t)}{4\pi |\mathbf{x} - \mathbf{x}'|}, \tag{5.1.5}$$

to give

$$\phi = \frac{1}{4\pi\varepsilon_0} \int \frac{[\rho]}{|\mathbf{x} - \mathbf{x}'|} \, d^3x',$$

$$\mathbf{A} = \frac{\mu_0}{4\pi} \int \frac{[\mathbf{j}]}{|\mathbf{x} - \mathbf{x}'|} \, d^3x',$$

(5.1.6)

where here and from now on bold square brackets indicate that a quantity is to be evaluated at the retarded time $t' = t - |\mathbf{x} - \mathbf{x}'|/c$.

Now consider a point charge of magnitude q, position $\mathbf{r}(t)$, and velocity $\mathbf{v}(t)$. Then

$$\rho = q\delta(\mathbf{x} - \mathbf{r}(t)), \qquad \mathbf{j} = q\mathbf{v}\delta(\mathbf{x} - \mathbf{r}(t)).$$

(5.1.7)

Substituting these values and performing the integrals we obtain

$$\phi = \frac{q}{4\pi\varepsilon_0} \left[\frac{1}{\kappa R} \right], \qquad \mathbf{A} = \frac{\mu_0 q}{4\pi} \left[\frac{v}{\kappa R} \right],$$

(5.1.8)

where $\mathbf{R} \equiv \mathbf{x} - \mathbf{r}$ is the vector from the charge to the field point and $\kappa \equiv 1 - \mathbf{R} \cdot \mathbf{v}/Rc$. These are the Lienard–Wiechert potentials.

The electric field may then be calculated from Eq. (5.1.3), taking care with derivatives of retarded quantities, to give

$$\mathbf{E} = \frac{q}{4\pi\varepsilon_0} \left[\frac{1}{\kappa^3 R^2} \left(\hat{\mathbf{R}} - \frac{\mathbf{v}}{c} \right) \left(1 - \frac{v^2}{c^2} \right) + \frac{1}{c\kappa^3 R} \hat{\mathbf{R}} \wedge \left\{ \left(\hat{\mathbf{R}} - \frac{\mathbf{v}}{c} \right) \wedge \frac{\dot{\mathbf{v}}}{c} \right\} \right],$$

(5.1.9)

where $\hat{\mathbf{R}} \equiv \mathbf{R}/R$.

The magnetic field may most easily be obtained as

$$\mathbf{B} = \frac{1}{c} [\hat{\mathbf{R}}] \wedge \mathbf{E}.$$

(5.1.10)

We shall be concerned primarily with the second term in Eq. (5.1.9), which is the radiation term, going as $1/R$. The first term is the near field, which we shall ignore. The radiated power is given by the Poynting vector

$$\mathbf{S} \equiv \mathbf{E} \wedge \mathbf{H} = \frac{1}{\mu_0 c} |E|^2 [\hat{\mathbf{R}}]$$

$$= \frac{q}{16\pi^2\varepsilon_0 c R^2} \left[\frac{1}{\kappa^6} \left| \hat{\mathbf{R}} \wedge \left\{ \left(\hat{\mathbf{R}} - \frac{\mathbf{v}}{c} \right) \wedge \frac{\dot{\mathbf{v}}}{c} \right\} \right|^2 \right] [\hat{\mathbf{R}}].$$

(5.1.11)

This is the energy radiated across unit area per unit time at the field point, not per unit time at the particle. Time at the particle is retarded time

$t' = t - R(t')/c$ so $dt = \kappa \, dt'$ and the energy emitted per unit time-at-particle is κ times the above.

5.1.2   *Frequency spectrum in the far field*

One often requires the frequency spectrum of the emitted radiation. This is obtained by writing the electric field as a Fourier integral

$$\mathbf{E}(t) = \int_{-\infty}^{\infty} e^{-i\omega t} \mathbf{E}(\omega) \, d\omega, \qquad (5.1.12)$$

so that

$$\mathbf{E}(\omega) = \frac{q}{4\pi\varepsilon_0 c} \frac{1}{2\pi} \int_{-\infty}^{\infty} e^{+i\omega t} \left[\frac{1}{\kappa^3 R} \hat{\mathbf{R}} \wedge \left\{ \left(\hat{\mathbf{R}} - \frac{\mathbf{v}}{c} \right) \wedge \frac{\dot{\mathbf{v}}}{c} \right\} \right] dt. \qquad (5.1.13)$$

We change the integration variable to retarded time t' and consider the far field ($R \gg r$) so that we can approximate $\hat{\mathbf{R}}$ as constant and the distance R as $x - \hat{\mathbf{R}} \cdot \mathbf{r}$ in the phase term and constant elsewhere. Then we get

$$\mathbf{E}(\omega) = \frac{q}{8\pi^2\varepsilon_0 cR} \int_{-\infty}^{\infty} \exp i\omega \left(t' - \frac{\hat{\mathbf{R}} \cdot \mathbf{r}(t')}{c} \right) \frac{1}{\kappa^2} \hat{\mathbf{R}} \wedge \left(\hat{\mathbf{R}} - \frac{\mathbf{v}}{c} \right) \wedge \left\{ \frac{\dot{\mathbf{v}}}{c} \right\} dt', \qquad (5.1.14)$$

which it is sometimes convenient to integrate by parts to obtain

$$\mathbf{E}(\omega) = \frac{-i\omega q}{8\pi^2\varepsilon_0 cR} \int_{-\infty}^{\infty} \hat{\mathbf{R}} \wedge \left(\hat{\mathbf{R}} \wedge \frac{\mathbf{v}}{c} \right) \exp i\omega \left(t' - \frac{\hat{\mathbf{R}} \cdot \mathbf{r}}{c} \right) dt'. \qquad (5.1.15)$$

[In these formulas we have omitted the unimportant additional phase term $\exp(i\omega x/c)$.]

The radiated energy $d^2W/d\Omega_s \, d\omega$ per unit solid angle (Ω_s) per unit angular frequency (ω) is then obtained by Parseval's theorem in the form

$$\int_{-\infty}^{\infty} |\mathbf{E}(t)|^2 \, dt = 4\pi \int_0^{\infty} |\mathbf{E}(\omega)|^2 \, d\omega, \qquad (5.1.16)$$

$$\frac{d^2W}{d\Omega_s \, d\omega} = \frac{q^2\omega^2}{16\pi^3\varepsilon_0 c} \left| \int_{-\infty}^{\infty} \hat{\mathbf{R}} \wedge \left(\hat{\mathbf{R}} \wedge \frac{\mathbf{v}}{c} \right) \exp i\omega \left(t' - \frac{\hat{\mathbf{R}} \cdot \mathbf{r}}{c} \right) dt' \right|^2. \qquad (5.1.17)$$

(Note that $d^2W/d\Omega_s \, d\omega$ is considered nonzero only for positive ω.)

These general formulas are the basis for our discussion of radiation by free electrons. Let us note, before moving on, some important general points.

First, radiation is emitted by a particle in free space when it is accelerated, as indicated by Eq. (5.1.9), for example. Now acceleration can occur as a result of the particle experiencing either a magnetic or an electric field. Most plasmas are permeated by some background magnetic field, and the resultant gyration of electrons gives rise to emission of what is called more or less interchangeably, cyclotron, synchrotron, or gyro radiation.

A constant electric field produces constant acceleration (unlike a magnetic field) resulting in very little radiation at high frequencies. The important sources of radiation by electrons accelerated by electric fields are therefore due to rapidly varying fields. When such a field is due to an incident electromagnetic wave, the process is one of scattering, whose treatment we defer to a later chapter. The other important source of varying electric fields in a plasma is the plasma particles themselves. An electron moving in the Coulomb field of an ion radiates, owing to its acceleration, by the process known as bremsstrahlung (German for braking radiation)

It is important to realize also that plasma is a refractive medium so that free space calculations represent a good approximation only for $\omega \gg \omega_p$. In particular, when the refractive index is greater than 1 it is possible for an energetic electron to travel faster than the local electromagnetic wave phase velocity. When this happens radiation is emitted, even though the energetic electron undergoes negligible acceleration; this is known as Čerenkov radiation. From the viewpoint of our free space treatment, one can think of the radiation as arising then from the acceleration of the background plasma electrons during the rapid passage of the energetic electron, rather than from the electron itself. This approach is cumbersome for actual calculation even though conceptually helpful, and so Čerenkov emission is usually treated by recognizing the dielectric properties of the plasma at the outset. Plasma dielectric effects can also be important for the other emission processes when $\omega \sim \omega_p$, in which case modifications to the free space calculation are important.

5.2 Cyclotron radiation
5.2.1 *Radiation by a single electron*

We consider the radiation from a single electron in a plasma sufficiently tenuous that the waves can be treated as being essentially in free space, so that the analysis of the previous section is applicable. Suppose there is a constant applied magnetic field $\mathbf{B}_0$ that we take to be in the z direction. Then the equation of motion of the electron is

$$\frac{d}{dt}(\gamma m_e \mathbf{v}) = -e(\mathbf{v} \wedge \mathbf{B}_0).$$

(5.2.1)

Note that it is essential to take into account the relativistic effects such as the mass increase by the factor $\gamma \equiv (1 - \beta^2)^{-1/2}$; we shall henceforth work in terms of $\boldsymbol{\beta} \equiv \mathbf{v}/c$ for convenience. The electron charge is $-e$ and we shall define

$$\omega_c \equiv \frac{\Omega}{\gamma} = \frac{eB_0}{m_e\gamma} \tag{5.2.2}$$

as the relativistically shifted cyclotron frequency, decreased by virtue of the mass increase.

The equation of motion is readily solved to obtain

$$\boldsymbol{\beta} = \beta_\perp \left(\hat{\mathbf{x}} \cos \omega_c t + \hat{\mathbf{y}} \sin \omega_c t\right) + \beta_\parallel \hat{\mathbf{z}},$$
$$\frac{\mathbf{r}}{c} = \frac{\beta_\perp}{\omega_c} \left(\hat{\mathbf{x}} \sin \omega_c t - \hat{\mathbf{y}} \cos \omega_c t\right) + \beta_\parallel t \hat{\mathbf{z}}. \tag{5.2.3}$$

The electron moves on a spiral path as illustrated in Fig. 5.1. We have ignored the constant of integration in $\mathbf{r}$ since it introduces only an irrelevant phase shift. It is convenient to choose axes such that the vector from the emission point to the observation point has no y component; that is,

$$\hat{\mathbf{R}} = (\sin\theta, 0, \cos\theta), \tag{5.2.4}$$

where θ is the angle between $\hat{\mathbf{R}}$ and $\hat{\mathbf{z}}$.

We now calculate the electric field of the radiation from the electron in the form of Eq. (5.1.15). For this calculation we need the phase factor

$$\exp i\omega\left(t - \frac{\hat{\mathbf{R}} \cdot \mathbf{r}}{c}\right) = \exp i\omega\left(t - \frac{\beta_\perp}{\omega_c}\sin\theta\sin\omega_c t - \beta_\parallel t\cos\theta\right), \tag{5.2.5}$$

where we note that the use of retarded time is implicit in Eq. (5.2.3). We

Fig. 5.1. Electron trajectory in a magnetic field.

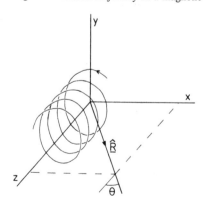

also require the vector determining the polarization,

$$\hat{\mathbf{R}} \wedge (\hat{\mathbf{R}} \wedge \boldsymbol{\beta}) = \hat{\mathbf{x}}\left(\beta_{\parallel}\cos\theta \sin\theta - \beta_{\perp}\cos^2\theta \cos\omega_c t \right)$$

$$- \hat{\mathbf{y}}\beta_{\perp} \sin\omega_c t$$

$$+ \hat{\mathbf{z}}\left(-\beta_{\parallel}\sin^2\theta + \beta_{\perp}\sin\theta\cos\theta\cos\omega_c t \right), \qquad (5.2.6)$$

after a little algebra.

In order to perform the required integral, we employ the Bessel function identity

$$e^{-i\xi\sin\phi} = \sum_{m=-\infty}^{\infty} e^{-im\phi} J_m(\xi) \qquad (5.2.7)$$

and two further identities that can be obtained immediately from it by differentiation:

$$-i\sin\phi\, e^{-i\xi\sin\phi} = \sum_{-\infty}^{\infty} e^{-im\phi} J_m'(\xi),$$

$$i\xi\cos\phi\, e^{-i\xi\sin\phi} = \sum_{-\infty}^{\infty} ime^{-im\phi} J_m(\xi). \qquad (5.2.8)$$

These enable the integrand to be expressed in the form

$$\hat{\mathbf{R}} \wedge (\hat{\mathbf{R}} \wedge \boldsymbol{\beta})\exp i\omega\left(t - \frac{\hat{\mathbf{R}} \cdot \mathbf{r}}{c} \right)$$

$$= \sum_{m=-\infty}^{\infty} \left\{ \hat{\mathbf{x}}\left(\beta_{\parallel}\cos\theta\sin\theta - \beta_{\perp}\cos\theta\frac{m}{\xi} \right) J_m(\xi) + \hat{\mathbf{y}}(-i\beta_{\perp})J_m'(\xi) \right.$$

$$\left. + \hat{\mathbf{z}}\left(-\beta_{\parallel}\sin^2\theta + \beta_{\perp}\sin\theta\cos\theta\frac{m}{\xi} \right) J_m(\xi) \right\}$$

$$\times \exp i\left[(1 - \beta_{\parallel}\cos\theta)\omega - m\omega_c \right]t, \qquad (5.2.9)$$

where

$$\xi = \frac{\omega}{\omega_c}\beta_{\perp}\sin\theta. \qquad (5.2.10)$$

All the t dependence is now in the exponential and the integration is trivial,

$$\int_{-\infty}^{\infty} e^{i\alpha t}\, dt = 2\pi\delta(\alpha). \qquad (5.2.11)$$

We therefore obtain the electric field as

$$\mathbf{E}(\omega) = \frac{+ei\omega}{4\pi\varepsilon_0 cR} \sum_{-\infty}^{\infty} \delta\left[(1 - \beta_{\parallel}\cos\theta)\omega - m\omega_c\right]$$

$$\times \left\{ \hat{\mathbf{x}}\left(-\frac{\cos\theta}{\sin\theta}(\cos\theta - \beta_{\parallel})\right)J_m(\xi) + \hat{\mathbf{y}}(-i\beta_{\perp})J_m'(\xi)\right.$$

$$\left. + \hat{\mathbf{z}}(\cos\theta - \beta_{\parallel})J_m(\xi)\right\},$$

(5.2.12)

where we have replaced m/ξ by $(1 - \beta_{\parallel}\cos\theta)/\beta_{\perp}\sin\theta$ because of the delta function.

We see, therefore, that the radiation consists of a series of discrete harmonics at frequency

$$\omega_m = m\omega_c/(1 - \beta_{\parallel}\cos\theta),$$

(5.2.13)

a result we might have anticipated because of the periodicity of the electron motion.

To calculate the power emitted, we obtain the instantaneous electric field

$$\mathbf{E}(t) = \int_{-\infty}^{\infty} e^{-i\omega t}\mathbf{E}(\omega)\,d\omega$$

$$= \sum_{1}^{\infty} \frac{ie\omega_m}{4\pi\varepsilon_0 cR} \frac{\mathbf{U}_m\exp(-i\omega_m t) - \mathbf{U}_m^*\exp(i\omega_m t)}{1 - \beta_{\parallel}\cos\theta}$$

$$= \sum_{1}^{\infty} \mathscr{R}e\left\{\frac{e\omega_m}{2\pi\varepsilon_0 cR} \frac{i\mathbf{U}_m\exp(-i\omega_m t)}{1 - \beta_{\parallel}\cos\theta}\right\} \equiv \mathscr{R}e\left\{\sum_{1}^{\infty}\mathbf{E}_m(t)\right\},$$

(5.2.14)

where we have combined the positive and negative summations using $J_m(\xi_m) = J_{-m}(-\xi_m)$, and dropped the $m = 0$ term, for the moment, under the assumption $1 - \beta_{\parallel}\cos\theta \neq 0$. $\mathbf{U}_m$ denotes the vector part of Eq. (5.2.12) with $\xi_m = m\beta_{\perp}\sin\theta/(1 - \beta_{\parallel}\cos\theta)$. The power is then the sum of the powers in the individual harmonics and, written per unit solid angle (Ω_s), is

$$\frac{dP}{d\Omega_s} = R^2\langle|\mathbf{S}|\rangle = R^2\sum_{1}^{\infty}\frac{\varepsilon_0 c|\mathbf{E}_m|^2}{2} = \frac{e^2\omega^2}{8\pi^2\varepsilon_0 c}\sum_{1}^{\infty}|\mathbf{U}_m|^2\frac{1}{(1 - \beta_{\parallel}\cos\theta)^2}$$

$$= \frac{e^2\omega^2}{8\pi^2\varepsilon_0 c}\sum_{1}^{\infty}\left\{\left(\frac{\cos\theta - \beta_{\parallel}}{\sin\theta}\right)^2 J_m^2(\xi_m) + \beta_{\perp}^2 J_m'^2(\xi_m)\right\}$$

$$\times \frac{1}{(1 - \beta_{\parallel}\cos\theta)^2}.$$

(5.2.15)

The spectral power density is obviously

$$\frac{d^2P}{d\omega\, d\Omega_s} = \frac{e^2\omega^2}{8\pi^2\varepsilon_0 c} \sum_1^\infty \left\{ \left(\frac{\cos\theta - \beta_\parallel}{\sin\theta}\right)^2 J_m^2(\xi) + \beta_\perp^2 J_m'^2(\xi) \right\}$$

$$\times \frac{\delta\left(\{1 - \beta_\parallel\cos\theta\}\omega - m\omega_c\right)}{1 - \beta_\parallel\cos\theta}. \tag{5.2.16}$$

Again we note that this is the energy radiated per unit time at the field (observation) point, not time at the particle. It is important to maintain this distinction because if we wish to know the rate of energy loss by the particle per unit solid angle per unit frequency the preceding expression must be multiplied by $1 - \beta_\parallel\cos\theta$. When we wish to calculate the emission from an assembly of particles, the power arising from any volume element is just the number of particles in that volume (of the particular velocity considered) $f\, d^3x\, d^3v$ times their rate of energy loss, that is, $(1 - \beta_\parallel\cos\theta)$ $(d^2P/d\omega\, d\Omega_s) f d^3x\, d^3v$. Thus the quantity $(d^2P/d\omega\, d\Omega_s)(1 - \beta_\parallel\cos\theta)$ is normally the power of interest. Expression (5.2.16) is often called the Schott–Trubnikov formula after two influential investigators of cyclotron emission (Schott 1912; Trubnikov 1958).

5.2.2 Plasma emissivity

In plasma that is sufficiently tenuous that the preceding free space treatment is appropriate, we can also assume that the electrons are completely uncorrelated so that to obtain the emissivity of the plasma we simply have to add all the intensities from the individual electrons. We define the emissivity $j(\omega, \theta)$ as the rate of emission of radiant energy from the plasma per unit volume per unit angular frequency per unit solid angle. Then if the plasma has an electron distribution function $f(\mathbf{v})\, d^3v = f(\beta_\perp, \beta_\parallel)\, d^3v$,

$$j(\omega, \theta) = c^3 \int \frac{d^2P}{d\omega\, d\Omega_s}(1 - \beta_\parallel\cos\theta) f(\beta_\perp, \beta_\parallel) 2\pi\beta_\perp\, d\beta_\perp\, d\beta_\parallel, \tag{5.2.17}$$

with $dP/d\omega\, d\Omega_s$ given by Eq. (5.2.16).

The result of this integration is to obtain emission lines of finite width because of the dependence of the resonance frequency $\omega_m = m\Omega(1 - \beta^2)^{1/2}/(1 - \beta_\parallel\cos\theta)$ upon the particle velocities. The two broadening mechanisms here are the *relativistic mass increase* $(1 - \beta^2)^{-1/2}$ and the *Doppler effect* $(1 - \beta_\parallel\cos\theta)^{-1}$. Additional broadening effects that are usually negligible are *natural broadening* (radiation broadening), due to the loss of energy by the electron as it radiates, causing an exponential decay of its energy, and *collision broadening*, due to collisions interrupting the wave train.

The relative importance of mass increase and Doppler effect is determined by the angle θ of the radiation emission to the magnetic field. Doppler shift will be larger if

$$\beta_\| \cos \theta > \beta^2 \qquad (5.2.18)$$

or, for thermal plasmas with thermal velocity v_t $[= (T_e/m_e)^{1/2}]$,

$$\cos \theta > v_t/c. \qquad (5.2.19)$$

Thus, for nonrelativistic plasmas ($v_t \ll c$) the broadening mechanism will be dominantly Doppler broadening except at angles of propagation very close to perpendicular.

5.2.3   *Nonrelativistic plasma*

If we can universally assume that for the plasma under consideration the velocities are small, $\beta \ll 1$, it is possible to obtain convenient nonrelativistic approximations for the emissivity. First, we note that the broadening effects will be small so that we can treat each harmonic separately and expect narrow lines close to the frequencies $\omega_m = m\Omega$. Second, we can adopt the approximation, valid for small argument,

$$J_m(\xi) \approx \frac{1}{m!} \left(\frac{\xi}{2} \right)^m. \qquad (5.2.20)$$

Third, we retain only lowest order terms in β elsewhere. Then

$$\left(\frac{\cos \theta - \beta_\|}{\sin \theta} \right)^2 J_m^2(\xi) + \beta_\perp^2 J_m'^2(\xi)$$

$$\approx \frac{m^{2(m-1)}}{(m-1)!^2} (\sin \theta)^{2(m-1)} (\cos^2 \theta + 1) \left(\frac{\beta_\perp}{2} \right)^{2m} \qquad (5.2.21)$$

and so

$$j_m(\omega) = \frac{e^2 \omega_m^2}{8\pi^2 \varepsilon_0 c} \frac{m^{2(m-1)}}{(m-1)!^2} (\sin \theta)^{2(m-1)} (\cos^2 \theta + 1)$$

$$\times c^3 \int \delta \big((1 - \beta_\| \cos \theta) \omega - m\omega_c \big)$$

$$\times \left(\frac{\beta_\perp}{2} \right)^{2m} f(\beta_\perp, \beta_\|) 2\pi \beta_\perp \, d\beta_\perp \, d\beta_\|. \qquad (5.2.22)$$

It is most convenient to deal with the total emissivity for a single harmonic, integrated over frequency in the vicinity of ω_m; that is,

$$j_m \equiv \int j_m(\omega)\, d\omega \qquad (5.2.23)$$

It is equal simply to the expression (5.2.22) for $j_m(\omega)$ without the delta function (to lowest order in β). If we take the electron distribution to be the Maxwellian

$$f = n_e \left(\frac{m_e}{2\pi T} \right)^{3/2} \exp\left(-\frac{m_e c^2 \left(\beta_\perp^2 + \beta_\parallel^2 \right)}{2T} \right), \qquad (5.2.24)$$

we can perform the velocity integrals, noting that

$$\int_0^\infty \beta_\perp^{2m} e^{-\beta_\perp^2/\alpha^2} 2\beta_\perp\, d\beta_\perp = \alpha^{2(m+1)} m!, \qquad (5.2.25)$$

so that

$$c^3 \int \left(\frac{\beta_\perp}{2} \right)^{2m} f 2\pi \beta_\perp\, d\beta_\perp\, d\beta_\parallel = \left(\frac{T}{2m_e c^2} \right)^m m!\, n_e \qquad (5.2.26)$$

and

$$
\begin{aligned}
j_m &= \int j_m(\omega)\, d\omega \\
&= \frac{e^2 \omega_m^2 n_e}{8\pi^2 \varepsilon_0 c} \frac{m^{2m-1}}{(m-1)!} (\sin\theta)^{2(m-1)} (\cos^2\theta + 1) \left(\frac{T}{2m_e c^2} \right)^m .
\end{aligned}
$$
$$(5.2.27)$$

The distribution of the emissivity within this line can be expressed as a shape function $\phi(\omega - \omega_m)$ such that

$$j_m(\omega) = j_m \phi(\omega - \omega_m), \qquad \int \phi(\omega)\, d\omega = 1. \qquad (5.2.28)$$

For angles that are not close to perpendicular ($\cos\theta \gg \beta$), Doppler broadening dominates, so we can calculate ϕ straightforwardly. The delta function in Eq. (5.2.22) becomes $\delta(\omega(1 - \beta_\parallel \cos\theta) - \omega_m)$, independent of $\beta_\perp$, so we can perform the $\beta_\perp$ integral as before and the $\beta_\parallel$ integral is then trivial because of the delta function. The result is

$$\phi(\omega - \omega_m) = \left(\frac{m_e c^2}{2\pi T} \right)^{1/2} \frac{\exp\left\{ -\left(m_e c^2/2T \right) \left((\omega - \omega_m)/\omega_m \cos\theta \right)^2 \right\}}{\omega_m \cos\theta}, \qquad (5.2.29)$$

a Gaussian line shape, as expected from Doppler broadening.

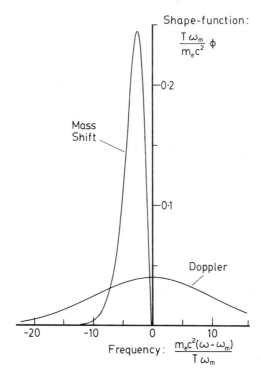

Fig. 5.2. Shape of the cyclotron emission line for the case of oblique and perpendicular propagation, giving Doppler- and mass-shift-broadened shapes, respectively.

When $\cos \theta \le \beta$, so that relativistic mass shift broadening is important, ϕ is more complicated and is asymmetric (see Exercise 5.1). It is also narrower, giving a line width of order $(T/m_e c^2)\omega_m$ rather than $\sim (T/m_e c^2)^{1/2}\cos \theta \omega_m$ for the Doppler-broadened case. This is illustrated in Fig. 5.2 where we see that for $\cos \theta \ll \beta$ the spread of the line is entirely downward in frequency owing to the relativistic mass increase. In either case, the frequency-integrated emissivity is correctly given by Eq. (5.2.27).

5.2.4   *Radiation transport, absorption, and emission*

It is a general principle that if a medium emits radiation it also absorbs radiation at the same frequency. The absorption coefficient $\alpha(\omega)$ is defined as the fractional rate of absorption of radiation per unit path length. Then, in a medium whose refractive index can be taken as unity, the intensity of radiation $I(\omega)$, which is the radiative power per unit area per unit solid angle per unit angular frequency, is governed by the equation

$$\frac{dI}{ds} = j(\omega) - I\alpha(\omega), \tag{5.2.30}$$

where s is the distance along the ray trajectory. The solution of this equation is straightforward:

$$I(s_2) = I(s_1)e^{(\tau_1 - \tau_2)} + \int_{s_1}^{s_2} j(\omega)e^{\tau - \tau_2} ds, \qquad (5.2.31)$$

where the "optical depth" is defined as

$$\tau \equiv \int^s \alpha(\omega) \, ds \qquad (5.2.32)$$

and s_1 and s_2 are two points on the ray. Suppose, then, that we consider a ray crossing a plasma slab, as illustrated in Fig. 5.3, in which we take j/α to be uniform. Then

$$I(s_2) = I(s_1)e^{-\tau_{21}} + \int_{\tau_1}^{\tau_2} (j/\alpha)e^{\tau - \tau_2} \, d\tau$$

$$= I(s_1)e^{-\tau_{21}} + (j/\alpha)[1 - e^{-\tau_{21}}], \qquad (5.2.33)$$

where $\tau_{21} \equiv \tau_2 - \tau_1$ is the total optical depth of the slab. This equation shows that the emergent intensity consists of some fraction of the incident intensity (first term) plus additional emitted intensity (second term).

When $\tau_{21} \gg 1$ the slab is said to be optically thick and we find simply

$$I(s_2) = j/\alpha. \qquad (5.2.34)$$

The slab absorbs all radiation (at this frequency) incident upon it.

It is a fundamental thermodynamic property that any body in thermodynamic equilibrium that is "black," meaning that it is perfectly absorbing,

Fig. 5.3. Intensity of radiation at a point 2 is determined by an integration of the transport equation from point 1 along the ray path (s).

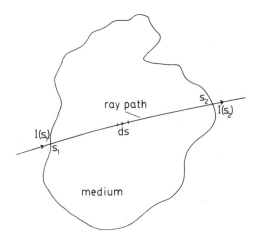

emits radiation with a unique blackbody intensity

$$I(\omega) = B(\omega) = \frac{\hbar\omega^3}{8\pi^3 c^2} \frac{1}{e^{\hbar\omega/T} - 1},$$

(5.2.35)

which for low frequency $\hbar\omega \ll T$ may be written

$$B(\omega) = \frac{\omega^2 T}{8\pi^3 c^2}.$$

(5.2.36)

In its first form this is Planck's radiation formula and the second form is called the Rayleigh–Jeans approximation or classical limit. Both expressions refer to a single wave polarization.

It is clear, then, that we must identify the emergent intensity $I(s_2)$ with $B(\omega)$ in the optically thick case so that

$$j/\alpha = B(\omega),$$

(5.2.37)

which is known as Kirchhoff's law. This allows us to deduce the absorption coefficient immediately from the emissivity or vice versa when the plasma is thermal. Since our treatment of emission has been entirely classical so far, it is appropriate to use the Rayleigh–Jeans formula. For cyclotron emission from laboratory plasmas this will always be an excellent approximation.

We can now write down the absorption coefficient for the mth harmonic for a tenuous nonrelativistic thermal plasma, using Eq. (5.2.27), as

$$\alpha_m(\omega) = \alpha_m \phi(\omega) = \left[\frac{j_m}{B(\omega_m)} \right] \phi(\omega)$$

$$= \frac{\pi}{2c} \cdot \frac{e^2 n_e}{m_e \varepsilon_0} \frac{m^{2m-1}}{(m-1)!} (\sin\theta)^{2(m-1)}$$

$$\times (\cos^2\theta + 1) \left(\frac{T}{2m_e c^2} \right)^{m-1} \phi(\omega).$$

(5.2.38)

In many cases the absorption at the lower (first and second) harmonics is strong enough that $\tau \gg 1$, that is, the plasma is optically thick. This proves to be of great advantage for diagnostic purposes because the intensity emitted by the plasma is then simply the blackbody level $\omega^2 T/8\pi^3 c^2$. A measurement of this intensity then provides a very direct measurement of the electron temperature, since the intensity is independent of other parameters such as density.

As a general point, true for any radiation measurement, the intensity at the plasma edge is measured by detecting the power per unit frequency crossing an area A in a solid angle Ω_s defined by some collimating optics.

This optics may consist of lenses, mirrors, or other focusing elements or merely of two apertures, but in any case the power observed is then $I(\omega)A\Omega_s$, assuming the collimation to be sufficient that $I(\omega)$ can be taken constant across A and Ω_s. The product $A\Omega_s$ is called the *étendue* of the optical system and determines its light gathering power. It is clear from our previous treatment of the radiation transport that one may place the collecting optics at any distance from the plasma and observe the same power provided only that there is no absorption in the intervening region and also that the plasma viewed fills the whole of the étendue. (This second requirement will always be violated if one goes too far away.)

In an optically thin case, $\tau \ll 1$, assuming no radiation to be incident on the other side of the plasma, the intensity emerging from the plasma is

$$I(s_2) = \int_{s_1}^{s_2} j\, ds \tag{5.2.39}$$

and the power detected via the collimating optics is

$$IA\Omega_s = \int jA\Omega_s\, ds. \tag{5.2.40}$$

Note that the étendue is simply a multiplying number and the fact that it can therefore be taken inside the integral is sometimes expressed by saying that the étendue is "conserved" through a lossless system.

The treatment given here of the radiation transport assumes, for simplicity, that the refraction in the plasma can be ignored. When this is not the case a more complicated treatment is required (Bekefi 1966), which recognizes that solid angle varies along the ray path due to refraction. The effect in an optically thin case is that the intensity is given by the ray integral, Eq. (5.2.39), of j/N_r^2 (rather than j), where N_r is the ray refractive index. N_r is equal to N for isotropic media but not generally otherwise. Although the blackbody intensity *within* the medium is also changed by refraction, the intensity observed *outside* the plasma in an optically thick case is still the usual blackbody level, Eq. (5.2.36), provided that the full étendue is accessible to the emitting plasma (see Exercise 5.2). We now turn to some related questions when plasma refraction is considered.

5.2.5 *Wave polarization and finite density effects*

Thus far, we have ignored the dielectric nature of the plasma and calculated the emission and absorption based on a single particle approach. However, now we must consider the modification of our results that occurs for finite density plasmas in which the dielectric effects are not negligible.

First, we know that the characteristic modes of wave propagation are not degenerate in a magnetized plasma but have specific polarization given by

Eq. (4.3.6). It is then often convenient to treat the two characteristic waves separately. This may be done within the framework of our tenuous plasma approximation as follows.

The direction of the electric vector for the characteristic waves in the low-density limit $X \ll 1$ is, in the coordinate system of Eq. (4.3.6) (that is, with z along $\hat{\mathbf{k}} \equiv \hat{\mathbf{R}}$, $B_x = 0$),

$$\hat{\mathbf{E}}_{\pm} = \frac{\left(i\left[\mp 1 + (F^2 + 1)^{1/2}\right]^{1/2}, \pm\left[\pm 1 + (F^2 + 1)^{1/2}\right]^{1/2}, 0 \right)}{\left[2F(F^2 + 1)^{1/2}\right]^{1/2}},$$

(5.2.41)

where Eq. (4.3.13) reduces to $F = 2\cos\theta / Y\sin^2\theta$. Now we project the vector $\mathbf{U}_m$, which determines the **E** vector of the radiation in Eq. (5.2.14), onto $\hat{\mathbf{E}}_{\pm}$ to determine what proportion of the radiation is in each mode.

For the nonrelativistic case

$$\mathbf{U}_m \propto (-i, \cos\theta, 0)$$

(5.2.42)

in the appropriate coordinates, and the proportion of power in each mode is then (see Exercise 5.3)

$$\frac{|\hat{\mathbf{E}}_{\pm} \cdot \mathbf{U}_m^*|^2}{|\mathbf{U}_m|^2}$$

$$= \frac{1}{2} \mp \frac{(\sin^4\theta/4m) + \cos^2\theta}{(\cos^2\theta + 1)(\cos^2\theta + (\sin^4\theta/4m^2))^{1/2}} = \eta_{\pm} \quad (\text{say})$$

(5.2.43)

(putting $Y = 1/m$). As usual, the upper sign here refers to the ordinary wave. Note that $\eta_+ + \eta_- = 1$, as must be the case since the ηs are just the fraction of power in each mode. Then we can write down the emission and absorption coefficients for each mode separately as

$$j_m^{\pm} = j_m \eta_m^{\pm}, \qquad \alpha_m^{\pm} = \alpha_m \eta_m^{\pm}.$$

(5.2.44)

Clearly, since the second term in Eq. (5.2.43) is positive definite, the power in the ordinary wave is always smaller than that in the extraordinary wave. Graphically, η_m^+ is as shown in Fig. 5.4, for the first few harmonics. It is always much less than 1.

It should be recalled, however, that we are working only to lowest order in $(T/m_e c^2)$, so at those angles (0 and $\pi/2$) where η_+ is small in this approximation, corrections of the order $T/m_e c^2$ must be invoked and cause η_+ to remain finite. In fact, as we shall see, our results so far are usually inapplicable at the fundamental ($m = 1$).

When we are dealing with a plasma in which the refractive index is significantly different from 1, we must abandon the tenuous plasma assumption and recognize that the wave propagation is no longer governed by the free space equations. A derivation based on single particle emission and the Lienard–Wiechert potentials is then no longer valid. Instead, the usual approach is to calculate the imaginary part of the refractive index and hence the absorption coefficient from a kinetic-theory treatment of the wave propagation problem. The derivation is far beyond our present scope. The interested reader is referred to a major review article (Bornatici et al. 1983) for the status of the ongoing theory. We shall discuss here the main physical issues and present the most important results.

As with our treatment of the line shape, it is necessary to distinguish oblique propagation from perpendicular propagation. In the latter case, $N \cos \theta < v_t/c$, relativistic effects dominate the line shape and make the calculations even more difficult.

1. Oblique propagation for harmonics higher than the first. The finite density modifications can be included in the absorption coefficient for nonrelativistic plasma by writing it as before,

$$\alpha_m^{\pm} = \alpha_m \eta_m^{\pm}, \tag{5.2.45}$$

with α_m given by Eq. (5.2.38), but now the coefficients $\eta_m^{\pm}$ are modified as follows:

$$\eta_m^{\pm} = \frac{N^{2m-3}[m-1]^2 \left[1 - ((m+1)/m)(m\Omega/\omega_p)^2(1-N^2)\right]^2}{(1+\cos^2\theta)(a_m^2 + b_m^2)^{1/2}}, \tag{5.2.46}$$

Fig. 5.4. Fraction of cyclotron emission (and absorption) in the ordinary wave as a function of propagation angle in the tenuous plasma approximation.

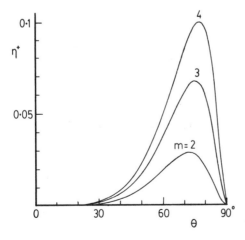

where

$$a_m^2 = \left\{ 1 + \frac{\left[1 - \left(\omega_p/m\Omega \right)^2 \right] N^2 \cos^2\theta}{\left[1 - \left(\omega_p/m\Omega \right)^2 - N^2 \sin^2\theta \right]^2} \right.$$

$$\left. \times m^2 \left[1 - \frac{m^2 - 1}{m^2} \left(\frac{m\Omega}{\omega_p} \right)^2 \left(1 - N^2 \right) \right] \right\}^2 \sin^2\theta,$$

$$(5.2.47)$$

$$b_m^2 = \left\{ 1 + \frac{1 - \left(\omega_p/m\Omega \right)^2}{1 - \left(\omega_p/m\Omega \right)^2 - N^2 \sin^2\theta} \right.$$

$$\left. \times m^2 \left[1 - \frac{m^2 - 1}{m^2} \left(\frac{m\Omega}{\omega_p} \right)^2 \left(1 - N^2 \right) \right] \right\}^2 \cos^2\theta,$$

and N is the (real part of the) refractive index given by the Appleton–Hartree relation

$$N_{\pm}^2 = 1 - \frac{\left(\dfrac{\omega_p}{m\Omega} \right)^2 \left[1 - \left(\dfrac{\omega_p}{m\Omega} \right)^2 \right]}{1 - \left(\dfrac{\omega_p}{m\Omega} \right)^2 - \dfrac{\sin^2\theta}{2m^2} \pm \left[\left(\dfrac{\sin^2\theta}{2m^2} \right)^2 - \left(1 - \left(\dfrac{\omega_p}{m\Omega} \right)^2 \right) \cos^2\theta \Big/ m^2 \right]^{1/2}}.$$

$$(5.2.48)$$

Evaluation of these formulas [e.g., Bornatici et al. (1980)] shows that they result in an absorption generally lower than in the tenuous plasma limit by a factor that for $\omega_p \leq \Omega$ is rarely smaller than 0.8. Thus, in many cases the previous tenuous plasma calculation is adequate, particularly when dealing with an optically thick harmonic, for which the precise value of α is unimportant.

2. Oblique propagation for the first harmonic. Here the situation is very different. The dielectric effects are strong when $(\omega_p/\Omega)^2 \geq v_t^2/c^2$; that is, even for rather low densities $(\omega_p/\Omega) \ll 1$. The most important effect is that the polarization of the extraordinary wave becomes nearly circular, rotating in the opposite direction to the electron gyration. As a consequence its coupling to the electrons, and hence absorption, is much weaker than calculated in the tenuous plasma limit. Also, as noted earlier, the ordinary wave possesses an absorption that is of order $(v_t/c)^2$, which was ignored in treating the two modes simultaneously. This can cause the ordinary wave to be the more strongly absorbed mode. The line shapes obtained are no

longer Gaussian, but the absorption can still be expressed as

$$\alpha_1^{\pm}(\omega) = \alpha_1^{\pm}\phi^{\pm}(\omega), \tag{5.2.49}$$

with

$$\int \phi^{\pm} \, d\omega = 1. \tag{5.2.50}$$

It is found that for large (but still oblique) angles

$$\left(\sin^4\theta \gg 4\left[1 - \omega_p^2/\Omega^2\right]^2 \cos^2\theta\right)$$

the absorption integrated over the line can be written

$$\alpha_1^+ = \frac{\pi}{2c}\omega_p^2 N_+ \frac{(1 + 2\cos^2\theta)^2 \sin^4\theta}{(1 + \cos^2\theta)^3}\left(\frac{T}{m_ec^2}\right) \tag{5.2.51}$$

and

$$\alpha_1^- = \frac{\pi}{2c}\left(1 + \frac{\omega_p^2}{\Omega^2}\right)\Omega^2\left(\frac{\Omega}{\omega_p}\right)^2 N_-^5 \cos^2\theta\left(\frac{T}{m_ec^2}\right) \tag{5.2.52}$$

for the ordinary and extraordinary modes, respectively. It should be noticed that the extraordinary mode absorption scales approximately inversely with density rather than proportional to density as in the tenuous plasma limit. For general angles no convenient analytic expressions are as yet available.

3. Perpendicular propagation $N\cos\theta < v_t/c$ for $m \geq 2$ in the extraordinary mode. Absorption integrated over the resonance is the same as given for oblique propagation [Eq. (5.2.46)], which reduces at $\theta = \pi/2$ to

$$\eta_m^- = \left[\frac{(1-X)^2 m^2 - 1}{(1-X)m^2 - 1}\right]^{m-3/2}\left[1 + \frac{mX}{(1-X)m^2 - 1}\right]^2, \tag{5.2.53}$$

with $X = \omega_p^2/m^2\Omega^2$. (Strictly speaking, there are some additional corrections for $m = 2$, but these are usually negligible.)

4. Perpendicular propagation for $m \geq 2$ in the ordinary mode. To lowest order in $(v_t/c)^2$ the absorption coefficient is

$$\alpha_m^+ = \frac{\pi}{c}\frac{e^2 n_e}{\varepsilon_0 m_e}\frac{m^{2m-1}}{(m-1)!}\left[1 - \frac{\omega_p^2}{m^2\Omega^2}\right]^{m-1/2}\left(\frac{T}{2m_ec^2}\right)^m \tag{5.2.54}$$

(rather than zero as in the tenuous plasma limit, which really does not treat this mode consistently).

5. Perpendicular propagation for $m = 1$. The ordinary wave has absorption given by the same formula as for oblique propagation [Eq. (5.2.51)]. The extraordinary mode has weaker absorption $\propto (T/m_ec^2)^2$; a simple analytic expression is not available.

In all these finite density cases the emissivity may be obtained from the absorption by application of Kirchhoff's law (suitably modified to account for the refractive index).

The refractive-index effects are most noticeable in their effect upon the cyclotron radiation when there exists a cutoff region where $N^2 < 0$ either at the point of emission or in the line of sight between the emission and observation points.

The presence of cutoffs may be determined from the Appleton–Hartree formula. For perpendicular propagation the ordinary wave is cut off if $\omega < \omega_p$ and the extraordinary wave is cut off if $\omega_H < \omega < \omega_R$ or $\omega < \omega_L$, where

$$\omega_R = \Omega \left[1 + \left(1 + 4\omega_p^2/\Omega^2 \right)^{1/2} \right] \Big/ 2, \tag{5.2.55}$$

$$\omega_L = \Omega \left[-1 + \left(1 + 4\omega_p^2/\Omega^2 \right)^{1/2} \right] \Big/ 2, \tag{5.2.56}$$

and

$$\omega_H^2 = \Omega^2 + \omega_p^2. \tag{5.2.57}$$

Because the wave is evanescent when cut off, the transmission through or emission from a cutoff layer is generally very small.

This constitutes a fairly serious limitation to the use of cyclotron radiation for diagnostics since emission only at $\omega \geq \omega_p$ will be useful. For high-density plasmas in which $\omega_p > \Omega$ the lowest harmonics are likely, therefore, to be unusable. Fortunately, many fusion plasmas such as tokamaks and stellarators satisfy $\omega_p \lesssim \Omega$ for most operating conditions so that cyclotron emission can be used.

5.2.6   *Spatially varying magnetic field*

In most diagnostic applications the magnetic field is not independent of position within the plasma. As a result, the emission of cyclotron radiation does not appear as a series of narrow harmonic lines; rather, these harmonics are broadened by the variation of the magnetic field along the line of sight. Often the inhomogeneity of the field causes broadening much greater than that due to Doppler and relativistic effects.

Consider then a case such as that illustrated in Fig. 5.5, in which the magnetic field varies slowly enough that the wavelength λ is much less than L (the scale length of variation) so that a WKBJ approach is possible. For a specific frequency ω_0, the cyclotron absorption and emission at the mth harmonic is appreciable only in a narrow resonance region where

$$[m\Omega(s) - \omega_0] \ll \omega_0. \tag{5.2.58}$$

(s denotes distance along the ray.) If the resonant layer is small enough that we can approximate the gradient of $m\Omega$ as constant through it, we can calculate the total optical depth through the resonance layer at frequency ω_0 as

$$\tau_m \equiv \int \alpha_m(\omega_0)\,ds = \int \alpha_m(\omega_0)\left|\frac{d(m\Omega)}{ds}\right|^{-1} d(m\Omega)$$

$$= \alpha_m(s)\left|\frac{d\Omega}{ds}\right|^{-1}\int \phi(\omega_0 - m\Omega)\,d\Omega. \qquad (5.2.59)$$

In the final expression here the α_m is the frequency-integrated absorption and ϕ is the line structure factor. We have assumed that α_m is effectively constant over the resonance region so that it can be taken outside the integral, leaving behind only ϕ, which is a strong function of Ω. Now because ϕ is, by presumption, narrow,

$$\int \phi(\omega_0 - m\Omega)\,d\Omega = \frac{1}{m}\int \phi(\omega_0 - m\Omega)\,d\omega_0 = \frac{1}{m} \qquad (5.2.60)$$

by the normalization of ϕ. Therefore,

$$\tau_m = \frac{\alpha_m(s)}{m|d\Omega/ds|} = \frac{L\alpha_m(s)}{m\Omega}, \qquad (5.2.61)$$

where $L \equiv \Omega/|d\Omega/ds|$. Thus, the precise shape of ϕ is unimportant in calculating the total optical depth of plasmas whose resonance is narrow. The plasma parameters to be used in $\alpha_m(s)$ are, of course, those at the resonance layer where $m\Omega(s) = \omega_0$.

Fig. 5.5. In a spatially varying field the cyclotron resonance at a certain point (s_0) has a certain narrow spectral shape $\phi(\omega, s_0)$. Equivalently, at a certain frequency ω_0 the resonance has a narrow spatial shape $\phi(\omega_0, s)$.

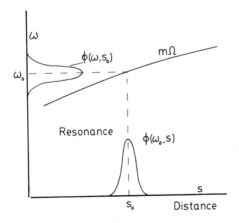

If this plasma is viewed from the vacuum, then, provided there is no radiation incident upon the plasma from outside, the intensity observed will be

$$I(\omega_0) = \frac{\omega_0^2 T(s)}{8\pi^3 c^2}(1 - e^{-\tau_m}),$$ (5.2.62)

assuming the frequency ω_0 to be resonant at only one position and harmonic.

This equation illustrates the potential power of cyclotron emission diagnostics. The intensity observed is a function only of the local plasma parameters $T(s)$, $\tau_m(s)$ at the resonant layer, not, as with many other diagnostics, a chordal average of the parameters. Therefore, if, as is often the case, one knows quite accurately the spatial distribution of Ω ($|B|$ is known), then excellent spatial resolution is possible.

5.2.7   *Diagnostic applications in thermal plasmas*

The most successful applications of cyclotron emission for diagnosis of laboratory plasmas have been in toroidal plasma confinement experiments such as tokamaks. In these plasmas the magnetic field is dominantly toroidal and accurately known (for tokamaks proportional to $1/R$, where R is the major radius). A typical plot of the relevant frequencies is then as shown in Fig. 5.6. In addition to the first three harmonics of the cyclotron frequency, we show the location of the cutoffs ω_p and ω_R for the ordinary and extraordinary waves, respectively; these depend on plasma density of course.

Fig. 5.6. Variation of cyclotron harmonic frequencies and cutoffs, etcetera, in a toroidal magnetic field (tokamak).

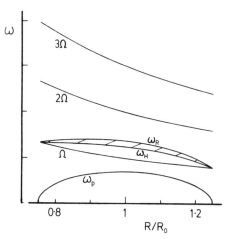

In Fig. 5.7 we show a typical emission spectrum obtained observing such a plasma from the low magnetic field (large R) side along a major radius. In the upper part of the figure we see spectra of intensity I versus frequency ν ($= \omega/2\pi$) for the characteristic modes when $\omega_p < \omega_c$. The parameters of the plasma here are such that the extraordinary mode second harmonic and ordinary mode first harmonic are optically thick ($\tau \gg 1$). Thus, the intensity in the second harmonic (e mode) is just the blackbody level $I(\omega) = \omega^2 T/8\pi^3 c^2$ at the point of resonance. The width and shape of the emission then corresponds to the width and shape of the temperature profile along the line of sight. The second harmonic o mode is optically thin and according to Eq. (5.2.62) should have an intensity much less than

Fig. 5.7. Typical cyclotron emission spectra from a tokamak plasma when (a) $\omega_p < \Omega$, (b) $\omega_p > \Omega$ (Hutchinson and Komm 1977).

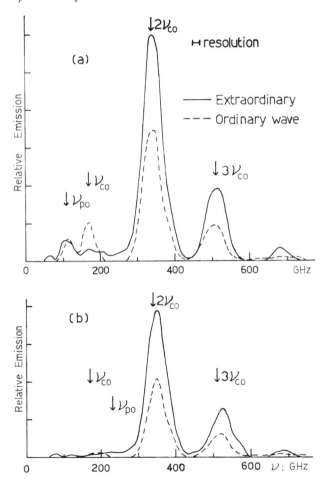

that observed. This discrepancy is resolved by taking into account reflections from the metal walls of the tokamak. Multiple reflections also increase the observed intensity of the optically thin third harmonic. At the fundamental, negligible emission is observed in the *e* mode because there is a cutoff region directly in the line of sight. An additional feature occurs near $\omega_p/2\pi$ that is attributable to Čerenkov emission.

The lower figure shows a higher density plasma in which $\omega_p > \omega_c$. In this case the *o*-wave first harmonic is also obscured because of the cutoff at $\omega = \omega_p$.

From an optically thick harmonic such as the extraordinary mode second harmonic we can obtain the temperature profile by inverting the emission formula

$$T_e = I\left(\frac{m\Omega_0 R_0}{R}\right) \cdot 8\pi^3 c^2 \left(\frac{R}{m\Omega_0 R_0}\right)^2, \qquad (5.2.63)$$

where Ω_0 is the cyclotron frequency at radius R_0. Figure 5.8 shows an example of the type of diagnostic information that can be obtained. From spectra measured every ~ 15 ms, the evolution of the temperature profile throughout the plasma pulse is obtained. The profile starts "hollow" during the early phases and evolves to a centrally peaked shape during the main part of the discharge.

In principle, it should be possible to deduce the density from an optically thin harmonic, once the temperature has been obtained from the optically thick harmonic, by deducing the optical depth from the observed intensity. The point is that the optical depth is a strong function of density as well as temperature. To date, however, no really convincing quantitative density

Fig. 5.8. Evolution of the temperature profile in Alcator C tokamak as determined from optically thick second harmonic cyclotron emission.

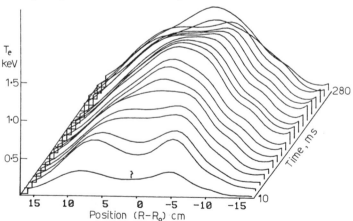

measurements have been made because multiple reflections from the walls enhance the observed intensity in an uncontrolled and difficult-to-calculate manner.

For low enough densities another attractive possibility is to use the polarization of the emission to determine the direction of the magnetic field inside the plasma at the point of emission. For tokamaks this would provide vital information on the poloidal magnetic field. Two problems make this application difficult. One is the multiple reflection problem that reduces the observed degree of polarization. The other is much more fundamental in that if the plasma density is large enough to cause strong birefringence, the waves do not preserve their emitted polarization as they travel out of the plasma. Instead, they remain ordinary or extraordinary waves with appropriate polarization relative to the magnetic field direction at the point of propagation not at the point of emission. The result is that they provide a measurement only of the field direction at the plasma edge rather than at the point of emission. At low densities (or high harmonics) this polarization rotation effect is small so that the desired information may still be retrievable, although no experiment has successfully demonstrated this measurement yet.

5.2.8 *Nonthermal plasmas*

When the plasma distribution function deviates significantly from Maxwellian many of the preceding results are altered. Kirchhoff's law no longer holds and so emission and absorption coefficients must be calculated separately. If the deviations involve very high-energy electrons, then the $\beta \ll 1$ approximation must be dropped. The emission intensity increases with perpendicular electron energy so that even a relatively small proportion of electrons possessing high energies can dominate the emission. On the one hand, this sensitivity requires that one be cautious about interpreting radiation as representative of the bulk electron temperature. On the other, cyclotron emission gives early warning of the presence of energetic electrons and provides a diagnostic of their distribution that, in principle, is very powerful but has yet to be fully exploited in practice.

First some qualitative points concerning the effects on temperature measurements. When the deviation from a Maxwellian distribution consists of the presence of a high-energy tail to the distribution function, one anticipates that observing an optically thick harmonic at right angles to the field from the low-field side, an intensity reflecting the temperature of the low-energy bulk part of the distribution will still be observed. The reason for this is that in the relativistic regime at perpendicular propagation the line shape is asymmetric, each electron resonating with the wave at a frequency downshifted from $m\Omega$ by the mass increase factor $(1 - \beta^2)^{1/2}$. Therefore, consider a specific frequency ω. The energy of particles that resonate with that frequency is given by

$$\frac{\omega}{m\Omega(R)} = (1 - \beta^2)^{1/2}.$$

(5.2.64)

So, taking $\Omega = \Omega_0 R_0 / R$ we have

$$\beta^2 = 1 - \left(\frac{\omega R}{m\Omega_0 R_0}\right)^2,$$

(5.2.65)

which is plotted in Fig. 5.9. For a wave propagating from left to right toward the low-field side, its last point of resonance is at $R = m\Omega_0 R_0/\omega$ at which point it is resonating only with $\beta^2 \sim 0$ particles. So, provided the optical depth is large, the radiation emerging will "remember" only its interaction with these last low-energy bulk electrons, since the nonthermal radiation from the high-energy tail at $R < m\Omega_0 R_0/\omega$ is absorbed by the thermal electrons and cannot be observed. (Naturally this is true only up to a point. If the nonthermal emission is very strong and the optical depth not very great, the small proportion of radiation that penetrates the layer may be sufficient to distort the measurements.)

By the same token if the radiation is observed from the high-field side, the contribution from the nonthermal tail will tend to dominate. Radiation from optically thin harmonics will not show this asymmetry since the emission will take the form of integrals, like equation (5.2.17), over the whole distribution function.

Fig. 5.9. At perpendicular propagation the high-β components are resonant at smaller major radius R (higher B) than the thermal bulk of the electron distribution.

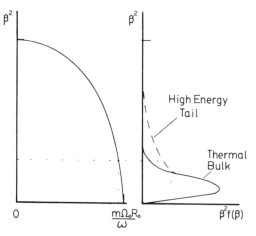

Since the emission from the mth harmonic goes approximately as $J_m^2(\omega\beta_\perp \sin\theta/\omega_c)$, which is $\sim \beta_\perp^{2m}$ (for $\beta \ll 1$), it is clear that higher harmonics will be more strongly enhanced by high-energy tails than low harmonics. Moreover, the broadening of the resonances due to Doppler and relativistic shifts can easily lead to overlap of the different harmonics if the tail is relativistic. As a result one can obtain a rather broad continuous emission spectrum extending to high harmonics, such as is illustrated in Fig. 5.10. From such broad spectra it is difficult to obtain more than a very approximate estimate of nonthermal tail temperatures because the intensity observed tends to be a complicated integral over $\Omega, \theta, \beta_\parallel, \beta_\perp$.

Current attempts to realize the potential of cyclotron emission for diagnosing highly nonthermal electron distributions seek to simplify the problem by observing along a line of constant magnetic field (i.e., vertically in a toroidal plasma) at fixed angle to the field (usually 90°). In such a situation there is a simple one-to-one relationship between frequency and total electron energy, provided harmonic overlap is negligible. As a result, one may potentially be able to diagnose the number density and velocity anisotropy of electrons as a function of energy (Hutchinson and Kato 1986) at the expense of returning to a chord-averaged measurement. A vital practical requirement is to avoid multiple reflections, which would otherwise confuse the desired simple relationship, using some kind of "viewing dump."

Fig. 5.10. A typical cyclotron emission spectrum when significant nonthermal components exist in the electron distribution.

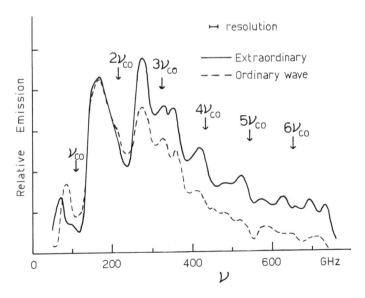

Measurements of cyclotron absorption (of a transmitted beam) are also of interest in diagnosing nonthermal distributions (and thermal too, although this proves usually to be much less useful than the emission). An absorption measurement has the advantage that it is much less affected by the multiple reflection problem. However, it tends to have smaller dynamic range. That is, it is hard to measure optical depths that are very large or very small.

5.2.9   Čerenkov emission

When relativistic particles are present in a finite density plasma, it is possible for Čerenkov radiation to occur. This requires that the refractive index be greater than 1 in order that the $m = 0$ ("cyclotron") resonance condition, which is the Čerenkov resonance condition

$$\omega = k_{\parallel}v_{\parallel} = \omega N_{\parallel}v_{\parallel}/c = \omega N \cos\theta v_{\parallel}/c, \tag{5.2.66}$$

can be satisfied. When the magnetic field is zero the only mode satisfying this condition is the electrostatic longitudinal wave. Thus electromagnetic emission requires conversion of this electrostatic wave into an electromagnetic wave, able to propagate in vacuum. In a magnetized plasma the two characteristic waves of the Appleton–Hartree dispersion relation [Eq. (5.2.48)] have frequency bands in which $N > 1$, as illustrated by Fig. 5.11. These branches are sometimes called the slow extraordinary (z mode) and slow ordinary (whistler) branches, "slow" implying $v_{\mathrm{phase}} < c$. Direct Čerenkov emission is possible into these modes.

Fig. 5.11. The Appleton–Hartree dispersion relation for the case when $\Omega = 1.5\omega_p$. For any propagation angle other than 0 or $\pi/2$ there are two ordinary and two extraordinary branches. Certain sections of these are slow.

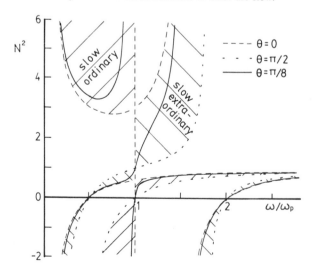

In straight magnetic fields it is difficult for this emission to be observed because the radiation tends still to be trapped inside the plasma. Usually the electron density is approximately uniform along the magnetic field even though it may vary across it. Because of this uniformity (i.e., the ignorability of the parallel coordinate) as the wave propagates, $N_{\parallel}$ is conserved. Now, by virtue of the emission resonance condition, $N_{\parallel}$ is greater than 1 and no propagating (nonevanescent) wave in free space ($N = 1$) can have $N_{\parallel} > 1$. Therefore, in straight magnetic fields the Čerenkov emission is again unable to propagate out into free space without some type of complicated mode conversion occurring. The radiation is normally trapped inside the plasma.

If there are abrupt "ends" to the plasma along the field lines or if the magnetic field is substantially curved, as will usually be the case in laboratory plasmas, escape of the radiation is possible. It turns out that the slow extraordinary wave escapes due to curvature rather easily and is quite often seen. In Fig. 5.12 the feature extending in frequency from ω_p upward to the lower-frequency end of the first cyclotron harmonic arises by Čerenkov emission from a high parallel energy runaway tail on the distribution function. As is clear qualitatively, it has the expected characteristics of the slow extraordinary wave (extending upward in frequency from ω_p), even though its observed degree of polarization is low, because of multiple reflections in the plasma chamber scrambling the polarization before its

Fig. 5.12. Emission spectrum with a distinctive feature between ω_p and Ω, arising from Čerenkov emission (Hutchinson and Komm 1977).

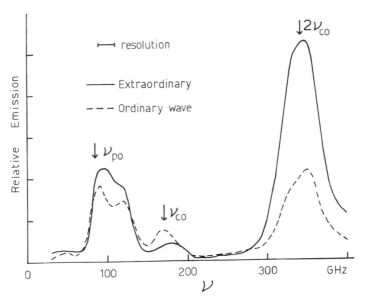

final exit through the port. Moreover, detailed theoretical calculations (Swartz et al. 1981) can reproduce in reasonable detail the line shape. The intensity of the emission is approximately proportional to the total number of electrons with parallel velocity greater than $c/N\cos\theta$. So this Čerenkov emission gives information about the high-energy tail density.

The slow ordinary wave is much more effectively trapped by electron density maxima. It is evanescent in the perpendicular direction for $\omega_p < \omega$. (Note the opposite inequality from the fast ordinary wave.) Because of this, toroidal plasmas tend to form excellent high Q cavities trapping the radiation. An intriguing consequence is that under some conditions, when these waves experience gain because of an unstable distribution, the whole plasma can act as a maser and generate coherent radiation (Gandy et al. 1985). When this occurs, the radiation frequency tends to be fractionally below the peak plasma frequency (by an amount that depends on plasma conditions but is often as little as 1%). Thus, the emission frequency may provide a highly accurate measurement of peak density. On the other hand, the "population inversion" giving rise to gain in this maser mechanism, consists of a positive slope $df_\parallel/dv_\parallel$ in the parallel distribution function. Hence the mere observation of the maser action indicates the presence of such a distribution, useful information in its own right.

Čerenkov emission is far from universal in the way that ($m \neq 0$) cyclotron emission is, and thus it is not really a routine diagnostic. Nevertheless, it does provide useful information about the relativistic components of the distribution function in nonthermal plasmas.

5.3   Radiation from electron–ion encounters

The radiation that occurs when a free electron is accelerated in the electric field of a charged particle is called bremsstrahlung. In fact, for collisions with a positively charged particle such as an ion, the radiative event can take one of two forms. It may be a free–free transition when the final state of the electron is also free (total energy greater than zero) or it may be a free–bound transition in which the electron is captured by the ion into a bound final state (total energy less than zero). When it is necessary to distinguish these two types, the free–bound transitions are often called recombination radiation, though we shall sometimes use the term bremsstrahlung to include both free–free and free–bound radiation.

Electron–electron collisions will be ignored because their contribution to radiation is generally small unless their velocities are very relativistic. This is easy to understand, since in a binary collision between identical particles there is no net acceleration of the center of mass or center of charge. Thus, to lowest order, the fields from the two particles exactly cancel.

Unlike the case of cyclotron radiation, in most practical situations a classical (nonquantum) treatment of bremsstrahlung is inadequate; we shall therefore be obliged to use quantum-mechanical formulas in our discussions. However, since it is usual to express the emission in terms of the classical formula multiplied by a correction factor, which takes into account the quantum effects, let us begin with a brief discussion of the classical problem.

5.3.1 Classical bremsstrahlung

The trajectory of an electron in the Coulomb field of an (assumed stationary) ion of charge Ze is as illustrated in Fig. 5.13. Assuming the radiative loss of energy to be negligible, the electron path is a conic section with the ion at the focus. In particular, for a free electron the path is a hyperbola given by the formula

$$r = \frac{b^2}{b_{90}(1 + \varepsilon \cos \theta)},$$ (5.3.1)

where b is the impact parameter, b_{90} is the impact parameter for $90°$ scattering, and ε is the eccentricity

$$b_{90} = \frac{Ze^2}{4\pi\varepsilon_0 mv_1^2}, \qquad \varepsilon = \left[1 + \left(\frac{b}{b_{90}}\right)^2\right]^{1/2}.$$ (5.3.2)

Here v_1 is the incident electron velocity and we shall drop the e suffix on the electron mass for brevity.

The radiation from such an encounter is given by our previous equations (5.1.13) and (5.1.17), which we write in the form integrated over all solid angles in the nonrelativistic limit:

$$\frac{dW}{d\omega} = \frac{e^2}{6\pi^2\varepsilon_0 c^3} \left| \int_{-\infty}^{\infty} \dot{\mathbf{v}} e^{i\omega t} \, dt \right|^2.$$ (5.3.3)

If we wished to obtain the polarization and angular dependence of the radiation we could use the form $\hat{\mathbf{R}} \wedge (\hat{\mathbf{R}} \wedge \dot{\mathbf{v}})$, but since for an isotropic

Fig. 5.13. Trajectory of an electron–ion collision.

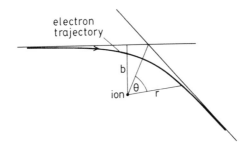

electron distribution the integration over angles will need to be performed, we allow it straight away. In order to calculate $dW/d\omega$, we need to substitute $\dot{v}$ as determined from the orbit and the energy equation,

$$\frac{1}{2}m\left[\dot{r}^2 + \left(r\dot{\theta}\right)^2\right] - \frac{Ze^2}{4\pi\varepsilon_0 r} = \frac{1}{2}mv_1^2, \tag{5.3.4}$$

into the Fourier integrals of Eq. (5.3.3) and evaluate them. Then, in order to calculate the radiation from a single electron colliding with a random assembly of ions of density n_i, we multiply by $n_i v_1$ and integrate over impact parameters to get the power spectrum

$$\frac{dP}{d\omega} = n_i v_1 \int_0^\infty \frac{dW}{d\omega}(\omega, b)2\pi b\, db. \tag{5.3.5}$$

Both these integration stages can be performed analytically (with considerable knowledge of Hankel function integral identities). The details are given by Shkarofsky et al. (1966) [and Landau and Lifschitz (1951)] and the result is

$$\frac{dP}{d\omega} = \frac{dP}{d\omega}\bigg|_c G(u_{90}), \tag{5.3.6}$$

where u_{90} is a nondimensionalized frequency $u_{90} = i\omega b_{90}/v_1$, the frequency independent part is

$$\frac{dP}{d\omega}\bigg|_c = \frac{Z^2 e^6}{(4\pi\varepsilon_0)^3} \frac{16\pi}{3\sqrt{3}} \frac{n_i}{m^2 c^3 v_1}, \tag{5.3.7}$$

and

$$G(u) = \frac{\pi\sqrt{3}}{4} u H_u(u) H_u'(u). \tag{5.3.8}$$

$H_\nu(x)$ here is the Hankel function of the first kind and order ν and the prime denotes differentiation with respect to argument. This result was originally obtained by Kramers (1923), by whose name it is sometimes known.

The dimensionless factor G, as we shall see shortly, is generally a weakly varying quantity of order 1. The physical significance of the first factor $dP/d\omega|_c$ may be illuminated by writing it in terms of the classical electron radius r_e and the fine structure constant α:

$$r_e \equiv \frac{e^2}{4\pi\varepsilon_0 mc^2}, \qquad \alpha = \frac{e^2}{4\pi\varepsilon_0 \hbar c} \qquad \left(\approx \frac{1}{137}\right). \tag{5.3.9}$$

We find

$$\frac{dP}{\hbar\,d\omega}\bigg|_c = n_i v_1 \frac{8}{3\sqrt{3}}\alpha\frac{mc^2}{\frac{1}{2}mv_1^2}\pi(Zr_e)^2. \tag{5.3.10}$$

This may conveniently be expressed in terms of a differential cross section $d\sigma_c/d\omega$, defined by setting the number of photons emitted in the frequency range $d\omega$ per unit path length, in an ion density n_i, equal to $n_i\,d\sigma$. Then the power spectrum may be expressed as

$$\omega\frac{d\sigma_c}{d\omega} = \frac{16}{3\sqrt{3}}\alpha\frac{c^2}{v_1^2}\pi(Zr_e)^2. \tag{5.3.11}$$

The subscript c on σ_c indicates that it does not include the factor G needed to give the full expression for the cross section. On the right hand side we have a numerical factor $(16/3\sqrt{3})$, two dimensionless factors (α and c^2/v_1^2, the latter representing the ratio of electron rest energy to kinetic energy), and finally a cross section. This last term, $\pi(Zr_e)^2$, is the area of a disk of radius Zr_e. We may thus loosely think of the ion as presenting to a colliding electron an area corresponding to a radius Zr_e but corrected by the other factors. (Note, though, that α and c^2/v_1^2 are not of order 1.)

Turning to the factor G, its functional form is rather obscure. It is of interest, therefore, to obtain some approximations for it in the limiting cases of high and low frequency. We may do this by some approximate physical arguments as follows.

Radiation occurs mostly during the time when the electron is closest to the ion, when the acceleration is greatest. If the subscript zero denotes quantities evaluated at the instant of closest approach, the conservation of angular momentum gives

$$v_1 b = v_0 r_0. \tag{5.3.12}$$

The acceleration at this instant is

$$\dot{v}_0 = \frac{Ze^2}{4\pi\varepsilon_0 mr_0^2} \tag{5.3.13}$$

and the approximate time duration of the close approach is

$$\tau \approx 2r_0/v_0. \tag{5.3.14}$$

When we evaluate the Fourier transform of $\dot{v}$ in order to obtain $dW/d\omega$, its peak value will occur when the oscillation of the $e^{i\omega t}$ factor is synchronized with the variation in $\dot{v}$. For this particular frequency, ω_0 say, we shall then find that

$$\int \dot{v}\,e^{i\omega_0 t}\,dt \sim \dot{v}_0\tau \approx 2\frac{Ze^2}{4\pi\varepsilon_0 mv_1 b}, \tag{5.3.15}$$

using Eqs. (5.3.12)–(5.3.14). Therefore, the peak in the spectrum of the emitted energy is

$$\frac{dW}{d\omega}\bigg|_{\omega_0} \approx \frac{Z^2 e^6}{(4\pi\varepsilon_0)^3} \frac{8}{3m^2 c^3} \frac{1}{v_1^2 b^2}. \tag{5.3.16}$$

Now we must recognize two opposite limits of impact parameter. (1) If $b \gg b_{90}$, the electron hardly deviates in its course and we have approximately a *straight-line collision*. (2) If $b \ll b_{90}$, the electron trajectory is then approximately parabolic in shape; it approaches the ion closely and then returns in approximately the direction from which it came. We shall call this a *parabolic collision*.

Although the peak value of the energy spectrum, $dW/d\omega$, is given by Eq. (5.3.16) for both these cases, the shape of the spectrum is different. This may be understood by considering the time history of $\dot{v}$ for the two cases, which is sketched in Fig. 5.14. The two components of the acceleration $\dot{v}_x$ and $\dot{v}_y$, parallel and perpendicular to the direction of incidence, respectively, are shown. For a straight-line collision $\dot{v}_y$ is symmetric and $\dot{v}_x$ antisymmetric about $t = 0$, the instant of closest approach. Also $\dot{v}_y$ is always negative (downward in Fig. 5.13). The importance of this last fact is that in either case the spectrum at very low frequency $\omega \approx 0$ involves $|\int \dot{v}\, dt|$, the time integral of $\dot{v}$. In the straight-line collision this integral is approximately equal to $|\dot{v}_0 \tau|$ because $\dot{v}_y$ has constant sign ($\dot{v}_x$ integrates to zero).

On the other hand a parabolic collision has $\dot{v}_y$ antisymmetric and $\dot{v}_x$ symmetric. Also $\dot{v}_x$ has both positive and negative excursions as the

Fig. 5.14. (*a*) Accelerations and (*b*) corresponding power spectra for straight-line and parabolic collisions.

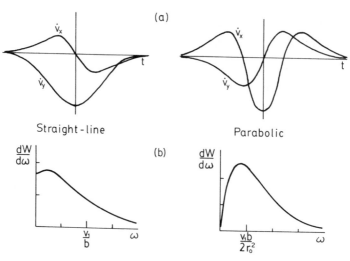

electron first gains energy in the Coulomb field (far in excess of its initial energy), passes the ion, experiencing rapid reversal of its v_x velocity, and then decelerates again as it retreats away from the ion. When we consider $\omega \approx 0$ for this collision, the positive and negative parts of $\dot{v}_x$ almost exactly cancel, and so

$$\left| \int \dot{v} \, dt \right| \ll |\dot{v}_0 \tau|. \tag{5.3.17}$$

Put another way, $\int \dot{v} \, dt$ is equal to approximately $2v_1$, but this is much less than the peak velocity $v_0 \approx |\dot{v}_0 \tau|$. What we have therefore shown is that

$$\left. \frac{dW}{d\omega} \right|_{\omega=0} \approx \left. \frac{dW}{d\omega} \right|_{\omega=\omega_0} \qquad \text{for a straight-line collision,} \tag{5.3.18}$$

but

$$\left. \frac{dW}{d\omega} \right|_{\omega=0} \ll \left. \frac{dW}{d\omega} \right|_{\omega=\omega_0} \qquad \text{for a parabolic collision.} \tag{5.3.19}$$

The actual shapes of the energy spectra in these two cases are shown in Fig. 5.14(b). They have an extent in ω of approximately

$$1/\tau \approx (v_1/2b), \quad \text{straight-line collision,} \tag{5.3.20}$$

$$1/\tau \approx (v_1/2b)(2b_{90}/b)^2, \quad \text{parabolic collision.} \tag{5.3.21}$$

When we come to integrate over impact parameters for a given frequency ω, the integral will be dominated by straight-line or parabolic collisions according to whether $\omega b_{90}/v_1$ is less than or greater than 1. Straight-line collisions thus correspond to low frequencies $\omega \ll v_1/b_{90}$ and parabolic collisions correspond to high frequencies $\omega \gg v_1/b_{90}$.

We may obtain an approximate value for the impact parameter integral by taking $dW/d\omega$ to be given by Eq. (5.3.16) over a range from b_{min} to b_{max}. Then one finds straight away that

$$\frac{dP}{d\omega} = \left. \frac{dP}{d\omega} \right|_c G, \tag{5.3.22}$$

with $dP/d\omega|_c$ given by Eq. (5.3.7) and

$$G = \frac{\sqrt{3}}{\pi} \ln \left| \frac{b_{max}}{b_{min}} \right|. \tag{5.3.23}$$

For low frequencies we must take $b_{max} \approx v_1/\omega$ because of the width of the $dW/d\omega$ spectrum and $b_{min} \approx b_{90}$ so that the straight-line approximation still holds. This leads to

$$G = \frac{\sqrt{3}}{\pi} \ln \left| \frac{2v_1}{\zeta \omega b_{90}} \right|, \tag{5.3.24}$$

where the additional factor $2/\zeta$ is inserted in the argument of the logarithm to conform our result to the exact limit obtainable from Kramer's formula [Eq (5.3.8)]. ζ is the reciprocal of Euler's constant ($\zeta = 1.78$) so the correction is negligible for most practical purposes.

For high frequencies b_{max}/b_{min} is a constant of order unity arising from the width of $dW/d\omega$. No singular lower limit occurs because $dW/d\omega|_{\omega=0} \ll dW/d\omega|_{\omega=\omega_0}$. Therefore, the logarithmic term is a constant of order unity. Again, detailed treatment of the asymptotic Hankel function limits provides the exact limiting value, namely

$$G = 1. \qquad (5.3.25)$$

Of course, it was in anticipation of this result that the division of the power spectrum into $dP/d\omega|_c$ and G was chosen in the precise manner of Eq. (5.3.8). A direct proof that $G = 1$ for parabolic collisions is given by Miyamoto (1980), though it still involves some tricky integrals.

Figure 5.15 shows the frequency dependence of the factor G, determining the spectral power density for a classical collision. G is called the Gaunt factor (Gaunt 1930) and modifications of the emission due to quantum effects are generally collected into corrections to G, as we shall see in the following section. Note, however, that at low frequencies (generally the only region where a classical treatment may be satisfactory) the classical Gaunt factor is greater than 1.

It should be noticed, too, that these classical results reveal an ultraviolet catastrophe; that is, the integral of $dP/d\omega$ to high frequencies becomes infinitely large. This emphasizes the importance of a quantum-mechanical treatment.

Fig. 5.15. The Gaunt factor for classical electron–ion bremsstrahlung.

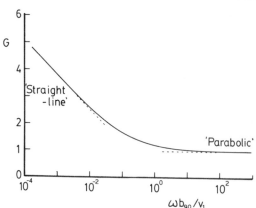

5.3.2 *Quantum-mechanical bremsstrahlung*

Semiclassically speaking, the quantum effects that enter are of two types: (a) the wave nature of the electron and (b) the particle nature of the electromagnetic radiation.

The de Broglie wave number of the electron is

$$k_h = mv/\hbar \qquad (5.3.26)$$

and the wave nature of the electron will be ignorable only if $k_h b \gg 1$. Now in the high-frequency ($b \ll b_{90}$) regime the requirement upon the photon size becomes more important than the de Broglie wavelength (see Exercise 5.4). Therefore, we need to consider the restriction on $k_h b$ only in the low-frequency regime $b > b_{90}$ and can write the electron requirement for the classical limit as

$$\eta_1 \equiv k_h b_{90} \gg 1. \qquad (5.3.27)$$

We note that

$$\eta_1^{-2} = \frac{\frac{1}{2} m v_1^2}{Z^2 R_y} \qquad (5.2.28)$$

is the initial particle energy in units of the ionization potential of the ground state of the atom formed by the electron and ion. R_y is the Rydberg energy

$$R_y = \frac{m}{2} \left(\frac{e^2}{4\pi\varepsilon_0 \hbar} \right)^2 = 13.6 \text{ eV}. \qquad (5.3.29)$$

The electromagnetic radiation quantum effects will be negligible only if the photon energy is much smaller than the initial particle energy,

$$\hbar\omega \ll \tfrac{1}{2} m v_1^2. \qquad (5.2.30)$$

It is helpful, then, to consider the bremsstrahlung coefficient, which is a function of ω and v_1, in terms of the initial electron energy and the photon energy, both measured in units of $Z^2 R_y$. Figure 5.16 shows this domain. The region of applicability of the classical Kramer's formula is indicated by K. No free–free emission can occur for frequencies above the photon limit ($\hbar\omega = \tfrac{1}{2} m v_1^2$), which appears as the diagonal right hand boundary. Above the line $m v_1^2 / 2 Z^2 R_y = 1$ the classical coefficient is inapplicable and a proper quantum-mechanical treatment is required.

A full nonrelativistic quantum analysis has been performed by Sommerfeld (1934), who calculated the transition probabilities between initial and final velocities v_1, v_2 and the corresponding "quantum numbers"

η_1, η_2, to obtain the emission coefficient for photons of energy

$$\hbar\omega = \tfrac{1}{2}m\left(v_1^2 - v_2^2\right) = Z^2 R_y\left(\eta_1^{-2} - \eta_2^{-2}\right). \qquad (5.3.31)$$

His result, in terms of a Gaunt factor, is

$$G = \frac{\pi\sqrt{3x}}{\left[\exp(2\pi\eta_1) - 1\right]\left[1 - \exp(-2\pi\eta_2)\right]} \frac{d}{dx}\left\{\left|F(i\eta_1, i\eta_2, 1; x)\right|^2\right\}, \qquad (5.3.32)$$

where F is the (ordinary) hypergeometric function and $x = -4\eta_1\eta_2/(\eta_1 - \eta_2)^2$. Again, despite its closed analytic form, the presence of the hypergeometric function makes this formula difficult to use.

A useful approximation may be obtained very simply by a semiclassical argument as follows. In the classical formula for straight-line collisions the logarithm $\ln(b_{max}/b_{min})$ allows the minimum impact parameter to be b_{90}. However, quantum limitations indicate that the classical formula for $dW/d\omega$ can apply only down to an impact parameter $\sim k_h^{-1}$. If we replace b_{min} with this value, the logarithm becomes $\ln(2mv_1^2/\hbar\omega)$. (An additional factor 2 in the argument is inserted to give agreement with other derivations.) Actually it may be more appropriate to use the mean velocity $\tfrac{1}{2}(v_1 + v_2)$ (although the difference will be small), in which case

$$G = \frac{\sqrt{3}}{\pi}\ln\left[\frac{m(v_1 + v_2)^2}{2\hbar\omega}\right] = \frac{\sqrt{3}}{\pi}\ln\left(\frac{v_1 + v_2}{v_1 - v_2}\right) = \frac{\sqrt{3}}{\pi}\ln\left(\frac{\eta_2 + \eta_1}{\eta_2 - \eta_1}\right). \qquad (5.3.33)$$

Fig. 5.16. The domains of applicability of approximations to bremsstrahlung: (K) classical Kramer's, (B) Born, (E) Elwert, (R) relativistic corrections needed [after Brussard and van de Hulst (1962)].

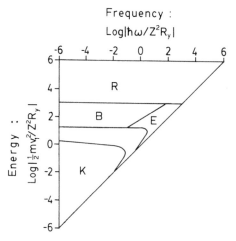

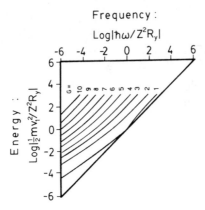

Fig. 5.17. Contour plot of the Gaunt factor [after Brussard and van de Hulst (1962)].

This result may also be obtained as the first Born approximation of the quantum-mechanical scattering treatment (Heitler 1944) and so is usually called the Born approximation. A modified form, due to Elwert (1939), with wider applicability is

$$G = \frac{\sqrt{3}}{\pi} \frac{\eta_2 [1 - \exp(-2\pi\eta_1)]}{\eta_1 [1 - \exp(-2\pi\eta_2)]} \ln\left(\frac{\eta_2 + \eta_1}{\eta_2 - \eta_1} \right), \qquad (5.3.34)$$

called the Elwert or Born–Elwert expression. All these formulas are valid in the region $\frac{1}{2} m v_1^2 \gg \hbar\omega, Z^2 R_y$ although the Elwert form provides good accuracy up to frequencies close to $\frac{1}{2} m v_1^2 / \hbar$. The regions B and E on Fig. 5.16 indicate where the Born and Elwert formulas are accurate to 1%. In the region R relativistic corrections become important.

Numerical values of the exact Gaunt factor are indicated in Fig. 5.17, which is a contour plot from which the Gaunt factor may conveniently be read.

5.3.3 Integration over velocities

The radiation per unit volume observed from a plasma with electron distribution function $f(\mathbf{v})$ must be calculated by integrating the power per electron over the distribution

$$\int \frac{dP}{d\omega} f d^3 \mathbf{v}. \qquad (5.3.35)$$

Assuming f to be isotropic we can write this as

$$4\pi j(\omega) = \left(\frac{dP}{d\omega} \bigg|_{v_1} \right)_c \int_0^\infty \frac{G(\omega, v)}{v} 4\pi v^2 f \, dv, \qquad (5.3.36)$$

where

$$\left(\frac{dP}{d\omega}\bigg|_{v_1}\right)_c = \frac{Z^2 e^6}{(4\pi\varepsilon_0)^3} \frac{16\pi n_i}{3\sqrt{3}m^2 c^3} \tag{5.3.37}$$

is a coefficient independent of velocity.

For a plasma in thermal equilibrium at temperature T, f is Maxwellian and the integral may be written

$$\left(\frac{2m}{\pi T}\right)^{1/2} n_e \int_0^\infty G(\omega, E) e^{-E/T} \frac{dE}{T}$$

$$= \left(\frac{2m}{\pi T}\right)^{1/2} n_e \left[\int_0^\infty G(\omega, E' + \hbar\omega) e^{-E'/T} \frac{dE'}{T}\right] e^{-\hbar\omega/T}, \tag{5.3.38}$$

where we have put $E = \frac{1}{2}mv^2 = E' + \hbar\omega$ and recognized that $G(\omega, E) = 0$ for $E < \hbar\omega$. The quantity

$$\bar{g}(\omega, T) \equiv \int_0^\infty G(\omega, E' + \hbar\omega) e^{-E'/T} \frac{dE'}{T} \tag{5.3.39}$$

is then the Maxwell-averaged Gaunt factor, written in lowercase to indicate that the factor $e^{-\hbar\omega/T}$ has been taken out of the G integral. The spectral power emitted into 4π sr per unit frequency is then

$$4\pi j(\omega) = n_e n_i Z^2 \left(\frac{e^2}{4\pi\varepsilon_0}\right)^3 \frac{16\pi}{3\sqrt{3}m^2 c^3} \left(\frac{2m}{\pi T}\right)^{1/2} e^{-\hbar\omega/T} \bar{g}. \tag{5.3.40}$$

Various approximations to $\bar{g}$ can be obtained from the approximate expressions for G, which are applicable under various conditions of temperature and frequency. Rather than derive them one by one we shall summarize the most important results by Fig. 5.18, which shows as solid lines the exact (nonrelativistic) Maxwell-averaged Gaunt factor as a function of $\hbar\omega/T$ for various values of $T/Z^2 R_y$, calculated numerically by Karzas and Latter (1961). Also shown are various broken lines corresponding to the following:

(a) Low-frequency semiclassical (Kramers) approximation

$$\bar{g} = \frac{\sqrt{3}}{\pi} \ln \left| \left(\frac{2T}{\varsigma m}\right)^{3/2} \frac{2m}{\varsigma\omega} \left(\frac{4\pi\varepsilon_0}{Ze^2}\right) \right|, \tag{5.3.41}$$

valid at low frequency for $T \ll Z^2 R_y$.

(b) Born approximation, valid for $T \gg Z^2 R_y$,

$$\bar{g} = \frac{\sqrt{3}}{\pi} K_0 \left(\frac{\hbar\omega}{2T} \right) \exp\left(\frac{\hbar\omega}{2T} \right) \qquad (5.3.42)$$

[see Shkarofsky et al. (1966)], where K_0 is the modified Hankel function. This reduces at low frequency to:

(c) Low-frequency Born approximation

$$\bar{g} = \frac{\sqrt{3}}{\pi} \ln \left| \frac{4T}{\zeta \hbar \omega} \right| . \qquad (5.3.43)$$

Perhaps the most important thing to notice about these results is that for a very wide range of temperatures the Gaunt factor lies between about 0.3 and 2 for $0.1 \le \hbar\omega/T \le 10$. The result is that, in this frequency range, virtually all the temperature and frequency dependence is contained in the $T^{-1/2}\exp(-\hbar\omega/T)$ factor of Eq. (5.3.40), compared to which $\bar{g}$ is almost unity.

5.3.4 *Recombination radiation contribution*

Before going further we must discuss the contribution that arises from free–bound transitions, since under some circumstances these can significantly change the radiation intensity emitted.

In his (1923) paper, Kramers' main concern was actually not to calculate bremsstrahlung but to estimate the cross section for absorption of electromagnetic radiation by bound–free electron transitions (the inverse of recombination). In the absence of a proper quantum theory at that time, his method was to estimate the free–bound transition probability from the

Fig. 5.18. Maxwell-averaged Gaunt factor. The solid line gives the exact (nonrelativistic) value. Other approximations are also shown [after Shkarofsky et al. (1966)].

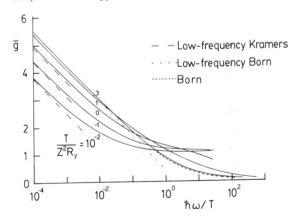

correspondence principle by comparison with the classical formula for radiation by an electron in a hyperbolic orbit, which we have previously discussed. By this means one arrives at a semiclassical estimate of recombination radiation. More complete quantum treatments usually express their results, just as for free–free transitions, as a Gaunt factor times Kramers' result.

The semiclassical approach is as follows. Consider a collision at velocity v_1 and impact parameter $b \ll b_{90}$; that is, a parabolic collision. The spectral energy density is $dW(v, b)/d\omega$ as in Fig. 5.14. Now for photon energy $\hbar\omega < \frac{1}{2}mv^2$, the final state of the electron is free so we expect the classical formula to hold (approximately). On the other hand if $\hbar\omega > \frac{1}{2}mv^2$ then the final electron state is bound and must take up a discrete spectrum of energies:

$$W_n = -\frac{Z^2 e^4 m}{2(4\pi\varepsilon_0\hbar)^2}\frac{1}{n^2} = -R_y\frac{Z^2}{n^2}, \qquad (5.3.44)$$

where n is the principal quantum number. The radiation spectrum thus consists of discrete lines at

$$\hbar\omega = \frac{1}{2}mv^2 + \frac{Z^2 R_y}{n^2}. \qquad (5.3.45)$$

The correspondence principle indicates that we should assign the classical power to these lines instead of to the continuum. This process is illustrated in Fig. 5.19. The energy to be put into the nth line is that in the frequency

Fig. 5.19. Schematic illustration of the application of the correspondence principle to estimate recombination.

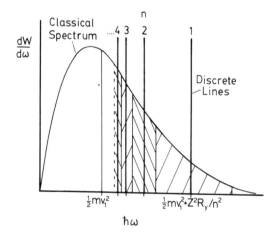

range

$$\frac{1}{2}mv^2 + \frac{ZR_y^2}{\left(n + \frac{1}{2}\right)^2} \leq \hbar\omega \leq \frac{1}{2}mv^2 + \frac{ZR_y^2}{\left(n - \frac{1}{2}\right)^2}, \qquad (5.3.46)$$

illustrated by the numbered (shaded) regions, so that

$$\Delta\omega_n \approx \frac{2Z^2R_y}{\hbar n^3} \qquad (5.3.47)$$

and the energy in the nth line is

$$\frac{dW}{d\omega}(b, v)\,\Delta\omega_n, \qquad (5.3.48)$$

where $dW/d\omega$ is the classical value. Note that this process has the effect of terminating the power spectrum at energy $R_y/(\frac{1}{2})^2$, so preventing an ultraviolet catastrophe.

The integral over impact parameters is now performed as before. This is unaffected by $\Delta\omega_n$ and so immediately we have the total power in the nth recombination line per electron of velocity v in a plasma of ion density n_i as

$$P_n = \left.\frac{dP}{d\omega}\right|_c \Delta\omega_n\, G_n = \left[\frac{Z^2 e^6}{(4\pi\varepsilon_0)^3}\frac{16\pi}{3\sqrt{3}}\frac{n_i}{m^2 c^3 v}\right]\frac{2Z^2 R_y}{\hbar n^3} G_n. \qquad (5.3.49)$$

The Gaunt factor G_n we shall expect to be close to 1, provided $\frac{1}{2}mv^2 \leq Z^2 R_y$.

To calculate the recombination radiation from an electron distribution $f(v)$ as a function of ω for level n, no integral is required; we simply put $v = [2(\hbar\omega - Z^2 R_y/n^2)/m]^{1/2}$ and then

$$4\pi j_n(\omega) = P_n 4\pi v^2 f(v)\left|\frac{dv}{d\omega}\right| = P_n 4\pi v f(v)\frac{\hbar}{m}. \qquad (5.3.50)$$

For a Maxwellian distribution this may be written

$$4\pi j_n(\omega) = n_e n_i Z^2 \left(\frac{e^2}{4\pi\varepsilon_0}\right)^3 \frac{16\pi}{3\sqrt{3}m^2 c^3}\left(\frac{2m}{\pi T}\right)^{1/2}$$

$$\times e^{-\hbar\omega/T}\left[\frac{Z^2 R_y}{T}\frac{2}{n^3}G_n e^{Z^2 R_y/n^2 T}\right]. \qquad (5.3.51)$$

This formula is identical to the bremsstrahlung equation (5.3.40), except for the term in square brackets that replaces $\bar{g}$, but it holds only for $\hbar\omega \geq Z^2 R_y/n^2$; otherwise, j_n is zero. No recombination occurs with frequency less than $Z^2 R_y/n^2\hbar$. Quantum-mechanical calculations (Karzas and Latter

1961) indicate that the Gaunt factors G_n are indeed within 20% of 1 for $\hbar\omega < 10Z^2R_y$. At higher frequency they decrease approximately logarithmically.

The relative importance of recombination radiation is determined by

$$\frac{Z^2R_y}{T}\frac{2}{n^3}\frac{G_n}{\bar{g}}\exp\left(\frac{Z^2R_y}{n^2T}\right). \qquad (5.3.52)$$

At low frequencies $\hbar\omega \ll Z^2R_y$, only very high n states contribute, $n^2 > Z^2R_y/\hbar\omega$. In view of the n^{-3} dependence, it is clear that recombination rapidly becomes negligible in this limit. On the other hand, when $\hbar\omega \geq Z^2R_y$ all ns can contribute and, for $Z^2R_y \geq T$, the recombination will dominate. If $Z^2R_y \ll T$, recombination is negligible for all frequencies.

As an example, Fig. 5.20 shows the theoretical emission spectrum of a Maxwellian plasma in which $T = Z^2R_y$. Significant steps occur at the so-called recombination edges $\hbar\omega = Z^2R_y/n^2$, where the next level begins to participate.

In calculating radiation from electrons colliding with incompletely ionized species, such as high-Z ions with electrons still bound to the nucleus, the usual approach is to assume that for free–free transitions and free–bound transitions, except to the lowest electron shell, the previous treatment is adequate using for Z the ionic charge rather than the nuclear charge. In other words, perfect screening of the nucleus by the bound electrons is assumed. However, this assumption will be valid only for photon (and hence electron) energies that are not much larger than the ionization potential. For the lowest unfilled shell we modify the recombination formula, recognizing that if this shell contains some electrons so that only ξ holes are available instead of the usual $2n^2$, then the recombination radiation is less by a factor $\xi/2n^2$. Also, it is usual to use the exact value χ_i

Fig. 5.20. Emissivity of a plasma in which $T_e = Z^2R_y$, illustrating the addition of recombination to the free emission spectrum [after Brussard and van de Hulst (1962)].

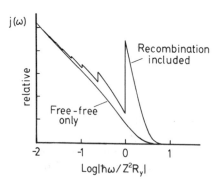

of the ionization potential of the recombined ion rather than the screened hydrogenic formula $Z^2 R_y/n^2$ for this level.

The total continuum radiation from collisions with such ions is then [cf. von Goeler (1978)]

$$
j(\omega) = n_e n_i Z^2 \left(\frac{e^2}{4\pi\varepsilon_0}\right)^3 \frac{4}{3\sqrt{3}m^2 c^3} \left(\frac{2m}{\pi T}\right)^{1/2} e^{-\hbar\omega/T}
$$

$$
\times \left[\bar{g}_{ff} + G_n \frac{\xi}{n^3} \frac{\chi_i}{T} e^{\chi_i/T} + \sum_{\nu=n+1}^{\infty} G_\nu \frac{Z^2 R_y}{\nu^2 T} \frac{2}{\nu} e^{Z^2 R_y/\nu^2 T}\right],
$$

(5.3.53)

the first term being the free–free contribution, the second recombination to the lowest unfilled shell (n), and the third to all other shells. It is often adequate to take $G_n = 1$ for $\hbar\omega \geq Z^2 R_y/n^2$, and, of course, $G_n = 0$ for $\hbar\omega < Z^2 R_y/n^2$. Evaluating the fundamental constants gives

$$
j(\omega) = 8.0 \times 10^{-55} n_e n_i Z^2 \left(\frac{T}{eV}\right)^{-1/2} e^{-\hbar\omega/T}
$$

$$
\times \left[\bar{g}_{ff} + G_n \frac{\xi}{n^3} \frac{\chi_i}{T} e^{\chi_i/T}\right.
$$

$$
\left. + \sum_{\nu=n+1}^{\infty} G_\nu \frac{Z^2 R_y}{\nu^2 T} \frac{2}{\nu} e^{Z^2 R_y/\nu^2 T}\right] \text{W m}^{-3} \text{ sr}^{-1} \text{ s}.
$$

(5.3.54)

5.3.5 Temperature measurement

Bremsstrahlung can be emitted over an exceptionally wide spectral range, from the plasma frequency ω_p, usually in the microwave region, right up to the frequency whose energy $\hbar\omega$ is of order the electron temperature or more, usually in the x-ray region. There are several possible uses of the emission for diagnostic purposes, using different spectral regions.

For $\hbar\omega \geq T$, as we have seen, the emission from both free–free and free–bound transitions has a strong (exponential) dependence upon the temperature. In this frequency range, the main diagnostic use of the emission is as a measurement of electron temperature.

Figure 5.21 shows a spectrum in the region 1–6 keV photon energy obtained from the Alcator tokamak. Only the relative spectral intensity is plotted (on a log scale) because the great advantage of this method is that it requires only the gradient of the spectrum, not its absolute intensity, to give the electron temperature. The instrument used to obtain this spectrum was an x-ray pulse height analyzer, which is simply a solid state detector in which each photon releases a charge pulse proportional to its energy. Pulses are then electronically sorted by size to provide the spectrum.

One can see that the spectrum falls off rapidly with photon energy as expected. Below ~ 2.5 keV a reasonably straight line is obtained on the log-linear plot. A line is fitted to these points (whose slight curvature is designed to account for the expected variation of T_e along the line of sight) representing the relative spectral shape for a plasma electron temperature of 900 eV. This corresponds to an exponential decay of intensity with characteristic (e^{-1}) spectral width 900 eV. At about 2.5–3 keV a sharp rise occurs in the spectrum. This is due to a combination of line (i.e., bound–bound) and recombination radiation from impurities in the plasma. The transitions involved are to the L shell of molybdenum and K shell of chlorine, both of which species are present as highly ionized impurities in the plasma. The feature is somewhat blurred by the presence of different ionization states that have different characteristic energies. Above this feature, from ~ 3–6 keV, the spectrum again shows the exponential decay anticipated but now at a higher absolute level. This may be in part because of the additional recombination contribution, although a nonthermal electron component may also be present.

Fig. 5.21. Typical emission spectrum from a tokamak plasma, also showing the effects of impurities [after Rice et al. (1978)].

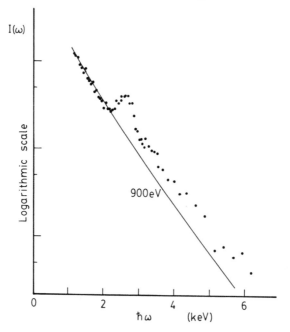

It is clear from such a spectrum that, even for a thermal plasma, considerable care must be exercised in estimating the electron temperature from the x-ray bremsstrahlung spectrum. If, for example, less spectral resolution were available, say just two broad windows at 1.5 and 2.7 keV, then an estimate of T_e from the ratio of the intensities there would be completely incorrect if a smooth exponential decay were assumed. It is generally required to have a fairly detailed spectrum such as that shown before any confidence can be gained that the measurement is not being distorted by unknown recombination edges and associated line radiation.

The continuum observed in such a spectrum, even below 2.5 keV, is mostly radiation, free–free and free–bound, arising from impurities such as oxygen and carbon in the (hydrogen) plasma. The reason for this is the dominance of the factor Z^2 in the bremsstrahlung coefficient and also the importance of the factor $\exp(Z^2 R_y / T)$ in recombination for any ion species in which $Z^2 R_y \sim T$. These tend to cause the radiation from impurities to exceed that from hydrogen for all but the cleanest plasmas. Fortunately, since the exponential dependence on ω is the same in all cases, this does not affect the logarithmic slope of the spectrum from which the temperature is deduced.

5.3.6   *Multiple species:* Z_{eff} *measurement*

Consider the practical case in which there are many different ion species i in the plasma of charge Z_i. The continuum radiation is then a sum of contributions of the form of Eq. (5.3.53) over all different species. When recombination is important, no general simplification of the radiation intensity is used, although quite often the extent to which the continuum exceeds hydrogen bremsstrahlung, at the particular temperature and electron density, is denoted by a single factor, the x-ray enhancement factor.

When recombination is negligible it is easy to see that, provided the Gaunt factor can be taken as equal for all species, the emission is proportional to

$$\sum_i n_e n_i Z_i^2 = n_e^2 Z_{eff}. \qquad (5.3.55)$$

Z_{eff} is then equal to the factor by which the bremsstrahlung exceeds that of hydrogen (at electron density n_e) and it is also given by the simple formula

$$Z_{eff} = \sum n_i Z_i^2 / \sum n_i Z_i, \qquad (5.3.56)$$

where quasineutrality has been invoked in writing $n_e = \sum n_i Z_i$. The quantity Z_{eff} is, therefore, a kind of mean ion charge in the plasma, weighted for each species by the proportion of electrons arising from that species. Clearly for a pure plasma of ionic charge Z, $Z_{eff} = Z$, while for a hydrogenic plasma (say) with impurities, Z_{eff} measures the degree of impurity in the plasma, an important quality for many purposes.

As indicated by Eq. (5.3.52), a sufficient condition for recombination to be negligible is $\hbar\omega \ll Z^2 R_y$. For singly ionized species this requirement is $\hbar\omega \ll 13.6$ eV, which is satisfied for visible radiation and longer wavelengths. Absolute calibration of the detector and spectrometer is necessary in order to deduce Z_{eff} from the theoretical emission coefficients given earlier. This is usually most easily accomplished in the visible, although measurements can also be made in the near and far infrared. In addition, independent estimates of the electron temperature and density are necessary in order to determine the factor $n_e^2 T^{-1/2}$ appearing in the bremsstrahlung coefficient. The exponential dependence $\exp(-\hbar\omega/T)$ may also be important, but for the purposes of Z_{eff} measurement it is usually best to choose $\hbar\omega \ll T$, so that this factor is unity.

Of course, if it is established that $Z_{\text{eff}} = 1$, one may use the bremsstrahlung to measure $n_e^2 T^{-1/2}$ and hence n_e (say) if $T_{(e)}$ is known. However, it is rarely, a priori, a reasonable assumption for hot plasmas that $Z_{\text{eff}} = 1$, so this approach is of less universal validity.

The most important precaution in performing the Z_{eff} measurement in the visible is to ensure that the intensity is measured in a spectral region free from strong impurity line radiation. This is often a difficult requirement to meet. Figure 5.22 illustrates a spectrum (from the Alcator tokamak) in a wavelength region of interest for bremsstrahlung purposes. As indicated, there are considerable numbers of lines in the region. A 3.0 nm wide

Fig. 5.22. Spectrum in the wavelength region suitable for Z_{eff} determination from visible bremsstrahlung [Foord et al. (1982)].

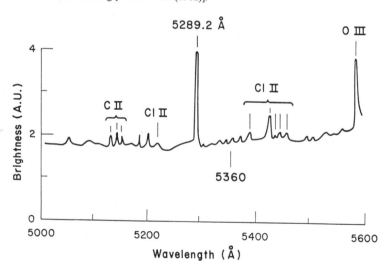

section is used at 536.0 nm for the continuum measurement, the line contribution being small enough there for these plasmas. In general a spectral survey such as this is required for any specific plasma to establish the freedom from line emission of a wavelength region of interest.

5.3.7   *Absorption: Blackbody level bremsstrahlung*

For a thermal plasma, Kirchhoff's law relates the absorption of radiation by the inverse of bremsstrahlung (collisional damping) to bremsstrahlung emission,

$$\alpha(\omega) = j(\omega)/B(\omega), \tag{5.3.57}$$

where the blackbody intensity is given by Eq. (5.2.35) or, in the Rayleigh–Jeans limit by Eq. (5.2.36). The ω^2 dependence of $B(\omega)$ for $\hbar\omega \ll T$ causes α to increase as ω decreases, so that absorption is more important at long wavelengths.

There is, of course, a low-frequency limit to the applicability of the bremsstrahlung emissivity we have calculated. This may generally be taken as the plasma frequency ω_p, below which waves do not propagate (in an unmagnetized plasma). For large dense plasmas, however, as the frequency is lowered, the optical depth due to bremsstrahlung processes may exceed 1 before the cutoff frequency is reached. In this case, it becomes possible to deduce the electron temperature from the blackbody emission intensity observed.

The principle is the same as with cyclotron radiation except that for bremsstrahlung there is no localized resonant interaction with the plasma. This is a considerable handicap since it means that in an inhomogeneous plasma the radiation temperature observed is that characteristic of the region of plasma that the wave traverses during its last optical depth of about unity before leaving the plasma. In other words, if the plasma has a total optical depth much greater than 1 we shall only "see" the edge of the plasma up to an optical depth ~ 1, not right into the plasma center (see Fig. 5.23). The radiation temperature measured will then be characteristic of the edge electron temperature. This point is rather familiar in astrophysics. When viewing a star, for example, we see radiation characteristic of its outer regions, the photosphere, and cannot directly observe the core temperature.

Of course, the advantage of viewing an optically thick plasma is that the radiation is then independent of n_e and Z_{eff} so that the interpretation is simpler. If one can choose a frequency such that the total optical depth is not much greater than 1, then the plasma center can be probed. Unfortunately, this choice does require knowledge of the emissivity and hence n_e and Z_{eff}. Thus a substantial part of the potential advantage is lost. For these reasons, optically thick bremsstrahlung emission tends not to provide very reliable temperature measurements for inhomogeneous plasmas.

In very high-density plasmas, such as occur in inertial confinement fusion or other laser-produced plasma experiments, it is sometimes fruitful to use inverse-bremsstrahlung absorption as a measurement of electron density. In such experiments the plasma to be diagnosed is generally "back-lighted" by some (x-ray) radiation source, itself often a laser-produced plasma. Viewing the plasma against this back-light one then observes absorption of the back-light corresponding to the form of the plasma under diagnosis. Obviously, since the source of the radiation is rather uncertain and the absorption is a function of T_e and Z as well as n_e, such measurements tend to be largely qualitative.

5.3.8   x-ray imaging

The spectral region in which the greatest bremsstrahlung power is emitted is in the vicinity of photon energy equal to the electron temperature. For much smaller energy, ω and hence $\Delta\omega$ are smaller and so, despite significant power per unit frequency, less total power is available. On the other hand, for $\hbar\omega \gg T_e$ the spectral intensity falls off exponentially. The radiation in this optimum spectral range $\hbar\omega \sim T_e$, which for hot plasmas falls in the soft x-ray region, tends to be sufficient and convenient for rapid time-resolved observations of its evolution to be made. Practically, this is facilitated by the availability of sensitive solid state detectors, usually semiconductor diodes of various specifically designed configurations, such

Fig. 5.23. The distance into which one can "see" at a certain frequency corresponds roughly to an optical depth of unity. This distance increases with ω for bremsstrahlung.

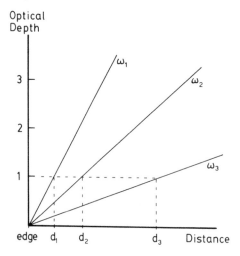

as surface barrier diodes, in which charge is liberated and collected, proportional to the total energy incident upon the detector (of photons above a certain energy). Thus, simple and convenient technology exists to view a plasma in the soft x-ray region.

A further important factor governing the usefulness of such an approach is that for $\hbar\omega > T_e$ the emission is a strong function of electron temperature as well as density. So, by employing a foil filter in front of the detector, chosen to pass photons above an energy of the order of T_e and to absorb (or reflect) them below that energy, the signal observed can be made a sensitive monitor of temperature. Naturally, the dependence of the emissivity on n_e^2 and also Z_{eff} remains; also recombination is often appreciable, if not dominant, so that it is very difficult to make any absolute deductions about T_e and n_e from the intensity observed. Instead the usual interest is to obtain *relative* variations of the complicated combination of plasma parameters that combine to determine the emissivity as a function of time and space. This enables one to see qualitatively, and sometimes quantitatively, the evolution of the shape of the plasma as a result of instabilities or other perturbations.

The approach typically adopted is illustrated in Fig. 5.24. The plasma is viewed, using an array of detectors, along a number of collimated chords. Each detector thus measures the mean emissivity along its chord. Ideally one would then want to deduce the entire spatial dependence of the

Fig. 5.24. Schematic illustration of the type of pinhole camera system used for x-ray imaging.

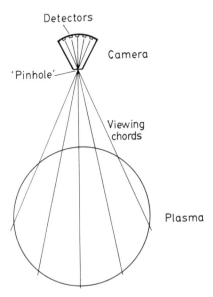

emissivity of the plasma by an inversion algorithm from the chordal measurements. Naturally, even when there is a symmetry to the plasma that allows one to concentrate on a single two-dimensional slice in attempting this reconstruction, the degree of spatial resolution achievable depends upon the number of views available. Moreover, under most conditions the reconstruction is not unique, in that a given chordal measurement set may arise from many different possible spatial configurations.

The problems involved in this reconstruction are essentially identical to those arising in computer aided tomography (Cormack 1964) (the medical CAT scanner), although the case we are considering is emission rather than absorption of x-rays. The reconstruction, therefore, draws upon the same mathematical techniques, although sometimes the plasma environment is less convenient and controllable than the patient in medical diagnostics. These inversion problems also arise in other plasma diagnostics that measure chordal averages such as interferometry or any optically thin radiation. Indeed these other diagnostic techniques can be used to the same ends as the soft x-ray bremsstrahlung when sufficient chord measurements are available.

The general principle of the approach to the inversion problem is to assume that the emissivity can be expanded as the sum of a finite set of known functions of space. For example, one possible choice would be to express the emissivity as a set of pixels covering some mesh. Each pixel then corresponds to one of the functions

$$f_i(x, y) = 1, \qquad x_i \le x < x_{i+1}, \ y_i \le y < y_{i+1},$$
$$= 0, \qquad \text{otherwise.} \tag{5.3.58}$$

The problem then is to obtain the coefficients of the functions that most satisfactorily reproduce the chordal observations. The meaning of "most satisfactorily" must be incorporated into the algorithm for determining the coefficients so as to produce a single answer even though many (or none) may be fully consistent with the data. Clearly, then, one cannot meaningfully determine more coefficients than one has observations. In general it is necessary, in fact, to employ significantly fewer coefficients than independent observations, and then by some optimization procedure such as least squares or maximum entropy [see, e.g., Gull and Daniell (1978)] to construct a fit to the data.

The quality of the results obtained will be strongly influenced by the basis function set adopted. The pixel example is actually a rather unsatisfactory choice for many situations because it employs none of the additional information that one has a priori, for example, that profiles tend to be smooth or that there are various approximate symmetries.

The obvious choice for an approximately cylindrical plasma is to take the azimuthal dependence to be represented by a sum of Fourier modes:

$\cos m\theta, \sin m\theta$. The Fourier sum must be truncated at an appropriate m number. For example, if $m = 0$ alone is allowed, the problem becomes a simple Abel inversion (see Section 4.4), that is, the plasma is assumed cylindrically symmetric.

In general, to proceed to higher m expansions requires more independent views of the plasma. This may be seen easily when we consider the plasma views to be sets of parallel chords. Figure 5.25 illustrates the point. If we have only one set of parallel chords at angle $\theta = 0$ [Fig. 5.25(a)], we can obtain a unique expansion of the form

$$f_0(r) + f_1(r)\sin\theta, \tag{5.3.59}$$

but cannot observe a $\cos\theta$ component nor distinguish a $\sin 2\theta$ component from $\sin\theta$. In effect, this single angle allows us to resolve only into odd and even components, which might be correctly expressed as $m = 0$ and $m = 1$ (sine), but might not. Figure 5.25(b), in which two views are available, enables an expansion of the form

$$f_0(r) + f_{1s}(r)\sin\theta + f_{1c}(r)\cos\theta + f_{2c}(r)\cos 2\theta \tag{5.3.60}$$

to be determined uniquely.

In some situations the plasma is cooperative and rotates poloidally at what can be taken as a uniform rate. This provides a single chordal array

Fig. 5.25. (a) Parallel views in one direction allow one to distinguish only even and odd modes. (b) Additional directions of view provide further mode resolution.

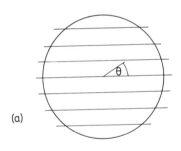

(a)

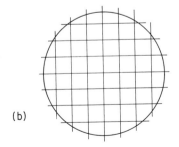

(b)

with, effectively, a large number of angular views of the plasma (just as does a rotation of the patient in a medical scanner). Then this Fourier expansion can be taken to high m numbers and works very successfully. In cases where the plasma cannot be taken as uniformly rotating, either a priori assumptions about the symmetry of the plasma must be invoked or the patience and funds of the experimenter tested by the number of independent simultaneous views he provides.

The deduction of the radial dependence of each Fourier mode involves a generalized form of the Abel integral equation. Because of its sensitivity to errors in the measurement, this inversion too is best solved by assuming an expansion in terms of a sum of special functions. The functions usually adopted, for the particular convenience of their inversion, are Zernicke polynomials.

An example of the type of reconstruction of plasma shape that can be achieved is given in Fig. 5.26. This is a contour plot of the x-ray emissivity. Under most conditions, because transport of energy and particles along the magnetic field is extremely rapid, one expects T_e, n_e, Z_{eff}, and, hence, the emissivity to be constant on a magnetic surface. Therefore, contours of emissivity should coincide with magnetic surfaces and a plot such as Fig. 5.26 can be considered as approximating a magnetic surface plot.

Fig. 5.26. Example of the plasma cross-section shape deduced from x-ray tomography [Granetz and Camacho (1985)].

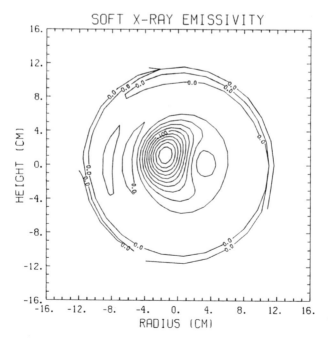

SOFT X-RAY EMISSIVITY

5.3.9 *Nonthermal emission*

When the electron distribution is not Maxwellian, bremsstrahlung emission can be very useful in diagnosing this fact. The situation of most frequent interest is one in which there is some high-energy tail on the distribution function. This shows up, naturally enough, as a tail on the bremsstrahlung spectrum. Thus, qualitatively at least, the bremsstrahlung immediately reveals the presence of such a tail. Bremsstrahlung measurements of x-ray continuum have long been used for this purpose. Figure 5.27 shows an example of a long high-energy tail on the x-ray spectrum from a nonthermal tokamak plasma.

To interpret quantitatively an observed spectrum such as Fig. 5.27 in terms of a distribution function, one immediately encounters a serious deconvolution problem, in that there is no one-to-one correspondence between photon energy and electron energy. In fact, an electron of velocity v_1 gives rise to a broad emission spectrum extending from $\omega = 0$ to $\omega = m v_1^2 / 2 \hbar$. The frequency dependence is contained in the Gaunt factor $G(\omega, v_1)$ (we ignore recombination in this section since we are mostly interested in high-energy electrons). As a reasonable approximation we may take $G = 1$ for $\hbar \omega < \frac{1}{2} m v_1^2$ and $G = 0$ for $\hbar \omega > \frac{1}{2} m v_1^2$. The Elwert expression [Eq. (5.3.34)] gives a more accurate form up to electron energy ~ 100 keV as Fig. 5.28 illustrates. For higher energies it underestimates the emission by a factor up to ~ 2 (Koch and Motz 1959). In general, though, an electron emits a whole spectrum of photon energies up to its kinetic energy.

As a result, the distribution function is related to the emissivity spectrum by an integral equation, which in an isotropic case is just Eq. (5.3.36); that

Fig. 5.27. An x-ray spectrum with a high-energy tail caused by radio frequency current drive [after Texter et al. (1986)].

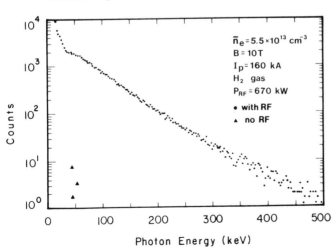

is

$$4\pi j(\omega) = \frac{Z^2 e^6}{(4\pi\varepsilon_0)^3} \frac{16\pi n_i}{3\sqrt{3} m^2 c^3} \int_0^\infty \frac{G(\omega, v)}{v} 4\pi v^2 f(v)\, dv. \quad (5.3.61)$$

This equation may, in principle, be solved for any functional form for G. It is particularly simple when the $G = 1$ approximation is adopted; then

$$f(v) = -\left[\frac{Z^2 e^6}{(4\pi\varepsilon_0)^3} \frac{16\pi n_i}{3\sqrt{3} m^2 c^3} \right]^{-1} \frac{m}{\hbar} \frac{dj}{d\omega}\bigg|_{\omega = mv^2/2\hbar} \quad (5.3.62)$$

(see Exercise 5.6). Using a more realistic expression for G leads to a much more complicated inversion integral, which requires a numerical solution.

Although Eq. (5.3.62) may appear to be a relatively tractable solution to the deconvolution problem, it is valid only for isotropic distribution functions. Unfortunately, most nonthermal distributions are highly anisotropic. They therefore demand consideration not only of the energy distribution of the electrons, but also the pitch angle distribution of the nonthermal components and the resulting anisotropy in the radiation.

The angular distribution of bremsstrahlung from a single electron may be understood qualitatively as arising from a dipole distribution of radiation,

Fig. 5.28. Bremsstrahlung spectrum at high energy (50 keV) where relativistic effects are important. The Elwert expression reasonably reproduces the experimental results (points) [after Koch and Motz (1959)].

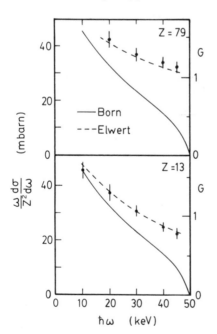

proportional to the square of the sine of the angle between the direction of emission and the dipole direction $\dot{v}$. Roughly speaking, for photon energy much less than the electron kinetic energy, the direction of $\dot{v}$ is arbitrary so that the emission is approximately isotropic. However, when $\hbar\omega \approx \frac{1}{2}mv_1^2$ the electron is decelerated almost to rest; so $\dot{v}$ is mostly in the direction of v_1. Thus photons near the high-frequency limit have a dipole distribution about the original velocity direction.

The preceding remarks refer to the angular distribution in the rest frame of the electron, or rather, ignoring the relativistic aberration [see any text on special relativity, e.g., Rindler (1960)]. In fact, however, this "headlight effect" in which emission is preferentially beamed in the direction of electron motion is usually the dominant effect upon the angular distribution. The relativistic aberration concentrates the radiation in the lab frame into a forward cone of half angle approximately $\sin^{-1}(1/\gamma)$ for $v_1 \sim c$, and at more modest energies ($\lesssim 100$ keV) a moderate enhancement of the emission occurs in the forward direction. This enhancement is sufficient in some cases to allow the anisotropy of the electron distribution to be investigated. Recent efforts in this respect are reviewed by von Goeler (1982).

Before leaving the topic of nonthermal bremsstrahlung, it should be noted that a major source of emission caused by high-energy electron tails is bremsstrahlung not from collisions with the plasma ions but from collisions with solid structures at the plasma edge. Such emission, often loosely referred to as hard x-ray radiation, is most properly called thick-target bremsstrahlung ("hard" and "soft" strictly refer to the energy–high or low–of the x-ray photons). This designation distinguishes it from our previous discussions by its most important characteristic, namely that electrons colliding with thick targets are slowed down by a series of collisions until they come to rest. During this process they may, at any stage, emit bremsstrahlung. The radiation spectrum is, therefore, characteristic not simply of the incoming electron energy, but of the average bremsstrahlung spectrum for all electron energies between the incident energy and zero. In other words, we may estimate the thick-target bremsstrahlung spectrum by averaging the previous (thin-target) spectrum over electron energies.

In the approximation where we take $G = 1$ (for $\hbar\omega \leq \frac{1}{2}mv_1^2$) this average is very simple. It leads to a spectrum proportional to $(\frac{1}{2}mv_1^2 - \hbar\omega)$, that is, a triangular shape as illustrated in Fig. 5.29.

Usually the thick-target bremsstrahlung is sufficiently complicated by these effects and the absorption of the radiation in the target itself that it gives only a rather rough indication of the electron tail energies. Nevertheless, because of its great intensity and considerable penetration power (since it consists of high-energy photons), the hard x-ray emission from thick-target

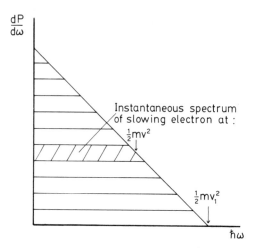

Fig. 5.29. The thick-target bremsstrahlung spectrum is composed of the sum of all the instantaneous (thin-target) spectra of the electron as it slows to rest in the target.

bremsstrahlung is usually very easy to detect and gives a simple indicator of the presence of nonthermal electrons. The fact that it arises from electrons leaving the plasma edge, rather than merely existing within the plasma, may sometimes be turned to advantage in indicating periods of rapid particle transport due, for example, to plasma instabilities.

Further reading

Radiation from an accelerated electron in free space is discussed in most books on electrodynamics, for example:

Jackson, J. D. (1962). *Classical Electrodynamics*. New York: Wiley.
Clemmow, P. C. and Dougherty, J. P. (1969). *Electrodynamics of Particles and Plasmas*. Reading, Mass.: Addison-Wesley.

Cyclotron radiation is discussed in these books and also, in the manner adopted in this chapter, by

Bekefi, G. (1966). *Radiation Processes in Plasmas*. New York: Wiley.
Boyd, T. J. M. and Sanderson, J. J. (1969). *Plasma Dynamics*. London: Nelson.

Bremsstrahlung is also discussed in some detail by Bekefi. More recent reviews of the diagnostic applications of free radiation in fusion research plasmas may be found in

von Goeler, S. (1978). In *Diagnostics for Fusion Experiments*, Proc. Int. School of Plasma Physics, Varenna, 1978. E. Sindoni and C. Wharton, eds. London: Pergamon.
von Goeler, S., et al. (1982). In *Diagnostics for Fusion Reactor Conditions*, Proc. Int. School of Plasma Physics, Varenna, 1982. P. E. Stott et al., eds. Brussels: Commission of E.E.C.
Costley, A. E. (1982). In *Diagnostics for Fusion Reactor Conditions*, Proc. Int. School of Plasma Physics, Varenna, 1982. P. E. Stott et al., eds. Brussels: Commission of E.E.C.

These reviews concentrate more on practical techniques of measurement.

Exercises

5.1 Show by integration over β and pitch angle that the shape function for cyclotron emission from a nonrelativistic ($T \ll m_e c^2$) Maxwellian distribution at perpendicular propagation ($\theta = \pi/2$) is

$$\phi(\omega) = \frac{\omega}{2\sqrt{\pi}(m\Omega)^2} \left[\frac{2m_e c^2}{T}\right]^{m+3/2} \frac{m!}{(2m+1)!} \left[1 - \frac{\omega^2}{m^2\Omega^2}\right]^{m+1/2}$$

$$\times \exp\left\{\frac{-m_e c^2}{2T}\left[1 - \frac{\omega^2}{m^2\Omega^2}\right]\right\}$$

for the mth harmonic.

5.2 When a blackbody emits thermal radiation in a medium of isotropic refractive index $N \ne 1$ the Rayleigh–Jeans formula is modified to give the intensity within the medium as

$$I_N(\omega) = \frac{N^2\omega^2 T}{8\pi^3 c^2}.$$

Nevertheless if the blackbody is observed from the vacuum (i.e., from an observation point outside the medium), the intensity observed must be the usual Rayleigh–Jeans formula (without the N). Otherwise, the second law of thermodynamics would be violated. Show quantitatively how this paradox is resolved.

5.3 Verify Eq. (5.2.43).

5.4 A tokamak with major radius $R = 1$ m and minor radius $a = 0.2$ m has magnetic field 5 T on axis, and temperature and density profiles that are parabolic [$\propto (1 - r^2/a^2)$], with central temperature 1 keV and central density 10^{20} m^{-3}.

   (a) Calculate the optical depth for a frequency corresponding to the second cyclotron harmonic on axis, in the tenuous plasma approximation.

   (b) Do the same for the third harmonic.

   (c) If one uses the second harmonic for temperature profile measurements, what is the maximum minor radius (r) at which the corresponding radiation can be assumed to be blackbody?

   (d) What is the answer to (c) if the central density is 10^{19} m^{-3} (but other parameters the same)?

   (e) Consider now what happens if the tokamak plasma is surrounded by a wall of average reflectivity 90%. Roughly what difference does this make to your answers (c) and (d)?

5.5 What will be the optical depth (in the tenuous plasma approximation) of the central cell of a tandem-mirror plasma of diameter D and uniform density temperature and magnetic field n_e, T, B as a function of frequency near the second harmonic, when the angle of propagation to the magnetic field is θ and $\cos\theta \gg (T/mc^2)^{1/2}$? Calculate this numerically for $D = 0.5$ m, $n_e = 10^{18}$ m^{-3}, $T = 200$ eV, $B = 1$ T, $\theta = 45°$, and $\omega = 2\Omega$ (exactly).

5.6 A certain reversed field pinch plasma has central magnetic field 0.3 T, density 2×10^{19} m^{-3}, and temperature 200 eV. Discuss (with numbers) the use of cyclotron emission as a diagnostic for this plasma.

5.7 For the case $b \ll b_{90}$ show that if $\hbar\omega \lesssim \frac{1}{2}mv_1^2$ then $mv_1 b/\hbar = mv_0 r_0/\hbar \gg 1$; that is, the photon size is more important than the de Broglie wavelength.

5.8 Show that Eq. (5.3.62) is the correct inversion of Eq. (5.3.61).

5.9 Calculate the optical depth due to bremsstrahlung processes, of a uniform hydrogen plasma 1 m across whose electron density is 10^{20} m^{-3} and electron temperature is 10 eV.
(a) At frequency $\nu = 200$ GHz.
(b) At frequency corresponding to wavelength $\lambda = 5000$ Å.

5.10 Oxygen is puffed into a plasma that is initially pure hydrogen. If the electron density doubles and the oxygen atoms can be assumed to be fully stripped (i.e., no bound electrons remain), then what is the final Z_{eff} of the plasma if the particles are perfectly confined (no recycling)?

5.11 A pure hydrogen plasma of uniform electron temperature and density ($T_e = 1$ keV and $n_e = 10^{20}$ m^{-3}) has a diameter 1 m. It is viewed through two collimating apertures of dimension 1 mm $\times$ 1 mm a distance 1 m apart, and afterward through a filter that absorbs photons with energy less than 1 keV but transmits those with higher energy.
(a) What is the étendue of the collimating system?
(b) What is the total bremsstrahlung radiation power passing through both apertures?
(c) What is the total power passing through the filter?
(d) How many photons per second pass through the filter? (This involves a troublesome integral. Use tables or else make a reasonable approximate estimate.)
(e) If the temperature changes by a small fraction F ($= \Delta T_e/T_e$), by what fraction does the power passing through the filter change?
(f) If photons are all perfectly detected by the detector, what is the standard deviation of the number of photons detected during a time t?
(g) Hence, calculate the fractional temperature resolution (i.e., what $\Delta T/T$ can be detected above the photon noise) during a time interval of 1 ms.

6   Electromagnetic radiation from bound electrons

Atoms and ions of the working gas and trace impurities emit radiation when transitions of electrons occur between the various energy levels of the atomic system. The radiation is in the form of narrow spectral lines, unlike the continua of free-electron radiation such as bremsstrahlung. It was, of course, the study of these spectral lines that originally led to the formation of the quantum theory of atoms.

Because of the enormous complexity of the spectra of multielectron atoms it would be inappropriate here to undertake an introduction to atomic structure and spectra. Many excellent textbooks exist [e.g., Thorne (1974) or, for a more complete treatment, Slater (1968)] that can provide this introduction at various levels of sophistication. Instead we shall assume that the energy level structure of any species of interest is known, because of experimental or theoretical spectroscopic research. Then we shall discuss those aspects of spectroscopy that more directly relate to our main theme, plasma diagnostics. It is fairly well justified for neutral atoms and for simpler atomic systems with only a few electrons, to assume that spectral structure is known (speaking of the scientific enterprise as a whole, not necessarily for the individual student!). It is not so well justified for highly ionized heavy ions that occur in hot plasmas. The spectra of such species are still a matter of active research; therefore, it should not automatically be assumed that all aspects of these spectra are fully understood. However, for those diagnostic purposes that we shall discuss, the gaps in our knowledge are not particularly important.

As a matter of terminology, we shall talk about atoms, but mean to also include ions that still possess bound electrons, except where a specific distinction is necessary and noted by the qualification "neutral."

Figure 6.1 shows an example of a photographic spectral plate of line radiation emitted by a magnetically confined hydrogen plasma. Position on the plate corresponds to the wavelength (or equivalently frequency) of the particular line. From the absolute wavelength and other circumstantial evidence, such as knowledge of the anticipated plasma composition, one can identify, by reference to tables of known spectral lines, the atomic species and transition that emits an observed line. It will be noticed that the majority of lines identified in this spectrum are not from hydrogen atoms, but from impurities: oxygen (O), nitrogen (N), nickel (Ni), and chromium (Cr). The ionization stage that gives rise to the line is denoted by a roman numeral; I means neutral, II means singly ionized, and so on.

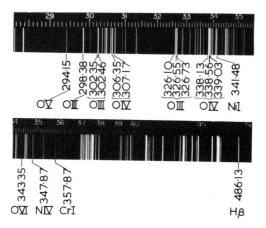

Fig. 6.1. Spectral plate or spectrogram of line radiation from a plasma. The wavelengths in nanometers and the emitting species are shown for some identified lines.

Even such a narrow spectral region as shown in Fig. 6.1 – in the near ultraviolet – illustrates the complexity of radiation spectra. Very rarely are such spectra fully understood in all their details. Nevertheless, extremely valuable and often quite accurate diagnostic information can be obtained from bound-electron radiation. To understand the methods used, we must first discuss the way in which the radiation comes about.

6.1   Radiative transitions: The Einstein coefficients

Consider just two energy levels (i, j) of a particular atomic species. Electrons can make three distinct types of radiative transitions between these levels. First, an atom with the electron in the upper level, energy E_i say, can decay spontaneously to the lower level E_j with the emission of a photon whose frequency is ν_{ij}:

$$h\nu_{ij} = E_i - E_j. \tag{6.1.1}$$

(We shall find it convenient in this chapter to work mostly in periodic frequency ν, rather than angular frequency ω.) The probability per unit time of this transition occurring is the spontaneous transition probability, denoted A_{ij}. Second, an atom with the electron in the lower level may absorb a photon by a transition to the upper level. If the energy density per unit frequency of electromagnetic radiation at the atom is $\rho(\nu)$, then the probability of absorption per unit time is written $B_{ji}\rho(\nu_{ij})$. Third, induced decay from E_i to E_j can occur due to the presence of the radiation. The probability of this event per unit time is written $B_{ij}\rho(\nu_{ij})$.

The numbers A_{ij}, B_{ij}, B_{ji} are called the Einstein coefficients for this transition. They are related by certain identities that can readily be deduced from fundamental thermodynamics as follows. Consider an assembly of atoms in the upper or lower energy levels of density n_i and n_j, respectively, residing in complete thermal equilibrium in a cavity of temperature T. Because of the thermal equilibrium we know immediately that the number of atoms in any quantum state is given by the Boltzmann distribution

$$N_i \propto \exp(-E_i/T),\tag{6.1.2}$$

and in particular,

$$N_i/N_j = \exp\left[-(E_i - E_j)/T\right] = \exp(-h\nu_{ij}/T).\tag{6.1.3}$$

Actually, the normal practice is to assume that the levels i, j consist not of single quantum states but of a number of degenerate states. The number of states g_i in the ith level is called the statistical weight of the level and we can immediately generalize Eqs. (6.1.2) and (6.1.3) to

$$N_i \propto g_i \exp(-E_i/T),\tag{6.1.4}$$

$$N_i/N_j = (g_i/g_j)\exp(-h\nu_{ij}/T).\tag{6.1.5}$$

Also, the radiation is in thermal equilibrium and so is given by the blackbody level

$$\rho(\nu) = 8\pi h\nu^3/\left[\exp(h\nu/T) - 1\right]c^3.\tag{6.1.6}$$

Now this will be an equilibrium situation only if the total rate at which atoms make transitions from i to j is equal to the rate at which they go from j to i. This is known as the principle of detailed balance and leads immediately to

$$\left(A_{ij} + B_{ij}\rho\right)N_i = B_{ji}\rho N_j.\tag{6.1.7}$$

Rewriting this as

$$\rho = \frac{A_{ij}}{(N_j/N_i)B_{ji} - B_{ij}} = \frac{A_{ij}}{(g_j/g_i)\exp(h\nu_{ij}/T)B_{ji} - B_{ij}},\tag{6.1.8}$$

we can see that the principle of detailed balance will be satisfied by the blackbody formula for ρ at all temperatures only if

$$A_{ij} = \frac{8\pi h\nu_{ij}^3}{c^3}B_{ij}\tag{6.1.9}$$

and

$$g_j B_{ji} = g_i B_{ij}.\tag{6.1.10}$$

Of course, since the coefficients are simply properties of the atom, these final relationships hold regardless of the actual situation of the atom, thermal equilibrium or not.

It is possible to obtain a general expression for the induced transition probability directly from a simple perturbation analysis of the system [e.g., Dirac (1958), Chap. 7]. For electric dipole transitions between single quantum states this is

$$B_{ij} = \frac{8\pi^3}{3h^2(4\pi\varepsilon_0)} S_{ij}, \tag{6.1.11}$$

where S_{ij}, which is often called in spectroscopy the line strength, is the square magnitude of the matrix element of the atomic dipole moment:

$$S_{ij} = |\langle i|\mathbf{D}|j\rangle|^2 = \left| \int \psi_i e\mathbf{r}\psi_j \, d^3\mathbf{r} \right|^2. \tag{6.1.12}$$

Here ψ_i refers to the wave function of the ith state and, strictly, for multiple electron atoms the integral should be summed over the position vectors $\mathbf{r}$ of all participating electrons. For levels consisting of degenerate states denoted m_i, m_j this must be generalized to

$$B_{ij} = \frac{1}{g_i} \frac{8\pi^3}{3h^2(4\pi\varepsilon_0)} S_{ij},$$

$$S_{ij} = \sum_{m_i, m_j} |\langle i, m_i|\mathbf{D}|j, m_i\rangle|^2. \tag{6.1.13}$$

Another way in which transition probabilities are sometimes expressed is in terms of the (absorption) oscillator strength f_{ji}. This is defined as the ratio of the number of classical oscillators to the number of lower state atoms required to give the same line-integrated absorption. Its relationship to the Einstein coefficients is

$$f_{ji} = \frac{m_e h(4\pi\varepsilon_0)}{\pi e^2} \nu_{ij} B_{ji}. \tag{6.1.14}$$

The usefulness of f_{ji} is that it is dimensionless, describing just the relative strength of the transition. For the strongest transitions its value approaches 1. (Strictly, the f sum rule is that the sum of all oscillator strengths of transitions from a given state is equal to the number of participating electrons.)

6.2   Types of equilibria

In order to predict the radiative behavior of any atomic species in a plasma we must know the expected population of its various possible states. In many cases the lifetimes of the plasma or atomic species are insufficient to guarantee that an equilibrium population has been reached. In these cases some kind of time-dependent calculation is necessary. However, some types of equilibria are attained in some plasmas and so it is of interest to understand these, both for their own uses and in order to determine whether a time-dependent calculation is necessary.

6.2.1 *Thermal equilibrium*

If the atoms adopt the Boltzmann distribution between all possible states and if also the radiation energy density corresponding to all possible transitions has the blackbody level of the system temperature, then the system is said to be in *complete thermal equilibrium*. Radiation is generally rather weakly coupled to the atoms. Therefore, this state is essentially never achieved in laboratory plasmas, though it may be approached in stellar interiors and it is useful (as in our discussion of the Einstein coefficients) as a theoretical construct for thought experiments. We therefore define a less complete form of thermal equilibrium, known as *local thermal equilibrium* (LTE). In this equilibrium the atoms adopt state populations given by the Boltzmann distribution but the radiation is not necessarily thermal. Unfortunately, even this condition is often too restrictive to be met by laboratory plasmas.

Before moving on to discuss the still less restrictive equilibrium models, let us explore somewhat further the consequences of the assumption of local or complete thermal equilibrium.

6.2.2 *Saha–Boltzmann population distribution*

Quantum statistical mechanics shows that in thermal equilibrium the population of any two (single) quantum states of a system are related by the Boltzmann factor

$$N_i/N_j = \exp\left[-\left(E_i - E_j\right)/T\right]. \tag{6.2.1}$$

Now when both i and j are bound states of the atom no complications enter. However, if state j (say) consists of a state in which the atom is ionized and thus the electron is free, a more careful enumeration of the quantum states is necessary in order to specify not just the distribution of states within the specific ionization state of the atom, but also the distribution among the various different possible ionization states of an atomic species (singly and multiply ionized).

Just as before, we must recognize the degeneracy of the states i and j. This arises first because of the statistical weights g_i, g_j, but second, in the case of ionization, because of the number of possible states of the free electron. The states are enumerated quite readily as follows.

Consider an assembly of electrons and atoms within a volume V. Specifically suppose that there are a total of N_i atoms in the specific level of ionization stage i, N_{i+1} in a specific level of ionization stage $i + 1$, and N_e electrons in the volume. We know that in thermal equilibrium the ratio of the probability of finding the system in this state to the probability of finding it in a state with exactly one further ionization (i.e., $N_i \rightarrow N_i - 1$, $N_{i+1} \rightarrow N_{i+1} + 1$, $N_e \rightarrow N_e + 1$), in which the electron liberated has speed v, is $\exp[-(\chi_i + \frac{1}{2}mv^2)/T]$ times the ratio of the number of possible quantum states for each case (χ_i is the energy difference between states i and $i + 1$, the ionization energy). The ratio of the number of possible distinguishable arrangements of $K + 1$ electrons among L states to that of K electrons among L states is $(L - K)/(K + 1) \approx L/K$ (for $L \gg K \gg 1$), the factor $1/(K + 1)$ arising because individual electrons are indistinguishable.

For a free electron the number of possible quantum states is

$$L = 2\frac{V}{(2\pi)^3}4\pi k^2 \, dk = 2\frac{V}{(2\pi)^3}4\pi\frac{m^3}{h^3}v^2 \, dv, \qquad (6.2.2)$$

where $2V/(2\pi)^3$ is the density of states in k space (allowing for two possible spin states) and $k = mv/\hbar$ is the de Broglie wave number of the electron. Also, the number of electrons in our velocity interval of interest is, from the Maxwellian distribution,

$$K = \left(\frac{m}{2\pi T}\right)^{3/2} N_e e^{-mv^2/2T} 4\pi v^2 \, dv. \qquad (6.2.3)$$

Substituting these values we find the ratio of the number of free electron states to be

$$\frac{L}{K} = 2\frac{V}{N_e}\frac{m^3}{h^3}\left(\frac{2\pi T}{m}\right)^{3/2} e^{mv^2/2T}. \qquad (6.2.4)$$

The translational states of the atom before and after ionization are approximately equal because the mass change is very small; therefore, the ratio of the number of indistinguishable ion states is

$$\frac{g_{i+1}L_{i+1}}{N_{i+1}}\left(\frac{g_iL_i}{N_i}\right)^{-1} = \frac{g_{i+1}N_i}{g_iN_{i+1}}, \qquad (6.2.5)$$

where $L_i \approx L_{i+1}$ is the number of translational states for a single ion. Hence, the total ratio of the number of states of the system is

$$\frac{N_ig_{i+1}}{N_{i+1}g_i}\frac{2V}{N_e}\frac{m^3}{h^3}\left(\frac{2\pi T}{m}\right)^{3/2} e^{mv^2/2T}. \qquad (6.2.6)$$

Now we multiply this by $\exp(-\chi_i - \frac{1}{2}mv^2)/T$ in order to obtain the ratio of the probabilities of the configurations and set this ratio equal to 1, since thermal equilibrium corresponds to the most probable state. Then we find

$$\frac{n_e n_{i+1}}{n_i} = \frac{g_{i+1}}{g_i} \left[\frac{2m^3}{h^3} \left(\frac{2\pi T}{m} \right)^{3/2} \right] e^{-\chi_i/T}, \tag{6.2.7}$$

where n refers to the number density (N/V) of a species.

This equation is a form of what is called the Saha equation and determines the relative populations of the ionization states in thermal equilibrium. The term in square brackets is roughly equal to the inverse of the volume of a cube of side length equal to the thermal de Broglie electron wavelength. It is usually much bigger than n_e, that is, there is on average much less than one electron in a de Broglie cube. One thus finds that the exponent, χ_i/T, can be appreciable but still allow a high degree of ionization (n_{i+1}/n_i). For example, for the hydrogen ground state, n_i versus the ion n_{i+1}, $\chi_i = 13.6$ eV so at $T_e = 1$ eV, $n_{i+1}/n_i = (4 \times 10^{21}$ m$^{-3})/n_e$ in thermal equilibrium. Thus modest density hydrogen plasmas $(n_e < 10^{21}$ m$^{-3})$ are almost fully ionized at 1 eV in local thermodynamic equilibrium.

6.2.3 *Nonthermal populations*

To achieve LTE generally requires high enough density for colli-sional transitions to dominate radiative transitions between all states; otherwise, the absence of thermal radiation will cause deviations from the thermal populations. A rule-of-thumb condition for collisional transitions to dominate may be written (McWhirter 1965) (see Exercises 6.1 and 6.2)

$$n_e \gg 10^{19}(T/e)^{1/2}(\Delta E/e)^3 \text{ m}^{-3}, \tag{6.2.8}$$

where T/e and $\Delta E/e$ are the temperature and energy level difference in eV.

It is rare in magnetic fusion plasmas, for example, for this criterion to be satisfied for low-lying levels, although it may be between high levels where ΔE is small. Therefore, more general equilibrium models must be used. For these, because the statistical thermodynamic arguments do not apply, it is necessary to use the rate coefficients for the relevant processes in order to calculate the populations.

The electron processes of importance are generally:

1. Radiative
   (a) Transitions between bound states.
   (b) Free–bound transitions: recombination/photoionization.
2. Collisional
   (a) Electron impact excitation/deexcitation.
   (b) Impact ionization/three-body recombination.
   (c) Dielectronic recombination/autoionization.

Radiative processes 1(a) have been introduced in Section 6.1 and 1(b), free–bound radiation, in the chapter on radiation by free electrons. The inverse of this recombination is photoionization.

The collisional processes 2(a) and 2(b) correspond to the radiative ones [1(a) and 1(b)] except that the transitions are induced by electron collisions and there is no spontaneous decay. Dielectronic recombination is a process by which an atom captures an electron into an upper energy level while using the electron's energy loss to excite another electron already in the atom to an upper level. Its inverse is autoionization. Of course, these apply only to multiple-electron species, not hydrogen. Dielectronic recombination should perhaps also be classed as radiative since the excess energy of the atom must be removed by a radiative transition subsequent to the capture; otherwise autoionization of the state simply breaks it up again.

The transition probabilities for radiative processes, the Einstein coefficients, have already been discussed. The collisional processes, like radiative recombination, are usually described by individual cross sections $\sigma_{ij}(v)$ (for any specific processes $i \to j$) defined by setting the number of such collisional events per unit path length of an electron of velocity v in a density n_i of "candidate" atoms equal to $\sigma_{ij} n_i$. Conversely, the rate at which any atom undergoes these collisions with electrons in the velocity interval $d^3\mathbf{v}$ is

$$\sigma_{ij}|v|f(\mathbf{v})\,d^3\mathbf{v}. \tag{6.2.9}$$

The total rate at which the atom undergoes these collisions with all types of electrons is

$$\int \sigma_{ij} f(\mathbf{v}) v\, d^3\mathbf{v} = n_e \langle \sigma_{ij} v \rangle, \tag{6.2.10}$$

where

$$\langle \sigma_{ij} v \rangle \equiv \int \sigma_{ij} v f\, d^3\mathbf{v} \bigg/ \int f\, d^3\mathbf{v} \tag{6.2.11}$$

is called the rate coefficient for this process. A rate coefficient exists no matter what the form of $f(\mathbf{v})$, but normally these are evaluated and tabulated for Maxwellian distributions. Usually, therefore, the rate coefficient of interest is the Maxwellian rate coefficient for temperature T.

Under thermodynamic equilibrium conditions the principle of detailed balance indicates that the inverse processes, excitation/deexcitation or impact ionization/three-body recombination, exactly balance one another so that, for example, there are just as many collisional deexcitations as collisional excitations. One can then derive relationships between the rate coefficients just as we did between the Einstein coefficients (see Exercise 6.3).

When we can assume neither complete nor local thermal equilibrium it is still possible to calculate the equilibrium populations of the atomic states provided all the rate coefficients are known. In principle we can write simultaneous equations for all states i that set the total rate of transitions (to and) from that state equal to zero:

$$0 = \sum_{j \neq i} \left[n_i A_{ij} - n_j A_{ji} + \left(n_i B_{ij} - n_j B_{ji} \right) \rho(\nu_{ij}) \right.$$

$$\left. + n_e n_i \langle \sigma_{ij} v \rangle - n_e n_j \langle \sigma_{ji} v \rangle \right], \qquad (6.2.12)$$

where the sum j is over all other states (including free-electron states) accessible to the state i by all types of transitions. For compactness of notation we include three-body recombination as a rate coefficient even though the coefficient is then proportional to n_e. To solve this system of equations, even for simple atoms, requires some sort of simplifying assumptions, since the number of equations is (effectively) infinite.

The approach normally adopted, even when all transition processes are retained in the equations is to truncate the system at some high level, ignoring transitions to or from all higher levels. This works rather well in many cases because the populations of the higher levels are often very small. The system is thus reduced to a finite matrix system that may be solved numerically given all the appropriate coefficients. In some situations further simplifications are possible as follows.

6.2.4   Coronal equilibrium

When the density is low, a rather simple model, which acquires its name from its applicability to the solar corona, may be used. In coronal equilibrium the fundamental approximation is that all upward transitions are collisional (since the radiation density is low) and all downward transitions are radiative (since the electron density is low).

To put the point more fully, upward radiative transitions (i.e., absorption) use up a photon that was presumably emitted by the same transition (downward) in an atom elsewhere in the plasma. Therefore, if the plasma is *optically thin*, so that most photons simply escape without being reabsorbed, then upward radiative transitions will be negligible compared with downward. On the other hand, the rate at which an excited state is depopulated by spontaneous downward transitions is independent of density. Therefore, if the electron density is low enough, the rate of collisional deexcitation becomes negligible compared with spontaneous emission.

Thus the only terms retained in Eq. (6.2.12) are the first two, spontaneous emission, and in the last, collisional excitation and recombination. Moreover, if an excited state can be populated by collisional excitation from the ground state, this process is the dominant one, so that the excited state population is determined by balancing this excitation against radiative deexcitation:

$$\frac{n_i}{n_1} = \frac{n_e\langle\sigma_{1i}v\rangle}{\Sigma_j A_{ij}}. \tag{6.2.13}$$

Ionization/recombination balance is governed by collisional ionization and recombination, and again essentially only the ground state is important because of its dominant population:

$$\frac{n_{k+1}}{n_k} = \frac{\langle\sigma_{k,k+1}v\rangle}{\Sigma_i\langle\sigma_{k+1,ki}v\rangle}, \tag{6.2.14}$$

where n_{k+1}, n_k are the ground state populations of ionization stages $k+1, k$, $\langle\sigma_{k,k+1}v\rangle$ is the ionization rate coefficient from the ground state, and $\langle\sigma_{k+1,ki}v\rangle$ is the total recombination rate coefficient to level i.

In actual fact, the spontaneous transition rate decreases quite rapidly as one goes to higher and higher states, both because of decreasing line strength S_{ij} and because of the v_{ij}^3 dependence of A_{ij} [Eq. (6.1.9)]. As a result, there exists a level, i', above which collisional transitions begin to dominate over radiative, even for low densities. Therefore, even when the low-lying states are in coronal equilibrium with the ground state, the higher states above level i' will be approximately in local thermal equilibrium with the free electron and ionized atom populations. That is, the Saha equation will be obeyed by the higher states. It is sometimes convenient, then, to extend the applicability of the coronal model by considering the upper levels ($> i'$) as belonging to the free-electron continuum and think of "ionization" and "recombination" as including all transitions between the lower levels ($< i'$) and the upper levels ($> i'$) as well as the continuum. One can then calculate generalized "ionization" and "recombination" coefficients, often denoted S and α. Then Eq. (6.2.14) becomes

$$\frac{n_{k+1}}{n_k} = \frac{S}{\alpha}. \tag{6.2.15}$$

Figure 6.2 illustrates the approach. In this case S and α may vary with n_e because i' is varying. Of course, at the lower limit of n_e, S and α become independent of n_e as in the primitive coronal model.

6.2.5 *Time-dependent situations*

Very often it is necessary to deal with situations in which the populations of states are not in equilibrium. In principle this is possible by changing Eq. (6.2.12) to allow for changes in population, setting the right

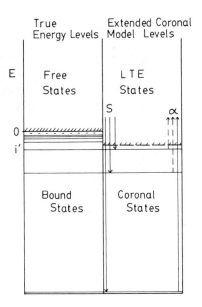

Fig. 6.2. Energy level diagram showing the relationship between the true energy levels (left) of the atomic system and the conceptual division (right) of the levels into states in coronal equilibrium with the ground state and states above level i' in LTE with the free-electron continuum.

hand side equal, not to zero, but to the time derivative of the population

$$\frac{dn_i}{dt} = -\sum_{j \neq i} \left[n_i A_{ij} - n_j A_{ji} + \left(n_i B_{ij} - n_j B_{ji} \right) \rho(\nu_{ij}) \right.$$

$$\left. + n_e n_i \langle \sigma_{ij} v \rangle - n_e n_j \langle \sigma_{ji} v \rangle \right]. \qquad (6.2.16)$$

Then we have a set of simultaneous differential equations to solve that is even more formidable than the equilibria.

The system can be considerably simplified in many situations by the assumption that the excited state populations are in equilibrium with the ground state even though the ground states of different ionization stages are not in equilibrium with one another. This assumption amounts to putting $dn_i/dt = 0$ except for the ground state in Eq. (6.2.16). Then, treating as before the high states $> i'$ as part of the continuum, the equations can be reduced to manageable proportions by using the coronal model or possibly a less restrictive model in which additional collisional processes (such as deexcitation) are included. [One such model is the collisional radiative model in which all processes except induced-radiative are included (McWhirter 1965).]

The reason for the success of this approach is that the time scale for relaxation to excited state equilibrium is very much shorter than that for relaxation to ionization stage equilibrium. Estimates for these time scales may be obtained as follows.

For excitation the appropriate time scale for equilibrium between two levels is the inverse of the fastest transition rate between them, generally A_{ij}. The order of magnitude of A_{ij} for transitions to the ground state may be estimated by noting that the line strength (S_{ij}) has order of magnitude $e^2 a_0^2/Z^2$, where $a_0 = 4\pi\varepsilon_0\hbar^2/me^2$ is the radius of the first Bohr orbit (for hydrogen). Also, the frequency is of order $Z^2 R_y/\hbar$ where R_y is the Rydberg energy $(m/2)(e^2/4\pi\varepsilon_0\hbar)^2$. Therefore, one can immediately calculate the order of magnitude of A_{ij} from the fundamental constants using Eqs. (6.1.9) and (6.1.11). One obtains

$$A_{ij} \approx 10^8 Z^4 \text{ s}^{-1}, \tag{6.2.17}$$

(for higher states A_{ij} tends to be less and so relaxation takes longer). The relaxation time is thus of order $10Z^{-4}$ ns, a time usually very short compared to typical transport and pulse times in magnetically confined plasmas.

On the other hand, for ionization equilibrium the rate of interest is the recombination rate. This rate is calculated in Section 6.3.1. Anticipating that result, we have (for $T < \chi_i$)

$$\langle\sigma_r v\rangle \approx 5.2 \times 10^{-20} Z(\chi_i/T)^{1/2} \text{ m}^3 \text{ s}^{-1}. \tag{6.2.18}$$

For multiple-electron species the highest ionization stage reached in equilibrium is such that

$$\chi_i/T \approx \ln\left|(18m_e c^2/Z^2\chi_i)/(\chi_i/T)\right|, \tag{6.2.19}$$

where χ_i is the ionization potential (see Exercise 6.5). This ratio thus ranges from about 11 for hydrogen down to near 1 for $Z \approx 20$ (where the approximations we are using break down). A simple way to approximate the appropriate Z^2 of the highest ionization stage reached, which we substitute in Eq. (6.2.18), is thus to take

$$Z^2 \approx \chi_i/R_y \approx 5T/R_y \tag{6.2.20}$$

(taking $\chi_i/T \approx 5$ as typical). We then find that as a rough approximation

$$\langle\sigma_r v\rangle \sim 2 \times 10^{-19}(T/R_y)^{1/2} \text{ m}^3 \text{ s}^{-1}. \tag{6.2.21}$$

For example, for a typical magnetically confined fusion research plasma, in which $T_e \approx 1$ keV and $n_e \approx 10^{20}$ m^{-3}, the radiative recombination rate is of order

$$n_e\langle\sigma_r v\rangle \sim 200 \text{ s}^{-1} \tag{6.2.22}$$

so that the time for relaxation to ionization stage equilibrium is ~ 5 ms, very much longer than the excitation equilibration time. (This estimate may require correction by an appreciable factor for some ions because dielectronic recombination becomes important. Also, under some circumstances, the total recombination rate is significantly enhanced by charge exchange – see Section 6.6.5. Nevertheless the point remains that the equilibration is slow.)

6.3   Rate coefficients for collisional processes

The determination of cross sections and rate coefficients for the various collisional processes that can occur to atoms in a plasma is the subject of virtually an entire subfield of physics. It would be impossible here to do justice to the extensive theoretical and experimental information on these rates. However, for many purposes, the most immediate need is for convenient general estimates of these rates rather than more rigorous results of lesser generality. This is true even if the estimates are subject to considerable errors. Besides, for all but the most common atomic species, the results of even the very best specialized quantum-mechanical theories are themselves subject to uncertainties that can be as great as a factor of 2.

The purpose of this section is to gather together some general estimates of rate coefficients for easy reference. While detailed calculations, of course, require a full quantum treatment, we shall see that the estimates we seek are most conveniently derived by semiclassical treatments, in the same spirit as our discussion of bremsstrahlung and recombination radiation.

In Fig. 6.3 are shown domains in the plane $(\Delta E, E)$ in which the different collisional processes we shall discuss occur. Here, E is the incident electron kinetic energy and ΔE its loss of energy during the collision. This diagram bears a strong relationship to Fig. 5.16, the E versus $\hbar\omega$ graph for bremsstrahlung. For *radiative recombination* the parallel is exact (since $\Delta E = \hbar\omega$). For this process, $\Delta E > E$ so that the final energy is negative (electron captured) and the energy change has discrete possible values $E - E_j$, where E_j is the jth energy level of the recombined atom (E_j is negative).

Collisional ionization can occur to a continuum of states provided that $\Delta E > \chi_i$ (so that the atom's electron gains enough energy to escape) and that $E > \Delta E$ (so that the incident electron remains free). *Collisional excitation* occurs with discrete ΔE ($= E_{ij}$) provided E is greater than the excitation energy ΔE. *Dielectronic recombination* can be thought of as an extension of this process to $\Delta E > E$ (so that the incident electron is captured). Its domain overlaps the radiative recombination domain, but it is a distinct process.

We now treat each of these processes in an appropriate semiclassical manner in order to obtain simple approximate estimates of their rate coefficients.

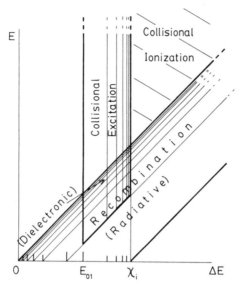

Fig. 6.3. Schematic representation of the regions in the plane $(\Delta E, E)$ in which the various collisional processes occur.

6.3.1   *Radiative recombination*

We already have an estimate of the power radiated by recombination [Eq. (5.3.49)]. We express that power in the form of a collision cross section like Eq. (5.3.11), so that for recombination to level n the cross section is

$$\sigma_{rn} = G_n \frac{16\alpha}{3\sqrt{3}} \pi (Zr_e)^2 \frac{c^2}{v_1^2} \frac{\Delta\omega_n}{\omega_n} \tag{6.3.1}$$

(here α is the fine structure constant).

Now we take $\hbar\Delta\omega_n = 2\chi_n/n$, as in Eq. (5.3.47), but writing χ_n for Z^2R_y/n^2, the ionization energy of the nth level of the recombined ion. Also $\hbar\omega_n = E + \chi_n$, where $E = \frac{1}{2}mv_1^2$. This gives our estimate for the cross section.

To obtain the rate coefficient we integrate over a Maxwellian electron distribution, ignoring any variation of G_n and merely replacing it by an average value $\bar{g}_n$:

$$\langle\sigma_{rn}v\rangle = \bar{g}_n \frac{16\alpha}{3\sqrt{3}} \pi (Zr_e)^2 c^2 \frac{2}{n} \left(\frac{m_e}{2\pi T}\right)^{3/2} \int_0^\infty \frac{\exp(-E/T)4\pi v \, dv}{1 + E/\chi_n}. \tag{6.3.2}$$

Using the identity $\alpha c = (2R_y/m_e)^{1/2}$ and making the substitution $s = (\chi_n$

$+ E)/T$, this becomes

$$\langle \sigma_{rn} v \rangle = \bar{g}_n \frac{64}{3} \left(\frac{\pi}{3} \right)^{1/2} r_e^2 c Z \left(\frac{Z^2 R_y}{n^2 T} \right)^{1/2} \frac{\chi_n}{T} \exp\left(\frac{\chi_n}{T} \right) \mathrm{Ei}\left(\frac{\chi_n}{T} \right),$$

$$(6.3.3)$$

where

$$\mathrm{Ei}(y) = \int_y^\infty \frac{\exp(-s)}{s} \, ds \qquad (6.3.4)$$

is the exponential integral. The distinction between $Z^2 R_y/n^2$ and χ_n is important only for low-lying states in nonhydrogenic ions, so usually it is ignored. However, the rate should be corrected by the factor $\xi/2n^2$ for a level that is partially filled, as in Eq. (5.3.53).

Evaluating the fundamental constants one gets

$$\langle \sigma_{rn} v \rangle = \bar{g}_n 5.2 \times 10^{-20} Z \left(\frac{\chi_n}{T} \right)^{3/2} \exp\left(\frac{\chi_n}{T} \right) \mathrm{Ei}\left(\frac{\chi_n}{T} \right) \, \mathrm{m}^3 \, \mathrm{s}^{-1}. \quad (6.3.5)$$

Note that the asymptotic forms are

$$\exp(x)\mathrm{Ei}(x) \to 1/x \qquad \text{as } x \to \infty,$$

$$\to -\ln x \quad \text{as } x \to 0. \qquad (6.3.6)$$

The former is generally the most useful for the lower levels, since the temperature is usually much smaller than the ionization energy in equilibrium.

It is often adequate to take $\bar{g}_n$ equal to 1. Other more complicated forms are available in the literature (Seaton 1959). The total recombination rate to all levels is sometimes required for the purpose of calculating ionization balance. A reasonable approximation to this sum over all levels is

$$\langle \sigma_r v \rangle \approx 5.2 \times 10^{-20} \frac{Z}{2} \left(\frac{Z^2 R_y}{T} \right)^{1/2}$$

$$\times \left[1 - \exp\left\{ \frac{-\chi_i}{T} \left[1 + \frac{1}{n_0} \left(\frac{\xi}{n_0^2} - 1 \right) \right] \right\} \right] \left[\left(\ln \frac{\chi_i}{T} \right)^2 + 2 \right]^{1/2},$$

$$(6.3.7)$$

where n_0 is the principal quantum number of the lowest incomplete shell of the ion and χ_i is the ionization potential of the recombined atom (see Exercise 6.6).

6.3.2 *Collisional ionization*

The most elementary classical approximate treatment of the collisional ionization process follows essentially that of Thomson (1912). We picture the bound electron as stationary in the atom and calculate the energy transferred to it (ΔE) in a collision by an incident electron of energy E. If ΔE is greater than the ionization potential χ_i, then ionization will occur. The ionization cross section is simply the area, integrated over all impact parameters b, for which $\Delta E > \chi_i$. Thomson showed that in a Coulomb collision between like particles,

$$\Delta E = E \left/ \left[1 + \left(\frac{4\pi\varepsilon_0}{e^2} \right)^2 E^2 b^2 \right] \right. . \tag{6.3.8}$$

This may readily be deduced from an orbit analysis, as in Section 5.2, but taking care to work in the center of mass frame with reduced mass, since we cannot take one of the particles to be fixed (Exercise 6.7). This equation shows that $\Delta E > \chi_i$ for all impact parameters less than a certain value b_c. Therefore, the cross section estimate is

$$\sigma_i = \pi b_c^2 = \pi \left(\frac{e^2}{4\pi\varepsilon_0} \right)^2 \frac{1}{E^2} \left(\frac{E}{\chi_i} - 1 \right). \tag{6.3.9}$$

Using the definition of the Bohr radius a_0, this is most conveniently written

$$\sigma_i = 4\pi a_0^2 \left(\frac{R_y}{E} \right)^2 \left(\frac{E}{\chi_i} - 1 \right). \tag{6.3.10}$$

This may be regarded as the classical collisional ionization cross section per active bound electron. Presumably if there are ξ electrons in the upper quantum level we must regard the cross section as applying to each and so multiply by ξ to get the cross section for the atom as a whole.

Although this cross section represents high-energy collisions quite well (once an appropriate Gaunt factor has been introduced), it overestimates quite seriously the cross section near threshold ($E \sim \chi_i$). In part this is because, even from a semiclassical viewpoint, the previous model omits several important physical features. In the first place, the bound electron's Bohr orbit is situated at a potential energy $-2\chi_i$ in the ion's Coulomb field. Presumably a slowly impinging colliding electron is accelerated in the ion's field so that its kinetic energy at the time of collision with the bound electron is $E + E_+$ (with $E_+ \sim 2\chi_i$). In the second place, if the colliding electron passes on to the atomic electron too much energy ($\Delta E > E$) it will itself become bound. There will then have been no net ionization.

A simple extension of the previous treatment, accounting for these two differences (Exercise 6.8), gives a cross section

$$\sigma_i = 4\pi a_0^2 \frac{R_y^2 (E/\chi_i - 1)}{E(E + E_+)}. \tag{6.3.11}$$

Notice the simple modification of one of the E factors in the denominator by the addition of the extra energy at collision. This form holds whatever value we take for this extra energy. Taking $E_+ = 2\chi_i$, the threshold cross section (as $E \to \chi_i$) is smaller by a factor of 3 in Eq. (6.3.11) than in the earlier form [Eq. (6.3.10)].

Rather than continue to attempt to refine further what is already a classical, and hence not fully satisfactory, method it is perhaps best simply to cite that experiments and quantum-mechanical calculations, taking account of the complexities of different atoms, indicate that a threshold cross section about 0.2 times the Thomson classical value represents a reasonable average for the different species (within a factor of about 2) (Seaton 1962). We shall obtain this factor if we use Eq. (6.3.11) with $E_+ = 4\chi_i$.

Our modified classical expression for the cross section is convenient because one can perform the integration analytically over a Maxwellian. The manipulations are elementary, giving the rate coefficient

$$\langle \sigma_i v \rangle = 4\pi a_0^2 \left(\frac{8R_y}{\pi m_e} \right)^{1/2} \left(\frac{R_y}{\chi_i} \right) \left(\frac{R_y}{T} \right)^{1/2} \exp\left(\frac{-\chi_i}{T} \right)$$

$$\times \left[1 - \left(\frac{\chi_i + E_+}{T} \right) \exp\left(\frac{\chi_i + E_+}{T} \right) \mathrm{Ei}\left(\frac{\chi_i + E_+}{T} \right) \right]. \tag{6.3.12}$$

The asymptotic limits of the square bracket factor here are

$$\begin{array}{ll} T/(\chi_i + E_+) & \text{as } T \to 0, \\ 1 & \text{as } T \to \infty. \end{array} \tag{6.3.13}$$

To avoid the exponential integral function, which is rather awkward to use, it is therefore convenient to replace this factor by the expression

$$[T/(\chi_i + E_+)][1 - \exp\{-(\chi_i + E_+)/T\}], \tag{6.3.14}$$

which has the same asymptotic limits and is quite a good approximation for all T, but is much easier to evaluate.

A final correcting factor arises because the cross section at high energies increases logarithmically above the simple classical estimate. This may be understood, as we shall explore more fully in the next section (for excitation), by regarding the ionization as a transition induced by the appropriate frequency component of the colliding electron's Coulomb field, experienced by the atom. Our bremsstrahlung treatment showed that this will involve a Gaunt factor approximately $(\sqrt{3}/\pi)\ln|4E/\Delta E|$ for $E \gg \Delta E$ [see Eq. (5.3.33)], the Born approximation. To include this factor exactly in the integration to obtain the rate coefficient is not possible analytically. Therefore, we continue in the semiempirical spirit of this section and write the average Gaunt factor as approximately

$$\bar{g} = 1 + \frac{\sqrt{3}}{\pi}\ln\left|1 + \frac{T}{\chi_i}\right|, \tag{6.3.15}$$

which has essentially the correct behavior at low and high temperatures. Our final semiempirical formula for collisional ionization is then

$$\langle \sigma_i v \rangle = \bar{g}4\pi a_0^2 \left(\frac{8T}{\pi m_e}\right)^{1/2} \frac{R_y^2}{\chi_i(\chi_i + E_+)}\exp\left(\frac{-\chi_i}{T}\right)$$

$$\times\left[1 - \exp\left\{-\frac{(\chi_i + E_+)}{T}\right\}\right] \tag{6.3.16}$$

or, evaluating the fundamental constants and putting $E_+ = 4\chi_i$,

$$\langle \sigma_i v \rangle = \bar{g}(1.7 \times 10^{-14})\left[\frac{R_y}{\chi_i}\right]^2\left[\frac{T}{R_y}\right]^{1/2}\exp\left(\frac{-\chi_i}{T}\right)$$

$$\times\left[1 - \exp\left(-\frac{5\chi_i}{T}\right)\right] \text{ m}^3 \text{ s}^{-1}. \tag{6.3.17}$$

This is, of course, the rate per atomic electron and needs to be summed over all significant electrons in the atom. Usually this requires only those in the uppermost level to be considered. However, occasionally lower levels must also be included, using the χ_i appropriate to that level. This expression, which is a form of what is called the exchange classical impact parameter approximation, will usually be correct to within a factor of about 2. Other similar expressions may be found elsewhere [e.g., Lotz (1967), Drawin (1968), Crandall (1983)].

6.3.3   *Collisional excitation*

The collisional excitation rates, which largely determine the excited state populations in coronal situations, require a treatment that recognizes the discrete quantum levels from the start. Transitions between these levels (whether up or down) may be treated via perturbation theory. The perturbation of interest is, of course, the electric field at the atom. In calculating rates for radiative transitions, such as induced emission or absorption, the electric field is that of the electromagnetic wave. For collisional excitation it is the field of the colliding electron that matters. A simple approach, therefore, is to calculate the appropriate frequency component of the collisional electric field and then obtain the rate of excitation by using the Einstein coefficient for that transition.

We adopt the same semiclassical treatment of the colliding electron as we used for calculating bremsstrahlung. In a collision with a given impact parameter b, the acceleration of the colliding electron is caused by the field of the ion (in our present case this is the atom to be excited). At any instant, the field of the electron at the atom (ion) is equal to $1/Z$ times the field of the atom at the electron. Therefore, the Fourier analysis of the electron's acceleration that we required for bremsstrahlung provides us also with the Fourier analysis of the electric field of the colliding electron experienced by the atom.

The total probability of a transition j to i being induced by the collision is the time integral of the transition rate

$$P_{ji} = \int_{-\infty}^{\infty} B_{ji} \rho(\nu_{ij}) \, dt, \tag{6.3.18}$$

where ρ is the energy density corresponding to the perturbing electric field (per unit frequency). This is related to the Fourier component of the electric field $E(\omega)$ by Parseval's theorem, which gives

$$\int_{-\infty}^{\infty} \rho(\nu) \, dt = 8\pi^2 \varepsilon_0 |E(\omega)|^2 \tag{6.3.19}$$

(see Exercise 6.9). Hence, the transition probability for a specific collision is

$$P_{ji} = 8\pi^2 \varepsilon_0 B_{ji} |E(\omega_{ij})|^2. \tag{6.3.20}$$

Recall, now, that the spectral power of bremsstrahlung radiation is given by Eq. (5.3.3), which may be written

$$\frac{dW}{d\omega} = \frac{e^2}{6\pi^2 \varepsilon_0 c^3} \frac{e^2}{m_e^2} |E'(\omega)|^2 (2\pi)^2, \tag{6.3.21}$$

where $E' = ZE$ is the field at the colliding electron. Thus P_{ji} is proportional to $dW/d\omega$. This means that the integration over impact parameters and subsequently over velocities proceeds exactly as in the bremsstrahlung calculation. In other words, we have already done the necessary mathematics; all that is required is to relate P_{ji} to the radiation $dW/d\omega$, using Eqs. (6.3.20) and (6.3.21):

$$P_{ji} = B_{ji} \left(\frac{4\pi\varepsilon_0}{e^2} \right)^2 \frac{3m_e^2 c^3}{4} \frac{1}{Z^2} \frac{dW}{d\omega}. \tag{6.3.22}$$

The impact parameter integration leads to a cross section for excitation,

$$\sigma_{ij} = B_{ij} \left(\frac{4\pi\varepsilon_0}{e^2} \right)^2 \frac{3m_e^2 c^3}{4} \frac{\hbar\omega}{Z^2} \frac{d\sigma_c}{d\omega} G$$

$$= B_{ji} \frac{8\pi}{\sqrt{3}} \frac{R_y a_0}{v_1^2} G$$

$$= \frac{8\pi}{\sqrt{3}} \pi a_0^2 \frac{2R_y^2}{E_{ij} m_e v_1^2} f_{ji} G \qquad (6.3.23)$$

[using Eq. (5.3.11) and identities for atomic parameters]. Then the velocity integration gives the rate coefficient

$$\langle \sigma_{ij} v \rangle = B_{ji} \left(\frac{4\pi\varepsilon_0}{e^2} \right)^2 \frac{3m_e^2 c^3}{4} \frac{4\pi j(\omega_{ij})}{n_e n_i Z^2}, \qquad (6.3.24)$$

where $j(\omega)$ is the bremsstrahlung emissivity given by Eq. (5.3.40). This may be written

$$\langle \sigma_{ij} v \rangle = B_{ji} \frac{8}{\sqrt{3}} (2\pi)^{1/2} R_y a_0 \left(\frac{m_e}{T} \right)^{1/2} \exp\left(\frac{-h\nu_{ij}}{T} \right) \bar{g}(\nu_{ij}) \quad (6.3.25)$$

or, in terms of the oscillator strength, via Eq. (6.1.14) as

$$\langle \sigma_{ij} v \rangle = 16 \left(\frac{\pi}{3} \right)^{1/2} \pi a_0^2 \alpha c \left(\frac{R_y}{T} \right)^{1/2} \exp\left(\frac{-E_{ij}}{T} \right) \frac{R_y}{E_{ij}} f_{ji} \bar{g}. \qquad (6.3.26)$$

Evaluating the fundamental constants, this becomes

$$\langle \sigma_{ij} v \rangle = 3.15 \times 10^{-13} f_{ji} \frac{R_y}{E_{ij}} \left(\frac{R_y}{T} \right)^{1/2} \exp\left(\frac{-E_{ij}}{T} \right) \bar{g} \ \text{m}^3 \ \text{s}^{-1}.$$

$$(6.3.27)$$

The remaining question is what to take for the Gaunt factor. Generally speaking, if all the significant collisions are sufficiently distant from the atom, so that the electric field of the colliding electron is indeed uniform at the atom, as this treatment is tacitly assuming, then the Gaunt factor we require is simply that which was given for bremsstrahlung. This will be the case for high-energy electrons $\frac{1}{2} m_e v_1^2 \gg E_{ij}$ and hence high temperatures, but not necessarily near the threshold or when $T \ll E_{ij}$. For a close collision, in which the colliding electron penetrates the bound-electron "cloud" of the atom, the exciting electric field is not uniform and the excitation probability is generally noticeably reduced.

Roughly speaking, the closest approach of the colliding electron in the dominant threshold collisions is greater or less than the atomic electron cloud dimensions according to whether χ_n/nE_{ij} is greater or less than 1. Therefore, for high principal quantum numbers (n) or excitations that are a large fraction of the ionization energy (χ_n), such as when $\Delta n \neq 0$ for the transition, there is a significant reduction of the Gaunt factor near threshold. Seaton (1962) suggests the average value $\bar{g} \approx 0.2$. On the other hand if n is fairly small and $E_{ij} \ll \chi_n$, for example when $\Delta n = 0$, the reduction due to cloud penetration is small and a $\bar{g}$ near 1 at threshold is appropriate [see, e.g., Bely (1966)]. Various empirical interpolation formulas for $\bar{g}$ have been given [e.g., Post (1977)], based usually on quantum-mechanical calculations. A simple form with correct asymptotic limits is

$$\bar{g} = g_t + \frac{\sqrt{3}}{\pi} \ln\left|1 + \frac{T}{E_{ij}}\right|, \qquad (6.3.28)$$

where g_t is the appropriate threshold value.

It should be noted that the approach we have adopted implicitly assumes the atom to be charged (partially ionized), since otherwise the colliding electron orbit is not hyperbolic but straight. The excitation cross section for a neutral atom tends to zero at threshold. Our treatment predicts a finite threshold value, qualitatively correct only for charged atoms.

6.3.4 Dielectronic recombination

When an electron collides with a charged atom with an energy slightly below the excitation threshold it may cause excitation and simultaneously be captured into a high bound state. Such dielectronic capture can reverse. The electron may be expelled again by autoionization as the atomic electron returns to its original level. In order to contribute to net recombination, a second process (stabilization) must intervene before autoionization occurs. In other words, a radiative transition, generally of the lower-level electron, must rid the atom of its excess energy before it breaks up of its own accord. Once the atom is stabilized, true dielectronic recombination (as opposed to simple capture) is completed.

We denote the capture rate coefficient by $\langle \sigma_{c,k} v \rangle$ (where k refers to the level to which the capture occurs and it is understood that we are discussing a single excitation $j \to i$ of the inner electron). The autoionization and radiative stabilization transition rates of the captured system are written A_a and A_r, respectively. Then the dielectronic recombination rate coefficient (including stabilization) is

$$\langle \sigma_{d,k} v \rangle = \langle \sigma_{c,k} v \rangle \frac{A_r}{A_a + A_r}. \qquad (6.3.29)$$

This expresses the fact that a proportion $A_r/(A_a + A_r)$ of captured states stabilize before autoionizing.

Since capture and autoionization are inverse processes they are related by detailed balance. No change of total energy of the system atom plus electron is involved; therefore, in thermal equilibrium the Saha–Boltzmann populations of captured (n_{z-1}) and ionized (n_z) states are related by

$$\frac{n_{z-1}g_z}{n_z g_{z-1}} \frac{1}{n_e} \frac{2}{h^3}(2\pi m_e T)^{3/2}\exp\left(\frac{m_e v^2}{2T}\right) = 1 \qquad (6.3.30)$$

[see Eq. (6.2.6)], where g_z is the statistical weight of the atomic state and v is the free-electron velocity, so that $\frac{1}{2}m_e v^2$ is the total net energy of the system. Setting the inverse rates equal in equilibrium we get

$$\langle\sigma_{c,k}v\rangle = A_a\frac{h^3}{2(2\pi m_e T)^{3/2}}\frac{g_{z-1}}{g_z}\exp\left(\frac{-m_e v^2}{2T}\right). \qquad (6.3.31)$$

Now we need to obtain an expression for $\langle\sigma_{c,k}v\rangle$ by extension of our excitation cross section σ_{ij} to values below threshold. For capture to all sublevels with principal quantum number n, we use the same correspondence principle as we used for radiative recombination, namely that the spread in final energies is $2Z^2R_y/n^3$. Thus the spread of initial velocities that recombine to the nth level is $\delta v = 2Z^2R_y/n^3 m_e v$. The rate coefficient for capture to the nth level is therefore

$$\langle\sigma_{c,n}v\rangle = \sigma_{ij}v\frac{f(v)}{n_e}4\pi v^2\,\delta v = \sigma_{ij}v^2\frac{8\pi Z^2R_y}{n^3 m_e}\frac{f(v)}{n_e}, \qquad (6.3.32)$$

where $f(v)$ is the (Maxwellian) distribution function and $\frac{1}{2}mv^2 + Z^2R_y/n^2$ is equal to E_{ij}, the excitation energy.

Notice, now, that σ_{ij} as given by Eq. (6.3.23) is proportional to B_{ji} and hence to A_{ij}, which will be equal to A_r (provided the captured electron is in a high orbit "out of the way" of the lower electron). We may therefore calculate A_a/A_r using Eqs. (6.3.32), (6.3.31), and (6.3.23). The algebra is straightforward and the result is

$$\frac{A_a}{A_r} = \frac{16}{\sqrt{3}}G\frac{1}{\alpha^3}\frac{Z^2R_y^3}{E_{ij}^3}\frac{1}{n^3} \qquad (6.3.33)$$

(α is the fine structure constant).

To obtain the total dielectric recombination rate coefficient we must sum $\langle\sigma_{d,n}v\rangle$ over all n:

$$\langle\sigma_d v\rangle = \sum_n\langle\sigma_{d,n}v\rangle = \frac{h^3}{2(2\pi m_e T)^{3/2}}\frac{g_i}{g_j}A_r$$

$$\times\sum_n\frac{\exp\left(-E_{ij}+Z^2R_y/n^2\right)/T}{1+(n/n')^3}, \qquad (6.3.34)$$

where n' is the value of n that makes $A_a/A_r = 1$. As may be seen by examining Eq. (6.3.33), n' is large provided $Z < \alpha^{-3/4}$ (recall $E_{ij} < Z^2 R_y$). Therefore, we may approximate the sum as an integral and also ignore the term $Z^2 R_y/n^2$ in the exponent (at least for all relevant T). We then get

$$\langle \sigma_d v \rangle = \frac{h^3}{2(2\pi m_e T)^{3/2}} \frac{g_i}{g_j} A_r \frac{2\pi}{3\sqrt{3}} n' \exp\left(-\frac{E_{ij}}{T}\right). \tag{6.3.35}$$

It is often convenient to express this in terms of the dimensionless oscillator strength f_{ji}. Substituting for n' explicitly and rearranging in terms of atomic constants we get

$$\langle \sigma_d v \rangle = \left(\frac{2G}{\sqrt{3}}\right)^{1/3} \left(\frac{2\pi}{3}\right)^{1/2} \frac{8\pi}{3} \pi a_0^2 \left(\frac{R_y}{m_e}\right)^{1/2} \alpha^2 f_{ji} Z^{2/3}$$

$$\times \frac{E_{ij}}{R_y} \left(\frac{R_y}{T}\right)^{3/2} \exp\left(\frac{-E_{ij}}{T}\right). \tag{6.3.36}$$

Fortunately, this expression is only very weakly dependent on the value adopted for the collisional cross section Gaunt factor G. For practical purposes, we may take $2G/\sqrt{3} = 1$ with negligible error compared to uncertainties involved in the overall approach. Then evaluating the fundamental constants one obtains

$$\langle \sigma_d v \rangle = 8.8 \times 10^{-18} f_{ji} Z^{2/3} \frac{E_{ij}}{R_y} \left(\frac{R_y}{T}\right)^{3/2} \exp\left(\frac{-E_{ij}}{T}\right) \text{ m}^3 \text{ s}^{-1}. \tag{6.3.37}$$

A somewhat different and more complicated expression, due to Burgess (1965), is widely used. It is based on empirical fits to a variety of calculations. The present expression has the merit of providing a more compact form with a clearer physical motivation; it gives values slightly smaller than the Burgess formula, by typically 30% at low Z and closer at higher Z. In either case, the total recombination rate must be obtained by summing over all relevant excitations $j \to i$; usually only one or two need be considered because of the exponential dependence on excitation energy. The required values of the oscillator strengths can usually be obtained from the extensive published tabulations [e.g., Wiese et al. (1966)]. For a more extensive review of dielectronic recombination processes see, for example, Seaton and Storey (1976).

The characteristic emission lines arising from the radiative stabilization of an atom formed by dielectronic capture have frequency approximately equal to the corresponding resonance line of the unrecombined atom. There is, however, a small frequency shift of the line due to the perturbing effect of the captured electron in an upper level. The shifted lines are called dielectronic satellites. The relative intensity of the satellite lines to the resonance line(s) can give a useful estimate of electron temperature relatively unaffected by the ionization stage equilibrium because they arise from collisions with the same species (i.e., the unrecombined atom) [see, e.g., Gabriel (1972)].

6.3.5   *Example: Carbon* V

As an example illustrating the general shape of the rate coefficients' variation with temperature, Fig. 6.4 shows the values obtained from the formulas of the previous sections for the four-times-ionized C V atom. C V is a heliumlike atom, having two bound electrons. Its ionization potential is 392 eV and the energy of its resonance transition from the ground state ($1s$-$2p$) is 308 eV. Figure 6.4 shows the rate coefficients for ionization and resonance excitation of the C V ion and also the radiative and dielectronic recombination rates to C V (from C VI). The dielectronic recombination involves the resonance excitation of the hydrogenlike core electron.

Fig. 6.4. Rate coefficients for collisional processes involving C V.

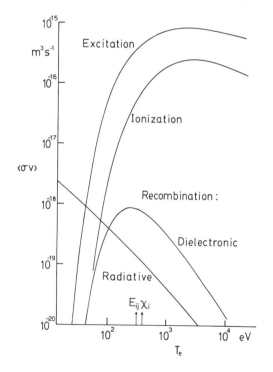

Notice that the temperature at which the C V would be 50% ionized, where $\langle \sigma_i v \rangle = \langle \sigma_r v \rangle$, is substantially below χ_i. Therefore, the threshold rates are the most important in practice because at higher temperatures C V is almost completely ionized up to C VI or beyond. For example, from the ratio of $\langle \sigma_i v \rangle$ to $\langle \sigma_d v \rangle$ at $T = 1$ keV we deduce an equilibrium ratio of C V to C VI of about 1.6×10^{-3}.

At the temperature at which the C V is 50% ionized (~ 75 eV) the dielectronic recombination is negligible. This is because for this ion the lowest excitation transition has an energy quite close to the ionization potential. For ions with more bound electrons, lower-energy resonance transitions often exist, so that the dielectronic recombination may not be negligible at the 50% temperature.

6.3.6   Charge-exchange recombination

There is a further process that we have so far ignored that can influence the state populations of species within the plasma, namely charge exchange. This is a mechanism by which a bound electron is transferred or exchanged between two atoms in a collision. We shall discuss its importance in more detail later when we deal with ionic processes; however, it arises in the present context because under certain circumstances it can have a noticeable, and from the diagnostic viewpoint helpful, influence on the state distribution of atomic impurities. Since this mechanism is not of importance in all plasmas and since detailed rate calculations have yet to be systematically summarized, we restrict our treatment to a few qualitative remarks.

The reason we are justified in ignoring the influence of charge exchange upon the excited states under normal circumstances is that for hot plasmas the number of neutrals (with which charge exchange could occur) is much less than the number of electrons (which are responsible for the competing collisional processes). An exception is radiative recombination, for which the cross section is so much smaller than other cross sections (including charge exchange) that it may not dominate over charge exchange. In some circumstances, therefore, the ionization stage of highly ionized impurities can be significantly depressed by charge-exchange recombination relative to our previous discussions (Hulse et al. 1980). Naturally this depends upon the density of neutrals, so charge-exchange effects are most likely to be important when neutral beams are deliberately injected into the plasma either for purposes of heating the plasma or specifically for diagnostics. Even without such beams, however, significant numbers of neutrals may be present, as we discuss further in Chapter 8.

The atomic states preferentially populated by charge exchange are those that have maximum overlap integral with the eigenfunction of the electron in the neutral (hydrogen) atom. For maximum resonance in the temporal frequency of the states the preference is to recombine to a state with binding energy $\sim Z^2 R_y/n^2$ equal to R_y, the ground state energy of hydrogen. Hence, $n \sim Z$ is preferred. However, the spatial overlap is maximized by having the orbital dimensions $\sim n^2 a_0/Z$ similar to the hydrogen ground state a_0. Thus spatial overlap prefers $n \sim Z^{1/2}$. Roughly speaking, the balancing of these two preferences leads to a population of levels dominantly with $n \sim Z^{3/4}$ (Olsen 1980).

Moreover, the recombination tends to occur to states with appreciable angular momentum, that is, the secondary quantum number ℓ will be significantly greater than 1. The importance of this is that the electric dipole transition selection rule $\Delta \ell = \pm 1$ then prevents this state from decaying by spontaneous emission directly to the ground state. Instead it must cascade down via many intermediate steps, each one emitting its characteristic line. Because the early steps of this cascade will have energy similar to the electron binding energy, the rich spectrum generated will include lines in the visible or near ultraviolet ($E \sim R_y$). These are often much more convenient for diagnostics (such as Doppler ion temperature measurements) than the extreme ultraviolet resonance lines that otherwise dominate the spectrum of these impurities.

A further diagnostic advantage may be gained by using a neutral beam crossed with the viewing chord. Insofar as the charge-exchange-produced excited states are localized to the beam, this then provides a well localized measurement of (for example) ion temperature or impurity density. In fact, it proves possible [e.g., Berezovskii et al. (1985)] under some conditions, using charge exchange between the majority ion (D^+) and a hydrogen beam, to localize the measurement of Balmer line emission to the interior of the plasma and hence to diagnose directly the majority ions. Thus, charge-exchange recombination with an active neutral beam is a method by which active perturbation of the state populations may be used to gain additional or improved diagnostic information.

In magnetic fusion research, diagnostic techniques based on charge-exchange recombination are the subject of considerable current development (Fonk, Darrow, and Jaehnig 1984).

6.4   Line broadening

The spectral lines emitted by bound–bound transitions do not have infinitesimal spectral width, but undergo several possible line broadening mechanisms that are extremely useful for diagnostics.

6.4.1   *Natural line broadening*

This arises because of the fact that the quantum states of an atom do not have a single energy, but a small spread in energy. The energy spread arises because perturbations of the atomic system due to interaction with the electromagnetic fields of virtual (or real) photons cause the quantum states to be only approximate eigenmodes of the system. A simpler way of saying the same thing in the language we have been using is that the lifetime of the atom in an upper state is finite owing to the spontaneous transitions to lower quantum states (induced transitions may also be important, but we shall suppose the radiation density to be negligible for our purposes). The effective spread in energy of a quantum state is given by the uncertainty principle

$$\Delta E \approx h/2\pi\tau, \tag{6.4.1}$$

where the lifetime τ is given by

$$\frac{2}{\tau} = \sum_j A_{ij}, \tag{6.4.2}$$

the sum of all possible spontaneous transitions. (The factor 2 here in our definition of τ is included to provide a standard form of the line profile.) The corresponding line broadening of the spectral lines of transitions from this broadened state i, due to the broadening of only this state, is simply $\Delta\nu = \Delta E/h \approx 1/2\pi\tau$.

The shape of the broadened line is determined by the shape of the energy broadening. This may be shown to be the Fourier transform of the square root of the exponentially decaying probability of the atom being in the ith state. The resulting spectral shape is

$$I(\nu) = I(\nu_0) \frac{1}{1 + \left[(\nu - \nu_0)2\pi\tau\right]^2}, \tag{6.4.3}$$

Fig. 6.5. The Lorentzian line shape typical of natural broadening: $I(\nu) = (1 + \Delta\nu^2)^{-1}$, with $\Delta\nu$ in units of $1/2\pi\tau$. The full width at half maximum is shown FWHM.

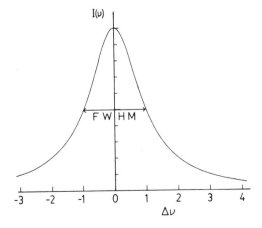

where I is the intensity and ν_0 the line center (note that if the lower level k is also broadened, for example, when it is not the ground state, the value to be used for $2/\tau$ is $\Sigma A_{ij} + \Sigma A_{kj}$). The line shape represented by Eq. (6.4.3) is called Lorentzian.

It is common to use as a measure of line width (regardless of the precise line shape) the full width at half maximum (FWHM). This is illustrated for a Lorentzian line shape in Fig. 6.5. Its relationship to the lifetime is

$$\Delta\nu_{1/2} = 1/\pi\tau. \tag{6.4.4}$$

Because $A_{ij} \propto \nu_{ij}^3$, natural broadening, though usually negligible in the visible, can become important for the extreme ultraviolet lines of highly ionized impurities.

6.4.2   *Doppler broadening*

This arises straightforwardly from the Doppler shift caused by thermal particle motion

$$\Delta\nu = \nu - \nu_0 = \nu_0 v/c, \tag{6.4.5}$$

where v is the particle velocity in the line of sight. For an atom velocity distribution $f(v)$, this gives a line shape

$$I(\nu) \propto f([\nu/\nu_0 - 1]c). \tag{6.4.6}$$

A Maxwellian velocity distribution gives rise to a Gaussian line profile

$$I(\nu) = I(\nu_0)\exp\left[-(\nu - \nu_0)^2 c^2/2v_{ta}^2\nu_0^2\right], \tag{6.4.7}$$

where $v_{ta}^2 = T_a/m_a$, the subscript a referring to the emitting atom. The FWHM is

$$\Delta\nu_{1/2} = \nu_0(v_{ta}/c)(2\ln 2)^{1/2}. \tag{6.4.8}$$

Of course, widths can also be expressed in terms of wavelengths λ. For small widths $\Delta\lambda/\lambda = \Delta\nu/\nu$.

6.4.3   *Pressure broadening*

This effect is also variously called collisional broadening or Stark broadening and arises from the influence of nearby particles upon the emitting atom. There are two main approaches to calculating the broadening, starting from opposite extremes [Traving (1968) provides a brief historical introduction as well as discussing the physical principles]. The *collisional* approach developed initially by Lorentz supposes that for most of the time the atom radiates undisturbed, but occasionally collisions occur that interrupt the wave train. If the mean time between collisions is τ, this approach assumes that the duration of the collision is $\ll \tau$. Then it is a general principle of Fourier transforms (mathematically equivalent to a form of the uncertainty principle used in natural broadening) that a sinusoidal wave of duration τ has frequency width of order

$$\Delta \nu \sim \frac{1}{\tau} \qquad (6.4.9)$$

(see Appendix 1). Taking into account the statistical nature of the time between collisions, one can show that the coherence of the wave falls off like $\exp(-t/\tau)$ (Poisson statistics; see Appendix 2). Fourier transforming this gives, as for natural broadening, a Lorentzian line shape [Eq. (6.4.3)], although the lifetime τ is now the collision time.

At the other extreme is the *quasistatic* approach developed by Holtzmark in which the atom is assumed to radiate in an environment that is effectively static during the period of emission. Any individual radiator thus experiences an instantaneous shift in wavelength and the average over all possible perturbations (and hence shifts) gives the line width and shape. The most important perturbing effect is generally the electric field of nearby particles. The atomic energy levels in an atom in an electric field are perturbed by the alteration of the form of the potential energy. States nearer the continuum are more strongly perturbed. The shift in a spectral line due to electric fields is called the Stark shift; hence the name Stark broadening.

To summarize: The impact or quasistatic approximations are applicable according as $\Delta \nu \ll t_p^{-1}$ or $\Delta \nu \gg t_p^{-1}$, where t_p is the duration of the perturbing interaction. Because of the difference in electron and ion velocities, the electron effects are usually best approximated by the impact approach while ion perturbations, for most of the line, are best modeled using the quasistatic approximation. It turns out too that the ion effects dominate the line width, at least for hydrogen lines.

Detailed calculations of Stark broadening are extremely complicated [see, e.g., Griem (1964) or Breene (1961)] and have been done in detail only for a few atoms. For hydrogen, good agreement between the theory and experiment has been obtained. We shall not attempt any outline of these theories. However, some important facts about the scaling of the broadening can be obtained by very simple arguments as follows.

In order to calculate the line shape in the quasistatic approximation we must first know how the line frequency is altered by a given electric field E. In the case of hydrogen, the Stark effect is linear, that is, $\Delta \nu \propto E$ (but note that for other atoms it is quadratic $\Delta \nu \propto E^2$ and much smaller); this is because of complicated quantum-mechanical effects upon which we shall not dwell. Also for hydrogen, because the Stark effect causes a symmetrical spread of initially degenerate lines, the effect causes no net shift of the line, unlike the situation with other atoms in which shifts as well as broadening of the lines occur.

Knowing the Stark effect on the line, we then must calculate the statistical distribution of electric field experienced by the atoms. For the case of ions as perturbers in a plasma, provided the atom is significantly closer to one ion than all the others, the electrical field E in the vicinity of that ion is $E \propto 1/r^2$, where r is the distance from that nearest ion. Also the proportion of space in that vicinity corresponding to r and hence the probability of experiencing E, say $P(E)\,dE$, is proportional to the volume of the spherical shell $4\pi r^2\,dr$ as illustrated in Fig. 6.6. Therefore, the distribution of intensity in the line is

$$I(\nu)\,d\nu \propto P(E)\,dE \propto -r^2\,dr. \tag{6.4.10}$$

Now $E \propto 1/r^2$ so $-r^2\,dr \propto E^{-5/2}\,dE$; also $\Delta\nu \propto E$ and hence,

$$I(\nu)\,d\nu \propto E^{-5/2}\,dE \propto (\Delta\nu)^{-5/2}\,d\nu. \tag{6.4.11}$$

Thus the shape of the line in the region where E and hence $\Delta\nu$ is large is $I(\nu) \propto (\Delta\nu)^{-5/2}$.

This functional form holds only up to a radius such that the electric field contributions from other ions are of the same order as that of the initial ion. This will occur when $\frac{4}{3}\pi r^3 \approx 1/n_i$ at $\Delta\nu \propto E \propto n_i^{2/3}$. At this point some kind of cutoff must be applied to $P(E)$, since obviously the relevant volume does not continue to increase. Thus the width of the line will be proportional to this cutoff, that is, to $n_i^{2/3}$. These scalings are illustrated in Fig. 6.7.

Fig. 6.6. The nearest neighbor approximation to Stark broadening. The probability that the nearest perturber is within a spherical shell dr at distance r is $\propto r^2\,dr$ up to the limit radius.

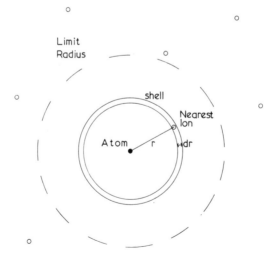

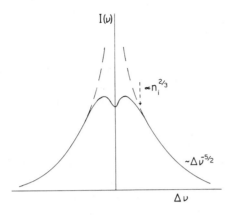

Fig. 6.7. Stark broadening line shape scalings.

Although this simple approach (often called the nearest neighbor treatment) does not give a full picture of the complexity of the problem nor account for important corrections due to electron effects (especially near $\Delta\nu = 0$), it does quite accurately give the density dependence of the line width. Full calculations give the constant of proportionality, which for the H_β line ($\lambda = 486.1$ nm) may be taken as given by

$$\text{FWHM } \Delta\lambda_{1/2} = \frac{\Delta\nu_{1/2}}{\nu_0}\lambda_0 = 0.040n_{20}^{2/3} \text{ nm}, \qquad (6.4.12)$$

where n_{20} is the ion density in units of 10^{20} m^{-3} ($n_{20} = n_i/10^{20}$) and λ, of course, is the wavelength. For H_α (656.2 nm) the Stark width is significantly narrower than for H_β, for example, by a factor ~ 10 at 10^{21} m^{-3} and 1 eV temperature.

6.4.4 Combinations of broadening effects

Sometimes both Doppler and other broadening effects and also, possibly, the finite resolution of the measuring instrument, are important. In general, when such a situation occurs, if there are two independent profiles, say $f_1(\Delta\nu)$ and $f_2(\Delta\nu)$, then the resulting observed profile is the convolution of the two profile functions:

$$f(\Delta\nu) = \int f_1(\Delta\nu - \Delta\nu')f_2(\Delta\nu')\,d(\Delta\nu'). \qquad (6.4.13)$$

Similarly, for further broadening functions f_3, and so forth, further convolutions are formed. Now for the Gaussian line shape, such as arises from Doppler broadening, the convolution of two profiles of width Δ_1 and Δ_2 gives a final width Δ given by

$$\Delta^2 = \Delta_1^2 + \Delta_2^2, \qquad (6.4.14)$$

while for the Lorentzian line shape

$$\Delta = \Delta_1 + \Delta_2. \tag{6.4.15}$$

In both these cases (separately) the convolved shape is another profile of the same form, Gaussian or Lorentzian (see Exercise 6.8).

A profile that is a convolution of Gaussian with Lorentzian is called a Voigt profile and its shape has been tabulated, for example, by Unsold (1955). These tabulated values are often used to obtain deconvolved widths corresponding to the Gaussian and Lorentzian parts, on the assumption that the non-Gaussian broadening may be approximated by the Lorentzian shape. Note, however, that as illustrated by our calculations indicating a wing shape $\propto \Delta\nu^{-5/2}$, the Stark profile will only approximately be represented by the Lorentzian. In high-temperature plasmas of moderate density typical of magnetic fusion, Stark broadening is usually negligible compared to Doppler broadening.

6.4.5   *Reabsorption: Optically thick lines*

Throughout most of this chapter we are tacitly assuming that the absorption of radiation due to induced upward transitions can be ignored. This will usually be a safe assumption for laboratory plasmas. Possible exceptions are for the strong resonance lines in very dense (or large) plasmas, in which reabsorption of the radiation may occur.

Such exceptional situations, in which the plasma is optically thick, must be treated by the radiation transport equation (5.4.6) in much the way we have discussed in Section 5.2.4. Additional complications arise when the radiation is sufficiently intense to change the population distributions, because then the excited state populations depend on conditions distant from the point in question. In other words the distribution is nonlocally determined. We shall not explore here these possible effects on the populations [see for example McWhirter (1965) for a discussion of them]. The most noticeable immediate effect on the radiation observed is a change of line shape, which we shall discuss briefly.

We can regard a ray passing through the plasma as gaining in intensity due to the additional emissivity of each volume element through which it passes. Mathematically, this is determined by the radiation transport equation, of course. Broadly speaking, the effect of reabsorption is to limit the intensity that can be reached to be less than or equal to the blackbody level corresponding to the effective temperature governing the atomic levels involved. This occurs because, as the blackbody level is approached, absorption balances emission and further increase in intensity is prevented. (Note that the effective temperature is given by the Boltzmann factor governing the ratio of the upper and lower state populations and may be completely different from, say, the local electron temperature.) Thus, we may regard the primary effect of reabsorption as being to place a "ceiling" at the blackbody level on the maximum radiation intensity.

Incidentally, this provides a simple way to check whether reabsorption is important. Calculate the intensity under the assumption that reabsorption is negligible by integrating the emissivity along the line of sight. If this is significantly less than the appropriate blackbody level, then reabsorption may indeed be ignored.

We can now understand what effect reabsorption will have on the line shape. Consider a line from a uniform plasma whose density (optical depth) we increase from a low value. Figure 6.8 illustrates the effect. Provided the maximum intensity is less than blackbody, no significant effect occurs (*a*). However, when the line center reaches close to the blackbody level it can increase no further (*b*), even though the lower intensity wings of the line are unaffected. (At the wing frequency the plasma is still optically thin.) The result is that the line shape observed is broadened. Further increase in optical depth causes further broadening of the observed line (*c*). In a uniform plasma the top of the line becomes flat at the blackbody level. When the plasma is nonuniform, for example if the effective temperature is lower at the plasma edge, inversion of the line may occur. That is, the line center intensity may decrease because it "sees" only the cooler plasma edge.

Thus, finite optical depth leads to a broadening of the observed line shape. Strictly speaking it is not a broadening mechanism itself; it merely serves to enhance the observed effect of broadening due to the other mechanisms in the line.

Fig. 6.8. The effect of reabsorption on line shape: (*a*) optically thin; (*b*) marginal; (*c*) optically thick.

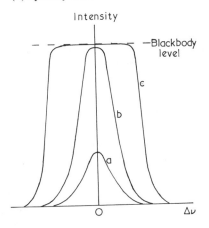

6.5 Applications

6.5.1 *Line intensities*

The most important application of measurements of the absolute line intensity is to estimate the density of an atomic species. Measurement of the line intensity for a line in which A_{ij} is known amounts to a rather direct measurement of the excited state density. Then, using an appropriate model to relate this to the ground state density (e.g., coronal equilibrium) the density of that species can be estimated.

For hydrogen, a very important parameter that can be deduced from such a measurement is the plasma source rate. The electron (and ion) density is determined by a balance between the divergence of the plasma particle diffusion flux and the rate at which plasma is replenished by ionization of neutrals. In order to estimate diffusion rates, usually quantified by a particle confinement time τ_p (equal to the ratio of the number of particles to the rate of loss of particles), it is convenient to measure the source rate. This is given by

$$n_N n_e \langle \sigma_i v \rangle, \tag{6.5.1}$$

where n_N is the neutral density and $\langle \sigma_i v \rangle$ is the rate coefficient for ionization. Because the excited state density depends on collisions – as does the ionization rate – in the coronal regime the ratio of ionization rate to line intensity is independent of density. Also, for $T_e > R_y$ the ratio is approximately independent of temperature because virtually all collision velocities are capable of ionizing or exciting the atom. Therefore, the ratio of, for example, H_α emission to ionization rate is constant. Figure 6.9 shows the ionizations per H_α photon calculated from a collisional radiative model. At low densities $n_e < 10^{18}$ m^{-3}, as expected, the ratio is approximately constant. The calculations extend to higher densities where the coronal equilibrium fails. Above $n_e \sim 10^{20}$ m^{-3} the ratio becomes proportional to n_e because LTE is approached and the excited state population becomes independent of n_e.

By the same principles, atomic species of higher charge can be diagnosed. These appear as impurities in, for example, hydrogen plasmas. In that case the impurity density itself is usually the parameter of interest. Again a model for the excited state population and knowledge of the important transition rates are required. If the total density of nuclei of a particular element is required rather than the density of a specific ionization stage, then it is important to know the ionization stage populations and also generally to make measurements on an abundant ionization stage of the element in question.

As noted in Section 6.3, in ionization equilibrium (which is not always attained) the dominant ionization stage reached is a strong function of

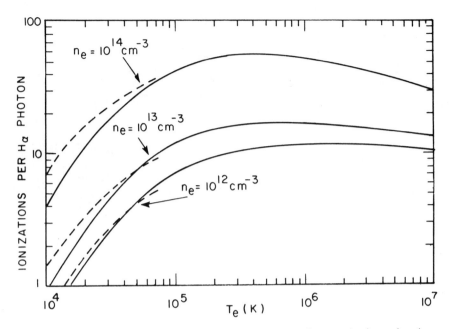

Fig. 6.9. The number of ionizations occurring per H_α photon emitted, as a function of electron temperature [after Johnson and Hinnov (1973)].

electron temperature. Examples of calculated populations as a function of temperature are shown in Figure 6.10. Ionization to the next higher level occurs (when collisional ionization exceeds radiative recombination) if $T \gtrsim \chi_i / \ln|(18mc^2/Z^2\chi_i)/(\chi_i/T)|$ [Eq. (6.2.19)]. This provides a qualitative understanding of the behavior of the fractional abundances in Fig. 6.10 (which are calculated using more accurate integrations of the rate coefficients than our rough approximation). For example, the significant increase in the ionization potential for heliumlike ions versus lithiumlike is the reason why there is quite a broad temperature range over which the heliumlike state is most abundant.

The ionization state of impurities can be used as a rather rough indication of electron temperature. This will rarely be better than a very crude estimate because it is usually rather difficult to measure the relative abundances with any great precision and also because the impurities are often not in ionization equilibrium because of the long time scales involved. On the other hand, the problem usually is most significant in the opposite way, namely that for a plasma of nonuniform temperature the different ionization stages dominate different parts of the plasma. For example, the lower ionization states will tend to occur in the outer cooler parts of the plasma. The effective volume from which radiation from a specific ioniza-

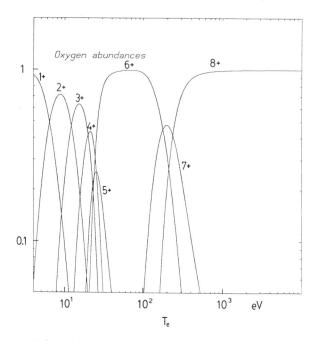

Fig. 6.10. Fractional abundances of different ionization stages of oxygen in coronal equilibrium as a function of electron temperature [after Piotrowicz and Carolan (1983)].

tion stage will occur will then often be much smaller than the total plasma volume viewed by the spectrometer. In the case of the lower ionization stages, emission will be concentrated in hollow shells near the periphery as illustrated schematically in Fig. 6.11.

The radial thickness of these shells will be determined by temperature gradient and also (when ionization relaxation times are longer than impurity transport confinement times) by impurity diffusion. Therefore, unless some kind of radial resolution of the emission profiles is possible, for example, by tomographic techniques or Abel inversion (see Section 5.3.8), a knowledge of the temperature profile and the corresponding ionization balance problem is required even to make a simple estimate of impurity density. The same is true for estimating neutral hydrogen density, since the hydrogen emission is dominated by the edge. This situation poses a rather severe problem for making estimates, for example, of neutral density in the plasma center of a hot plasma, since the emission there can be orders of magnitude less than at the edge.

A long established technique for estimating electron temperature and density is to use the ratios of intensities of lines from the same species. Taking the ratio cancels the direct dependence on the species density and

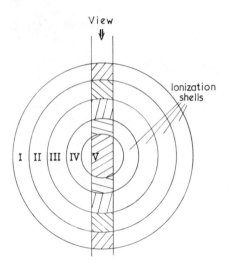

Fig. 6.11. In nonuniform plasmas the different ionization stages of a given species tend to adopt a distribution of concentric shells, with the lowest stage in the outermost (coolest) region.

leaves a parameter that depends mostly on the rates for excitation, and so forth. These rates depend on the temperature through the rate coefficients. The density dependence tends to cancel out, since most rates are proportional to n_e. Therefore, for density measurement, one must seek to employ lines whose intensity has additional dependence on density, for example, forbidden lines from atomic states whose population is influenced by collisional deexcitation (as well as excitation). These techniques rarely give accuracy better than $\sim 30\%$ because of uncertainties in the modeling and in the rate coefficients.

Line intensities are less often used nowadays for laboratory plasma diagnosis of temperature and electron density, and so forth, because of the availability of more satisfactory techniques that are often not much more difficult to perform. They retain their importance, of course, for astrophysical measurements where alternatives are not so readily available, and also in some short-lived laser-produced plasmas for which the alternatives prove very difficult.

6.5.2 *Doppler broadening*

Line widths prove to be of considerable importance and use for diagnosis of the majority particle species. In particular the line width due to Doppler broadening measures rather directly the temperature of the atomic species. When this species is charged, it acquires, by collisions with the majority ions, a temperature equal to the ion temperature. This equilibration occurs with a collision frequency [see, e.g., Schmidt (1969), p. 387]

$$\nu_\varepsilon \sim 10^{-12} \frac{\sqrt{\mu_1}}{\mu_2} \frac{n_1 Z_1^2 Z_2^2}{(T_1/e)^{3/2}} \, \text{s}^{-1}, \tag{6.5.2}$$

where we assume that the impurity species (2) is considerably heavier than the majority species (1) and where μ represents the mass in units of the proton mass (see Exercise 6.9). For many situations this is fast enough for full temperature equilibration to be assumed.

In many situations the other broadening mechanisms, natural broadening, and Stark broadening, are negligible compared to the Doppler width. If so, then a straightforward measurement of line width is sufficient to give ion temperature from any impurity line. If the other broadening mechanisms are not negligible, either because the transition probability is great enough to give significant natural line width (as can sometimes occur for high-energy lines from highly charged ions), or because high density leads to significant Stark broadening, then some kind of deconvolution of the different effects is necessary. This may be performed either by detailed analysis of the line shape, since the different mechanisms give different line shapes, or by auxiliary knowledge of the non-Doppler broadening. For example, if the density is known, the Stark effect may be estimated.

Of course, Doppler width measurement provides ion temperature information only on the region of the plasma in which the particular atomic species is present in abundance. Thus, in hot ($T_e > 10$ eV) hydrogen plasmas, broadening of hydrogen lines is rarely useful as a measure of anything but the plasma edge temperature (unless other effects intervene; see, e.g., Section 6.3.6) because essentially all the atomic radiation is from the cooler edge.

On the other hand, this feature can be used to advantage when the positions of the shells of various ionization stages of impurities are known, for then by observing the Doppler widths of lines from different species, spatially resolved information on the profile of the ion temperature can be obtained. Figure 6.12 shows an example.

In fusion plasmas the technical difficulties of performing Doppler line width measurements are considerable because the high electron temperatures cause the impurities in the discharge center to be very highly stripped and correspondingly to possess line spectra with dominantly very high-energy (short wavelength) photons, usually $h\nu \gtrsim T_e$. To obtain sufficient resolution in the extreme ultraviolet and soft x-ray spectral regions is difficult, but recent developments in detector technology make it quite practical. Figure 6.13 shows an example of a high resolution x-ray spectrum in the vicinity of 3.7 Å ($= 0.37$ nm) wavelength, about 3.15 keV photon energy. Three lines are visible: two are from the Lyman alpha line of

Fig. 6.12. Radial profile of ion temperature during heating by neutral beam injection in PLT tokamak. Temperatures at different radii are obtained from Doppler broadening of lines of different ion species [after Eubank et al. (1978)].

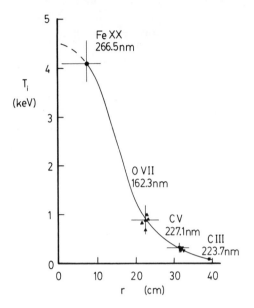

Fig. 6.13. High resolution soft x-ray spectrum from Alcator C tokamak showing two lines from hydrogenlike argon and a molybdenum line [after Marmar and Rice (1985)]. Doppler widths give ion temperatures of 1200 eV for each line (the Mo being narrower), thus confirming the thermal origin of the Doppler broadening.

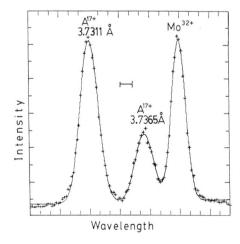

hydrogenlike argon A^{17+} and a third is from molybdenum Mo^{32+}. All three are predominantly excited in the center of this plasma ($T_e \sim 1500$ eV). The molybdenum line is noticeably narrower because molybdenum moves more slowly owing to its greater mass. The ion temperatures deduced from the line widths are consistent (~ 1200 eV) to within the experimental uncertainty.

Another possibility for Doppler broadening measurements in high-temperature plasmas is to use certain "forbidden lines," that is lines emitted by other than electric dipole radiation, arising from transitions within the ground configuration of some highly ionized medium- and high-Z elements. The lines emitted are often in the longer wavelength (near) ultraviolet, where measurement techniques can be easier.

Obviously, the line choices are manifold and at present are made in a more or less ad hoc manner. Considerable further developments for high-temperature plasmas may be anticipated.

6.5.3   *Ion flow velocity*

An important plasma parameter that can also be probed by the Doppler effect on spectral lines is the bulk flow velocity of the ions. Impurities will generally acquire a mean velocity equal to that of the ions in a time similar to that for temperature equilibration. If the mean velocity parallel to the line of sight is significant, this will lead to a shift of the emission line from its usual spectral position proportional to the ion flow velocity.

Generally speaking, the presence of broadening due to thermal ion motion will set a lower limit to the detectable line shift at some fraction of the thermal Doppler width, the fraction depending on just how accurately the center of the broadened line can be determined and hence on signal-to-noise levels (see Exercise 6.10).

Because the thermal velocity is inversely proportional to the square root of the mass of the species, heavier ions will generally have a smaller ratio of line width to line shift (assuming their mean velocity to be equal to the mean ion velocity) and so tend to provide better candidates for flow measurements.

It is sometimes difficult to measure the absolute wavelength with sufficient accuracy to provide line shift data. This can then be obtained by a measurement that observes both parallel and antiparallel to the flow, taking the difference in the line position for the two cases. An example of this type of measurement is illustrated in Fig. 6.14. These measurements were obtained during neutral beam heating on the tokamak PLT. Suckewer et al. (1979) used a rotating mirror to scan the plasma view from forward to backward along the toroidal direction and a more rapidly vibrating mirror

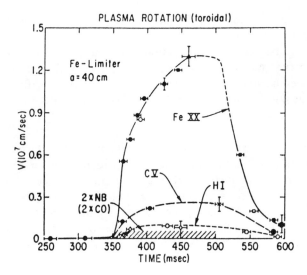

Fig. 6.14. Measurements of plasma flow velocity using the Doppler shift of lines from different impurity species. The injection of energetic neutral beams causes this tokamak plasma to rotate toroidally. Prior to injection the flow is too small to measure [after Suckewer et al. (1979)].

to scan the spectral line during this process. The neutral beams impart substantial rotation to the plasma when they are switched on. Measurements of lines from three different species, Fe xx, C v and H ɪ give the flow velocity at three radial positions $r \approx 0$, 30, and 40 cm (the edge), respectively.

In the presence of turbulent flows, caution must be exercised in the interpretation of line widths. It is possible for flow velocities to mimic the effect of thermal velocities and the experimenter may then be deceived into interpreting as ion temperature broadening that which is in reality the effect of simply coherent, but possibly fine scale, flow patterns. For just the reasons mentioned, heavy atomic species are more susceptible to this problem than light species. Thus, one way of distinguishing turbulent flow effects from true thermal broadening is to perform measurements on species of different mass. Agreement in the temperatures deduced from assuming thermal Doppler broadening then confirms the dominance of thermal velocity over flow velocity (see Fig. 6.13).

6.5.4 Stark widths

In lower-temperature higher-density plasmas it is possible for the Stark broadening (particularly in hydrogen) to become significant or even dominate the line width. If so, then the electron density can be deduced from a measurement of the line width.

The H_β line is the best analyzed transition from the point of view of Stark width diagnostics, and the formula Eq. (6.4.12) has been used as the basis for relatively convenient measurements. All the usual difficulties of spatially inhomogeneous plasmas apply to the interpretation of the results.

6.5.5   *Bolometry*

Under some circumstances, the energy loss from the plasma by radiation of all types (but notably line radiation) can be comparable to other losses by thermal conduction. This is a matter of considerable importance for fusion research, since the energy balance of the plasma is altered thereby. Moreover, this is by no means a purely pathological situation because, even when radiation losses in the plasma center are small, there is very often substantial radiation from the cooler outer regions of the plasma.

In principle, if all the impurity species responsible for the radiation were known, it should be possible to estimate the total energy loss by an elaborate sum over different states and transitions. The proportion of each ionic species would have to be estimated from line emission from that species. In practice, this procedure proves to be too cumbersome to be useful routinely and in those cases where it has been attempted it sometimes provides results that are not in agreement with other measurements (Hsuan et al. 1978).

An alternative and more direct measurement of radiation loss is to use a radiative bolometer. This is a detector specifically designed to respond to the total radiative energy flux with a spectral response as near as possible constant in the regions of major radiative loss (usually the ultraviolet). Most bolometers consist of an absorbing element designed to absorb all the incident energy, whose temperature rise, measured by some appropriately sensitive method, is then equal to the total energy flux divided by the bolometer's thermal capacity. Solid state detectors, such as surface barrier diodes, can also be used, but uncertainty in the number of electron–hole pairs generated per unit energy incident (i.e., the spectral response) makes their calibration less reliable.

It is usually most satisfactory to adopt some kind of imaging system using apertures and to employ sufficient chords that a spatial reconstruction of the emission profile by, for example, Abel inversion can be performed. This allows one to determine where the radiation is mostly coming from and is considerably more useful than, say, a single bolometer near the plasma edge.

Naturally, most bolometers are sensitive to energy loss by all neutral particles; in particular, fast charge-exchange neutrals will be detected as well as photons. This is usually perfectly satisfactory, since all direct energy losses are of interest.

6.6 Active diagnostics

There are several attractive possibilities for diagnostics in which active perturbation of the excited state populations is used to improve on the usefulness of emitted line radiation.

6.6.1 Resonant fluorescence

The basis of this approach is to irradiate some portion of the plasma with intense electromagnetic radiation at some resonant (line) frequency of an atom in the plasma. Generally, a tunable laser such as a dye laser is used to provide appropriate frequency (e.g., H_α) and sufficient intensity. The effect of the radiation is to induce transitions between the atomic levels of the transition chosen. If the radiation is intense enough, these induced transitions will dominate over all other (spontaneous and collisional) processes between these levels, and the result will be to equalize the occupancy of each quantum state of the two levels. In other words, the effective temperature describing the population difference of the two levels becomes infinite.

The alteration of the state population, generally an enhancement of the upper-level population, causes a change (usually an increase) in the observed radiation. When the radiation observed is from the same transition as that excited, the effect is of so-called fluorescent scattering, although this terminology may be misleading, since the radiation observed is from spontaneous transitions from the upper level, not those induced by the laser beam. The induced photons are emitted in the direction of the illuminating beam whereas the observed fluorescence is generally in a different direction and consists only of the spontaneous transitions of the enhanced upper-level population.

The observable fluorescence is limited by the extent to which the upper-level population can be enhanced. If the exciting radiation resonates with a line whose lower level is the ground state, then the degree of enhancement possible is usually very great because, for example, in coronal equilibrium the ground state population is usually much larger than the excited state population. For this reason excitation from the ground state appears most attractive. However, for hydrogen the longest wavelength line to the ground state is Ly_α (121.6 nm), which is in the ultraviolet where suitable lasers of sufficient power are not yet available.

Excitation of hydrogen by Balmer transitions (H_α, etc.) that are accessible to existing tunable lasers is possible. The population of the upper level is enhanced usually by a modest factor only (perhaps 2–3), because the initial populations of the $n = 2$ and $n = 3$ levels are not greatly different.

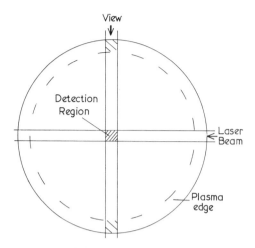

Fig. 6.15. Resonant fluorescence of atomic transitions with orthogonal viewing direction allows a very localized detection region.

The enhancement is observable, however (Razdobarin et al. 1979), and so use can be made of the technique.

The usefulness of resonant fluorescence for hydrogen arises mostly from the ability to localize the region of measurement. The use of crossed exciting and viewing beams, as illustrated in Fig. 6.15, means that fluorescence is observed from a well-defined spatial region. If the fluorescence can be detected, it can then give information on the neutral population in the plasma center, for example, which is normally unavailable through passive H_α observations because of the dominant emission from the plasma edge. In detecting the fluorescence, the edge radiation plays the role of a strong background noise signal that must be discriminated against as far as possible. This noise dictates the minimum detectable internal neutral density in most situations, despite various techniques to reduce it.

Another diagnostic of interest involves the application of fluorescence to impurity species. The principles are essentially the same except that many proposals envisage the use of atoms (for example, of lithium) injected in the form of an energetic neutral beam into the plasma. Lithium is of particular interest since it has lines to the ground state of long (visible) wavelength (670.8 nm).

6.6.2   *Zeeman splitting: Magnetic field measurements*

The splitting of emission lines by a magnetic field into three groups of components, the unshifted π component and the symmetrically shifted σ components, can in principle be used to diagnose the magnetic field even from the natural line radiation.

The magnitude of the field may be deduced from the magnitude of the splitting. Oversimplifying slightly, the shift of the σ components is proportional to B, the exact constant of proportionality being determined by the Landé g factor g_J (~ 1):

$$\Delta\nu = \frac{1}{2\pi} g_J \frac{\Omega}{2}. \tag{6.6.1}$$

In many configurations the magnitude of the field is of great interest, but sometimes it is primarily the direction that one wishes to measure. For example, in a tokamak the dominant toroidal magnetic field is rather well known but the smaller poloidal component is determined by the internal plasma current profile and is of great importance in determining MHD stability, for example. It can be measured by a measurement of the precise field direction. For this purpose the polarization dependence of the Zeeman effect is useful. When viewed perpendicular to the magnetic field, the π component is polarized with E vector parallel to the magnetic field and the σ components perpendicular. By a measurement of the angle of polarization of one of these components the magnetic field angle may, in principle, be measured.

Complications arise in a passive measurement because of the lack of spatial resolution along the line of sight or the dominance (for hydrogen) of the edge region emission. A solution that has been made to work (McCormick 1977 and 1986), though not really routinely, is to use an injected lithium beam, so as to localize the emission source. A further enhancement of this technique is to use resonance fluorescence to increase the emission intensity. The main difficulty in either case is in producing a lithium beam sufficiently energetic to penetrate the plasma but sufficiently monoenergetic as not to obscure the Zeeman splitting by Doppler broadening. Further discussion of the use of lithium beams is given in Section 8.2.4.

Because of the importance of nonperturbing measurements of internal magnetic fields [see the review of Peacock (1978)], these techniques are undergoing current research and development, despite their difficulty.

Further reading
Various reviews of bound-electron radiation processes in the plasma diagnostic context are found in:

McWhirter, R. P. (1965). In *Plasma Diagnostic Techniques*. R. H. Huddlestone and S. L. Leonard, eds., Chap. 5. New York: Academic.
Wiese, W. L. (1965). In *Plasma Diagnostic Techniques*. R. H. Huddlestone and S. L. Leonard, eds., Chap. 6. New York: Academic.
Griem, H. R. (1964). *Plasma Spectroscopy*. New York: McGraw-Hill.
Lochte-Holtgreven, W., ed. (1968). *Plasma Diagnostics*. Amsterdam: North-Holland.

Introductory texts on atomic physics and spectroscopy:

Thorne, A. P. (1974). *Spectrophysics*. London: Chapman and Hall.
Kuhn, H. G. (1969). *Atomic Spectra*. New York: Longmans.
Born, M. (1969). *Atomic Physics*. 8th ed. London: Blackie.

Quantum mechanics:

Dirac, P. A. M. (1958). *Principles of Quantum Mechanics*. 4th ed. London: Oxford.
Schiff, L. I. (1964). *Quantum Mechanics*. New York: McGraw-Hill.

Collision processes, Classical encyclopedic work:

Massey, H. S. W. and Burhop, E. H. S. (1952). *Electronic and Ionic Impact Phenomena*. London: Oxford.

In addition, collision phenomena involving ions typical of hot plasmas have recently been reviewed in:

Janev, R. K., Presnyakov, L. P., and Shevelko, V. P. (1985). *Physics of Highly Charged Ions*. New York: Springer.

Reference works on spectral lines and line strengths:

Wiese, W. L., Smith, M. W., and Glennon, B. M. (1966). *Atomic Transition Probabilities I-Hydrogen through Neon*. NSRDS-NBS4. Washington: U.S. Government Printing Office.
Wiese, W. L., Smith, M. W., and Miles, B. M. (1969). *Atomic Transition Probabilities II-Sodium through Calcium*. NSRDS-NBS22. Washington: U.S. Government Printing Office.
Striganov, A. R. and Sventitskii, N. S. (1968). *Tables of Spectral Lines of Neutral and Ionized Atoms*. New York: Plenum.

Exercises

6.1 Show that two levels i, j of an atom will be in local thermodynamic equilibrium in a plasma of negligible optical depth if

$$n_e \gg A_{ij}/\langle \sigma_{ij}v \rangle.$$

6.2 Using Eq. (6.3.25) for $\langle \sigma_{ij}v \rangle$, show that

$$\frac{A_{ij}}{\langle \sigma_{ij}v \rangle} = E_{ij}^3 T^{1/2} \frac{4\pi\varepsilon_0}{e^2} \frac{(6\pi)^{1/2}}{h^2c^3m_e^{1/2}} \exp\left[\frac{h\nu_{ij}}{T}\right].$$

What value of $h\nu_{ij}/T$ does the rule-of-thumb equation (6.2.8) correspond to?

6.3 Use the principle of detailed balance to deduce the relationship between the Maxwellian rate coefficients for collisional excitation/deexcitation, $\langle \sigma_{ij}v \rangle, \langle \sigma_{ji}v \rangle$, between energy levels E_i, E_j, with statistical weights g_i, g_j.

6.4 Calculate, using Eq. (6.2.8), the principal quantum level in a hydrogen atom above which LTE can be assumed when:
(a) $n_e = 10^{20}$ m^{-3}, $T_e = 1$ eV.
(b) $n_e = 10^{17}$ m^{-3}, $T_e = 1$ eV.

6.5 Suppose that, in coronal equilibrium, there are equal densities of atoms of ionization stages i and $i+1$. By equating the collisional ionization rate to the radiative recombination rate, using Eqs. (6.3.12) and (6.3.3) in the limit $\chi_i \gg T$, obtain the equation

$$\exp\left[\frac{\chi_i}{T}\right] = \frac{3\sqrt{3}}{80g\alpha} \frac{m_e c^2}{Z^2 \chi_i} \frac{T}{\chi_i}.$$

For $g = 0.5$ this gives Eq. (6.2.19).

6.6 The sum over levels of the recombination rate requires the sum

$$S = \sum_n \left[\frac{Z^2 R_y}{n^2 T}\right]^{1/2} \frac{\chi_n}{T} \exp\left[\frac{\chi_n}{T}\right] \mathrm{Ei}\left[\frac{\chi_n}{T}\right].$$

Perform this sum, maintaining the distinction between χ_n and $Z^2 R_y/n^2$ only in recombination to the lowest level n_0, when the result is sensitive to the value used. Consider separately the two cases representing the extremes of temperature:

(a) $\chi_i \gg T$. Let n' be the level at which $\chi_{n'} = T$. The sum above n' may be ignored, and below n' the approximation $\exp(x)\mathrm{Ei}(x) \approx 1/x$ $(x \gg 1)$ may be used. Show that

$$S = \sum_{n_0}^{n'} \left[\frac{Z^2 R_y}{T}\right]^{1/2} \frac{1}{n}$$

and evaluate this sum approximately as an integral over n to get:

$$S = \tfrac{1}{2}\left(Z^2 R_y/T\right)^{1/2} \ln|\chi_i/T|.$$

(b) $\chi_i \ll T$. Use $\exp(x)\mathrm{Ei}(x) \approx -\ln(x)$ $(x \ll 1)$ to obtain

$$S = \sum_n \left[\frac{Z^2 R_y}{n^2 T}\right]^{1/2} \left[\frac{\chi_n}{T}\right]\left[-\ln\left|\frac{\chi_n}{T}\right|\right].$$

Ignore the weak dependence of the logarithmic term on n and perform the sum for the higher levels $n > n_0$ approximately as an integral with lower limit $n_0 + \tfrac{1}{2}$. Since this sum is most significant when n_0 is substantially greater than 1, expand the term $(n_0 + \tfrac{1}{2})^{-2}$ as $\sim n_0^{-2}(1 - 1/n_0)$, to obtain

$$S_{n>n_0} = \left[\frac{Z^2 R_y}{T}\right]^{1/2}\left[\frac{\chi_i}{T}\right]\left|\ln\frac{\chi_i}{T}\right|\frac{1}{2}\left(1 - \frac{1}{n_0}\right).$$

Add on the term for the lowest level, including the hole correction factor $\xi/2n_0^2$,

$$S_{n_0} = \left(\xi/2n_0^2\right)\sqrt{\left(Z^2 R_y/n_0^2 T\right)}(\chi_i/T)|\ln(\chi_i/T)|,$$

to obtain

$$S \approx \left[\frac{Z^2 R_y}{T}\right]^{1/2}\left[\frac{\chi_i}{T}\right]\left|\ln\frac{\chi_i}{T}\right|\frac{1}{2}\left[1 + \frac{1}{n_0}\left(\frac{\xi}{n_0^2} - 1\right)\right].$$

Show that the empirical interpolation formula Eq. (6.3.7) correctly gives these values for the sum at high and low temperatures.

6.7 Derive Eq. (6.3.8) from an analysis of the Coulomb orbits.

6.8 Take the colliding electron's energy at collision to be $E + E_+$, due to the potential energy in the nuclear field. Ionization will occur if $\chi_i < \Delta E < E$, where ΔE must now be evaluated from Eq. (6.3.8), but substituting $E + E_+$ instead of E. Show that this leads to Eq. (6.3.11) for σ_i.

6.9 The EM wave energy density may be written $\rho(t) = \varepsilon_0 |E(t)|^2$ (twice the electric field energy density, because half the energy is in the magnetic field). Write Parseval's theorem for E, for finite time duration T and use it to obtain an expression for the mean value of $\rho(\nu)$ [$= 2\pi\rho(\omega)$]. Hence, prove Eq. (6.3.19) in the limit $T \to \infty$.

6.10 Show that convolutions of Gaussians and of Lorentzians separately lead to Gaussian or Lorentzian resultants, respectively, and prove Eqs. (6.4.14) and (6.4.15).

6.11 Consider a straight-line Coulomb collision between a stationary heavy impurity atom and a majority ion. Show that the momentum transfer is

$$2\frac{Z_i Z_a e^2}{4\pi\varepsilon_0 vb},$$

where Z_i and Z_a are ion and atom charges, v is ion velocity, and b is the impact parameter. Hence, show that the rate of gain of energy by the atom due to all collisions with a Maxwellian distribution of ions is

$$\frac{dW}{dt} = \left(\frac{Z_i Z_a e^2}{4\pi\varepsilon_0}\right)^2 \frac{8\pi n_i}{m_a} \left(\frac{m_i}{2\pi T_i}\right)^{1/2} \ln \Lambda,$$

where $\ln \Lambda$ is the Coulomb logarithm arising from applying cutoffs to the b integration. The effective collision frequency may be taken as $(dW/dt)/T_i$. Obtain Eq. (6.5.2).

6.12 Suppose one had a perfect spectrometer that gave an exact spectrum of all photons arriving at it. Because of photon statistics there would still be some noise in the spectrum. Calculate the minimum detectable line shift as a fraction of the (Gaussian) line width if the total number of detected photons is N. How is this result affected if the spectrometer has a finite resolution with a spectral shape that is Gaussian with width equal to the emission line width (but is perfect in other respects)?

7   Scattering of electromagnetic radiation

One of the most powerful methods of diagnosis is to use the scattering of electromagnetic radiation from the plasma. The attractiveness of this diagnostic derives from two main features. First, it is, for all practical purposes, a nonperturbing method, requiring only access of radiation to the plasma. Second, it offers the potential of determining detailed information about the distribution function of electrons and sometimes even of the ions too. These advantages are sufficient to offset the fact that the measurements are generally very difficult technically to perform. Electromagnetic wave scattering diagnostics are now widespread, especially in hot plasma experiments.

The process of electromagnetic wave scattering by charged (elementary) particles may be thought of as follows. An incident electromagnetic wave impinges on the particle. As a result of the electric and magnetic fields of the wave, the particle is accelerated. The charged particle undergoing acceleration emits electromagnetic radiation in all directions. This emitted radiation is the scattered wave.

Of course, this description is purely classical. From a quantum-mechanical viewpoint we might have described the process in terms of photons colliding with the particle and hence "bouncing off" in different directions. This would lead to an identical mathematical formulation provided there is negligible change in the mean particle momentum during collision with the photon. This will be the case provided that the photon mass is much smaller than the particle mass: $\hbar\omega \ll mc^2$. This classical limit of scattering by free charges is called Thomson scattering. On the other hand, when the photons are sufficiently energetic that their momentum cannot be ignored, the quantum-mechanical modifications lead to different results and the situation is called Compton scattering. The plasma applications of scattering tend to be limited to visible or longer wavelengths, whose photons have much less energy than the rest mass of an electron (~ 1 eV versus ~ 500 keV). Therefore, we shall deal only with the classical Thomson scattering case.

Because the ions are much heavier than the electrons, their acceleration and hence radiation is usually small enough to be negligible. Thus, it is the electrons that which do the scattering, at least from the microscopic point of view. This does not mean that the ions are irrelevant, as we shall see, but rather that in order to obtain information about them we shall have to measure some effect they have on the electrons.

7.1   Relativistic electron motion in electromagnetic fields

One finds that when electron thermal energies exceed a fraction of a percent of their rest mass (that is, if $T_e > 1$ keV) the relativistic corrections become significant in the scattering spectrum observed. It is important, therefore, to perform a truly relativistic calculation for scattering from hot plasmas. Thus, we cannot rely on our earlier nonrelativistic calculation of the electron motion.

Here we suppose that an electron of velocity $\mathbf{v} \equiv c\boldsymbol{\beta}$ moves in (time-varying) electric and magnetic fields $\mathbf{E}$ and $\mathbf{B}$. The equation of motion is

$$\frac{\partial}{\partial t}\left\{\frac{m_0 \mathbf{v}}{(1-\beta^2)^{1/2}}\right\} = -e(\mathbf{E} + \mathbf{v} \wedge \mathbf{B}). \tag{7.1.1}$$

The left hand side of this equation can be calculated explicitly to give

$$m_0 \gamma \dot{\boldsymbol{\beta}} + \gamma^3 m_0 \boldsymbol{\beta}(\boldsymbol{\beta} \cdot \dot{\boldsymbol{\beta}}) = -e\left(\frac{1}{c}\mathbf{E} + \boldsymbol{\beta} \wedge \mathbf{B}\right), \tag{7.1.2}$$

where m_0 is the electron rest mass, γ is the relativistic factor $(1 - \beta^2)^{-1/2}$, and overdots denote time derivative. Taking $\boldsymbol{\beta} \cdot$ in this equation we obtain

$$\dot{\boldsymbol{\beta}} \cdot \boldsymbol{\beta} = -\frac{e}{m_0 c \gamma^3}\boldsymbol{\beta} \cdot \mathbf{E}. \tag{7.1.3}$$

We then substitute this back into Eq. (7.1.2) to solve for $\dot{\boldsymbol{\beta}}$, obtaining

$$\dot{\boldsymbol{\beta}} = \frac{-e}{m_0 \gamma}\left\{\frac{\mathbf{E}}{c} - \left(\frac{\boldsymbol{\beta} \cdot \mathbf{E}}{c}\right)\boldsymbol{\beta} + \boldsymbol{\beta} \wedge \mathbf{B}\right\}. \tag{7.1.4}$$

We suppose that the wave fields are small enough that the value of $\boldsymbol{\beta}$ to be used in the right hand side of this formula is that of the unperturbed particle orbit. That is, we linearize the equations. Also, for now, we shall suppose that no applied constant $\mathbf{E}$ or $\mathbf{B}$ fields are present so that the unperturbed $\boldsymbol{\beta}$ is simply constant.

The electric and magnetic fields we shall suppose to be given by an incident transverse plane wave $\mathbf{E}_i$ propagating in the direction $\hat{\mathbf{i}}$, for which

$$\mathbf{B}_i = \frac{1}{c}\hat{\mathbf{i}} \wedge \mathbf{E}_i. \tag{7.1.5}$$

Inserting these fields gives

$$\dot{\boldsymbol{\beta}} = \frac{-e}{m_0 c \gamma}\left\{\mathbf{E}_i - (\boldsymbol{\beta} \cdot \mathbf{E}_i)\boldsymbol{\beta} + (\boldsymbol{\beta} \cdot \mathbf{E}_i)\hat{\mathbf{i}} - (\boldsymbol{\beta} \cdot \hat{\mathbf{i}})\mathbf{E}_i\right\}. \tag{7.1.6}$$

This is the electron acceleration we require.

Now we recall the formula for the radiated electric field from an accelerating charge Eq. (5.1.10), which we write

$$\mathbf{E}_s = \frac{e}{4\pi\varepsilon_0}\left[\frac{1}{\kappa^3 Rc}\{\hat{\mathbf{s}} \wedge (\dot{\boldsymbol{\beta}} \wedge \{\hat{\mathbf{s}} - \boldsymbol{\beta}\})\}\right]. \qquad (7.1.7)$$

We have changed our notation slightly by putting $\hat{\mathbf{s}}$ for $\hat{\mathbf{R}}$, consistent with the identification of $\hat{\mathbf{s}}$ as the direction in which the scattered radiation is detected. Figure 7.1 shows the general geometry. Remember that bold square brackets denote quantities to be evaluated at retarded time and $\kappa \equiv 1 - \hat{\mathbf{s}} \cdot \boldsymbol{\beta}$. The value for $\dot{\boldsymbol{\beta}}$ is substituted into this formula and gives the scattered electric field in terms of the incident field. A convenient form of the resulting expression that may be obtained by vector manipulation of the cross products in the formula (see Exercise 7.1) is

$$\begin{aligned}\mathbf{E}_s = \frac{e^2}{4\pi\varepsilon_0 m_0 c^2}\Bigg[&\frac{E_i(1-\beta^2)^{1/2}}{R(1-\beta_s)^3}\big\{-(1-\beta_i)(1-\beta_s)\hat{\mathbf{e}} - \beta_e(1-\beta_s)\hat{\mathbf{i}}\\ &+\big(\{1-\beta_i\}\hat{\mathbf{s}}\cdot\hat{\mathbf{e}} + \{\hat{\mathbf{s}}\cdot\hat{\mathbf{i}} - \beta_s\}\beta_e\big)\hat{\mathbf{s}}\\ &-\big(\{1-\beta_i\}\hat{\mathbf{s}}\cdot\hat{\mathbf{e}} - \{1-\hat{\mathbf{s}}\cdot\hat{\mathbf{i}}\}\beta_e\big)\boldsymbol{\beta}\big\}\Bigg],\end{aligned}$$
$$(7.1.8)$$

where $\hat{\mathbf{e}} \equiv \mathbf{E}_i/E_i$ and subscripts on β indicate components in the direction of the vector indicated (e.g., $\beta_s \equiv \boldsymbol{\beta} \cdot \hat{\mathbf{s}}$, etc.).

This is the general formula for relativistic Thomson scattering from a single electron. The multiplicative factor is simply the classical electron radius, defined as

$$r_e \equiv \frac{e^2}{4\pi\varepsilon_0 m_0 c^2}. \qquad (7.1.9)$$

If we consider a specific incident wave polarization such that the electric

Fig. 7.1. Vector diagram for general scattering geometry.

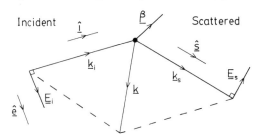

field is perpendicular to the scattering plane (i.e., the plane containing $\hat{\mathbf{s}}$ and $\hat{\mathbf{i}}$), then $\hat{\mathbf{e}} \cdot \hat{\mathbf{s}} = 0$ ($\hat{\mathbf{e}} \cdot \hat{\mathbf{i}} = 0$ always). Also suppose that we select, using a polarizer, for example, only that part of the scattered wave that has $\mathbf{E}$ parallel to $\hat{\mathbf{e}}$; then we have a rather simpler formula,

$$\hat{\mathbf{e}} \cdot \mathbf{E}_s = r_e \left[\frac{E_i (1 - \beta^2)^{1/2}}{R(1 - \beta_s)^3} \left\{ (1 - \hat{\mathbf{s}} \cdot \hat{\mathbf{i}}) \beta_e^2 - (1 - \beta_i)(1 - \beta_s) \right\} \right].$$

(7.1.10)

Figure 7.2 illustrates this more restricted scattering geometry. Equation (7.1.10) is the formula usually used for relativistic scattering from a single electron.

7.2   Incoherent Thomson scattering

7.2.1   *Nonrelativistic scattering: The dipole approximation*

In some cases the additional complexities of the relativistic treatment are unnecessary. The nonrelativistic formulas are particularly simple and bear a rather more elegant relationship to plasma parameters, so we derive them here. In the nonrelativistic limit, the initial particle velocity is small enough that the equation of motion becomes simply

$$\dot{\boldsymbol{\beta}} = \frac{-e}{m_0 c} \mathbf{E}_i$$

(7.2.1)

and the radiated (scattered) field is

$$\mathbf{E}_s = \frac{-e}{4\pi\varepsilon_0} \left[\frac{1}{Rc} \left(\hat{\mathbf{s}} \wedge (\hat{\mathbf{s}} \wedge \dot{\boldsymbol{\beta}}) \right) \right].$$

(7.2.2)

This latter equation is just the radiation field for an oscillating dipole of moment $\mathbf{p}$ such that $\mathbf{p} = -e\dot{\mathbf{v}}$; hence, the term dipole approximation for this nonrelativistic limit. The scattered electric field is therefore

$$\mathbf{E}_s = \left[\frac{r_e}{R} \hat{\mathbf{s}} \wedge (\hat{\mathbf{s}} \wedge \mathbf{E}_i) \right].$$

(7.2.3)

Fig. 7.2. Vector diagram for restricted scattering geometry.

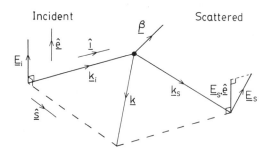

The power per unit solid angle (Ω_s) scattered in the direction $\hat{s}$ by a single electron is simply $R^2 c\varepsilon_0 |E_s|^2$, which is

$$\frac{dP}{d\Omega_s} = r_e^2 \sin^2\phi\, c\varepsilon_0 |E_i|^2, \tag{7.2.4}$$

where ϕ is the angle between $\hat{s}$ and E_i. This scattering is illustrated in Fig. 7.3, which shows a polar plot of the power scattered in different directions. It is convenient then to define the differential scattering cross section as the ratio of $dP/d\Omega_s$ to the incident power per unit area $c\varepsilon_0 |E_i|^2$: This cross section is then

$$\frac{d\sigma}{d\Omega_s} = r_e^2 \sin^2\phi. \tag{7.2.5}$$

The total Thomson scattering cross section is simply the integral of this expression over all solid angles. Noting $d\Omega_s = 2\pi \sin\phi\, d\phi$, this is easily evaluated to give

$$\sigma = \frac{8\pi}{3} r_e^2. \tag{7.2.6}$$

Naively, one may interpret this as indicating the effective "size" of the electron for scattering. The total power it scatters in all directions is equal to that power of the incident wave that would fall on a disk of area $\sigma = 8\pi r_e^2/3$.

7.2.2 Conditions for incoherent scattering

So far we have calculated the scattering from a single electron. However, we need to know what the scattering will be from a plasma consisting of many electrons. In order to calculate the total scattering, it is necessary to add up the electric field contributions from each of the electrons. To do this correctly requires information on the relative phase of each contribution as well as the amplitude.

Fig. 7.3. Polar plot of the angular distribution $\sin^2\phi$ of power in the dipole approximation.

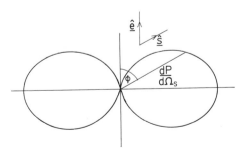

If one could assume that the phases of all contributions were completely uncorrelated then the prescription would be simple. For any such case of *incoherent* summation we know that the *powers* add. So we need simply to take the modulus squared of the electric field and sum over all electrons. For the nonrelativistic case, the total fraction of the incident power scattered over all angles per unit path length is then simply

$$n_e \sigma = \tfrac{8}{3}\pi n_e r_e^2. \tag{7.2.7}$$

The problem is, though, that we do not know a priori that the phases are purely random like this, because plasma is a medium supporting all kinds of collective effects in which electron positions and motions are correlated. Thus, in some cases the summation requires *coherent* addition, which will give a very different result.

First let us note that the scattered field from a single particle can be written quite generally as

$$\mathbf{E}_s(\mathbf{x}, t) = \left[\frac{r_e}{R}\, \boldsymbol{\Pi} \cdot \mathbf{E}_i \right], \tag{7.2.8}$$

where $\boldsymbol{\Pi}$ is a tensor polarization operator that, as we have seen, is equal to $\hat{\mathbf{s}} \wedge \hat{\mathbf{s}} \wedge \equiv \hat{\mathbf{s}}\hat{\mathbf{s}} - \mathbf{1}$ for the dipole approximation, but is more complicated in the fully relativistic case.

Now we are considering the far field where the distance R may be approximated as $x - \hat{\mathbf{s}} \cdot \mathbf{r}$, where $\mathbf{r}$ is the position of the particle relative to an origin in the scattering region and $\mathbf{x}$ is the position of observation of $\mathbf{E}_s$ as illustrated in Fig. 7.4. The difference between $\mathbf{R}/R$ and $\mathbf{x}/x$ is negligible; both are equal to $\hat{\mathbf{s}}$. The retarded time at which we must evaluate $\boldsymbol{\Pi} \cdot \mathbf{E}_i$ is then

$$t' = t - \frac{1}{c}(x - \hat{\mathbf{s}} \cdot \mathbf{r}'), \tag{7.2.9}$$

where primes again denote retarded quantities.

Fig. 7.4. Coordinate vectors for calculation of the wave phase.

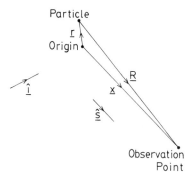

We shall be interested in the frequency spectrum of the field scattered during a finite time interval T (at $\mathbf{x}$), so we require the Fourier transform of $E_s(t)$:

$$\mathbf{E}_s(\omega_s) = \frac{1}{2\pi}\int_T \mathbf{E}_s(t)e^{i\omega_s t}\,dt = \frac{1}{2\pi}\int_T\left[\frac{r_e}{R}\,\mathbf{\Pi}\cdot\mathbf{E}_i\right]e^{i\omega_s t}\,dt. \quad (7.2.10)$$

Now we transform the integral to retarded time noting $dt = \kappa'\,dt'$ and ignore the difference between R and x except in the calculation of the retarded phase; we get

$$\mathbf{E}_s(\omega_s) = \frac{r_e}{2\pi x}\int_{T'}\mathbf{\Pi}'\cdot\mathbf{E}_i(r',t')e^{i\omega_s\{t'+(x-\hat{s}\cdot\mathbf{r}')/c\}}\kappa'\,dt'. \quad (7.2.11)$$

Defining $\mathbf{k}_s = \hat{s}\omega_s/c$, this is

$$\mathbf{E}_s(\omega_s) = \frac{r_e e^{i\mathbf{k}_s\cdot\mathbf{x}}}{2\pi x}\int_{T'}\kappa'\,\mathbf{\Pi}'\cdot\mathbf{E}_i e^{i(\omega_s t'-\mathbf{k}_s\cdot\mathbf{r}')}\,dt'. \quad (7.2.12)$$

If we take the input wave to be monochromatic,

$$\mathbf{E}_i(\mathbf{r},t) = \mathbf{E}_i\exp i(\mathbf{k}_i\cdot\mathbf{r}-\omega_i t), \quad (7.2.13)$$

then this becomes

$$\mathbf{E}_s(\omega_s) = \frac{r_e e^{i\mathbf{k}_s\cdot\mathbf{x}}}{2\pi x}\int_{T'}\kappa'\,\mathbf{\Pi}'\cdot\mathbf{E}_i e^{i(\omega t'-\mathbf{k}\cdot\mathbf{r}')}\,dt', \quad (7.2.14)$$

where $\omega = \omega_s - \omega_i$ and $\mathbf{k} = \mathbf{k}_s - \mathbf{k}_i$ are the so-called scattering frequency and k vector. In particular, the phase difference between scattered fields from two electrons at $\mathbf{r}_1$ and $\mathbf{r}_2$, respectively, is $\mathbf{k}\cdot(\mathbf{r}_1-\mathbf{r}_2)$.

Now, as we saw in Chapter 2, the Debye shielding effects of a plasma cause a test charge to be surrounded by shielding charges in a cloud of characteristic length λ_D. The combination of the charge and its shield is referred to as a dressed particle; consideration of the interaction of radiation with the combination of charge and shield enables us to see when particle correlations are important and when they are not, as follows.

A test electron has a shielding cloud (whose total charge is $+e$) consisting purely of electrons, or rather, the absence of them. This is because electrons move so much faster than ions (for comparable temperatures) that the ions "can't keep up" with the electron in order to contribute to its shielding. On the other hand, a thermal test ion is slow enough to allow the other ions to participate in the Debye shielding, so it will be surrounded by a cloud of roughly $-e/2$ total charge of electrons and $-e/2$ of (absence of) ions.

Now, if the phase difference between the scattering from an electron and from electrons in its shielding cloud is large, as will be the case if $k\lambda_D \gg 1$, then the random distribution of the electrons within the cloud will be sufficient to ensure that the scattered fields of electron and shielding cloud are incoherent. In this case, no correlation alterations to the power are necessary and the total scattered power is a simple sum of single-electron powers.

If, on the other hand $k\lambda_D \ll 1$, the contribution from test particle and cloud will add up coherently since there is negligible phase difference between them. For a test electron, this means that its scattering is almost exactly balanced by the (absence of) scattering from its electron shielding cloud. As a result the total scattering from test electrons is greatly reduced. For a test ion, though, there is scattering from the electrons in its shielding cloud, but direct ion scattering is negligible. Thus, in this case, the total scattered power from a uniform plasma is essentially just from the electrons shielding ions, which is approximately half that from uncorrelated electrons and is characteristic of the ion distribution function, since the scattering comes from the "$\frac{1}{2}$ electron" cloud shielding each ion and moving with it. This second situation is sometimes called ion Thomson scattering or perhaps more accurately collective scattering or coherent scattering.

We shall return to the problem of particle correlations and put these heuristic arguments on a more rigorous footing when we discuss collective scattering, but for now we proceed with a discussion of incoherent scattering.

7.2.3   *Incoherent Thomson scattering (B = 0)*

In the limit $k\lambda_D \gg 1$, when particle correlations can be ignored, the scattered power can be obtained as an incoherent sum of scattered powers from single electrons.

Consider a monochromatic input wave; take the wave amplitude $\mathbf{E}_i$ to be constant (in space) and also (for now) the particle velocity $\mathbf{v} = \dot{\mathbf{r}}$ to be effectively constant, because we assume no applied steady magnetic field. The time integration in Eq. (7.2.14) can then be performed to get

$$\mathbf{E}_s(\omega_s) = \frac{r_e e^{i\mathbf{k}_s \cdot \mathbf{x}}}{2\pi x} 2\pi\kappa \, \mathbf{\Pi} \cdot \mathbf{E}_i \, \delta(\mathbf{k} \cdot \mathbf{v} - \omega). \qquad (7.2.15)$$

Thus, the scattered field from this electron has a single frequency

$$\omega_s = \omega_i + \mathbf{k} \cdot \mathbf{v} = \omega_i + (\mathbf{k}_s - \mathbf{k}_i) \cdot \mathbf{v}, \qquad (7.2.16)$$

which may also be written $\omega_s = \omega_i(1 - \hat{\mathbf{i}} \cdot \boldsymbol{\beta})/(1 - \hat{\mathbf{s}} \cdot \boldsymbol{\beta}) \equiv \omega_d$. This is precisely the Doppler-shifted frequency of the input wave, the shift arising from a combination of the shift occuring at the electron due to its motion toward the source of the incident wave $(\mathbf{k}_i \cdot \mathbf{v})$ and the additional shift at the observation point due to the electron's motion toward it $(\mathbf{k}_s \cdot \mathbf{v})$.

The scattered power may be obtained easily in this monochromatic case by forming

$$E_s(t) = \int_{-\infty}^{\infty} E_s(\omega_s) e^{-i\omega_s t} \, d\omega_s = \frac{r_e e^{i(\mathbf{k}_s \cdot \mathbf{x} - \omega_d t)}}{x} \, \mathbf{\Pi} \cdot \mathbf{E}_i, \qquad (7.2.17)$$

using $|d(\mathbf{k} \cdot \mathbf{v} - \omega)/d\omega_s| = |1 - \hat{\mathbf{s}} \cdot \mathbf{v}/c| = \kappa$ so that $\kappa \delta(\mathbf{k} \cdot \mathbf{v} - \omega) = \delta(\omega_s - \omega_d)$. (Note that real part is implied.) Then the Poynting vector magnitude is

$$\langle S_s \rangle = \left\langle c\varepsilon_0 \big(\mathcal{R}e(E_s(t)) \big)^2 \right\rangle = \frac{1}{2} c\varepsilon_0 \left| \frac{r_e}{x} \, \mathbf{\Pi} \cdot \mathbf{E}_i \right|^2. \qquad (7.2.18)$$

Expressed as mean power per unit solid angle per unit frequency this is

$$\frac{d^2 P}{d\Omega_s \, d\omega_s} = r_e^2 |\mathbf{\Pi} \cdot \hat{\mathbf{e}}|^2 \langle S_i \rangle \, \delta(\omega_s - \omega_d), \qquad (7.2.19)$$

where $\langle S_i \rangle$ is the mean incident Poynting vector usually expressed as P_i/A, the total input power divided by the total input beam area.

Recall once more that this is energy per unit time-at-observer. If we want scattered energy per unit time-at-particle we must multiply this by the factor κ relating these two times. When we wish to calculate the total power from an assembly of electrons within a specified volume element $d^3\mathbf{x}$, the total power is equal to the number of particles in the element (with specified velocity) $f d^3\mathbf{v} d^3\mathbf{x}$ times the rate of scattering per unit time-at-particle. This is just the same issue as arose in Section 5.2.1 with cyclotron emission, although in Thomson scattering it has acquired (somewhat misleadingly) the specific name finite transit time effect.

Some early treatments erroneously omitted the additional κ factor until the point was thoroughly discussed by Pechacek and Trivelpiece (1967). A similar confusion also arose over the cyclotron emission formula and was resolved almost simultaneously (Scheuer 1968).

We are now able to write down the total incoherently scattered power spectrum from an assembly of electrons with distribution function f. It is

$$\frac{d^2 P}{d\Omega_s \, d\omega_s} = r_e^2 \int_V \langle S_i \rangle \int |\mathbf{\Pi} \cdot \hat{\mathbf{e}}|^2 \kappa f \kappa \, \delta(\mathbf{k} \cdot \mathbf{v} - \omega) \, d^3\mathbf{v} \, d^3\mathbf{r}, \qquad (7.2.20)$$

where V is the scattering volume from which the scattered radiation is detected, and we have again used $\delta(\omega_s - \omega_d) = \kappa \delta(\mathbf{k} \cdot \mathbf{v} - \omega)$. Recall here that, in general,

$$\mathbf{\Pi} \cdot \hat{\mathbf{e}} = \frac{(1 - \beta^2)^{1/2}}{\kappa^3}$$

$$\times \hat{\mathbf{s}} \wedge \left\{ [\hat{\mathbf{s}} - \boldsymbol{\beta}] \wedge \left[\hat{\mathbf{e}} - (\boldsymbol{\beta} \cdot \hat{\mathbf{e}})\boldsymbol{\beta} + (\boldsymbol{\beta} \cdot \hat{\mathbf{e}})\hat{\mathbf{i}} - (\boldsymbol{\beta} \cdot \hat{\mathbf{i}})\hat{\mathbf{e}} \right] \right\} \qquad (7.2.21)$$

from Eqs. (7.1.6), (7.1.7), and (7.2.8).

The power spectrum Eq. (7.2.20) is sometimes expressed as a mean differential scattering cross section per electron by dividing by $\int\langle S_i\rangle n_e\,d^3\mathbf{r}$ so that

$$\frac{\omega_s}{\omega_i}\frac{d^2\sigma_p}{d\Omega_s\,d\omega_s}=r_e^2\int|\mathbf{\Pi}\cdot\hat{\mathbf{e}}|^2\frac{f}{n_e}\kappa^2\delta(\mathbf{k}\cdot\mathbf{v}-\omega)\,d^3\mathbf{v}\qquad(7.2.22)$$

provided n_e is uniform in the scattering volume V. The extra factor ω_s/ω_i is required to compensate for the fact that the scattered photons have somewhat different photon energy from the incident. This σ is the cross section for *photon* scattering. Some authors use an expression without the factor ω_s/ω_i; their cross-section is then for *energy* scattering.

In the *nonrelativistic* dipole approximation these expressions become particularly simple because $\kappa^2|\mathbf{\Pi}\cdot\hat{\mathbf{e}}|^2$ reduces to $\hat{\mathbf{s}}\wedge(\hat{\mathbf{s}}\wedge\hat{\mathbf{e}})$ a quantity independent of $\mathbf{v}$. The velocity integral can then be performed trivially to obtain

$$\frac{d^2P}{d\Omega_s\,d\omega_s}=\left[r_e^2\int_V\langle S_i\rangle\,d^3\mathbf{r}|\hat{\mathbf{s}}\wedge(\hat{\mathbf{s}}\wedge\hat{\mathbf{e}})|^2\right]f_k\left(\frac{\omega}{k}\right)\frac{1}{k},\qquad(7.2.23)$$

where $f_k(v)$ is the one-dimensional velocity distribution in the $\mathbf{k}$ direction:

$$f_k(v_k)\equiv\int f(\mathbf{v}_\perp,v_k)\,d^2\mathbf{v}_\perp.\qquad(7.2.24)$$

($\perp$ denotes perpendicular to $\mathbf{k}$.) For a Maxwellian distribution this is

$$f_k=n_e\left(\frac{m_e}{2\pi T_e}\right)^{1/2}\exp\left(-\frac{m_ev_k^2}{2T_e}\right).\qquad(7.2.25)$$

The potential power of this result is clear. The frequency spectrum (for fixed scattering geometry) is directly proportional to the velocity distribution function, giving, in principle, complete information on the electron distribution in one dimension along $\mathbf{k}$.

A typical scattering spectrum is shown schematically for this case in Fig. 7.5. The spectral shape is proportional to the distribution function.

Unfortunately, it is rather rare for the results of a practical scattering experiment to be sufficiently accurate as to provide detailed information on the precise shape of the distribution function. Signal-to-noise limitations usually require that a specifically chosen curve shape (Maxwellian) be fitted to the spectrum obtained, thus in effect measuring T_e from the width and n_e from the height, again providing only moments of the distribution function.

For high-temperature plasmas it is essential to retain the *relativistic* treatment. The main effect of relativistic corrections arises from the relativ-

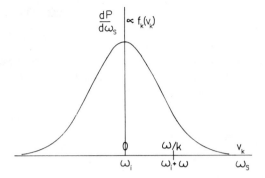

Fig. 7.5. The scattered spectrum in the dipole approximation is directly proportional to the one-dimensional velocity distribution with $v_k = (\omega_s - \omega_i)/k$.

istic aberration or "headlight effect," whereby a relativistic particle emitting radiation isotropically in its rest frame preferentially beams this radiation in the forward direction in the frame of a stationary observer. In the present context this means that we observe preferentially greater scattering intensity from electrons moving toward the observation point than from those moving away. Now the Doppler shift for scattering from electrons moving toward the observer is upward in frequency, that is, toward the blue (rather than red for the optical spectrum). Thus scattering from a relativistic plasma shows enhancement of the blue (high-frequency) side of the spectrum. This blue shift is important even for temperatures of only a few keV, especially when, as is often the case, the spectrum is measured only on one side (usually blue) of the incident frequency. An interpretation of the single-sided blue spectrum ignoring the relativistic blue shift would then tend to overestimate the width of the spectrum and hence the temperature.

These factors are automatically accounted for in our relativistic formulas. It is inconvenient to maintain the full generality of the $\mathbf{\Pi} \cdot \hat{\mathbf{e}}$ term, so we use the simpler form of Eq. (7.1.10) when $\mathbf{E}_i$ is perpendicular to $\hat{\mathbf{s}}$ and $\hat{\mathbf{i}}$ and we select only the $\hat{\mathbf{e}} \cdot \mathbf{E}_s$ component. The scattered power is then

$$\frac{d^2 P}{d\Omega_s \, d\omega_s} = r_e^2 \int_V \langle S_i \rangle \, d^3\mathbf{r} \int |\hat{\mathbf{e}} \cdot \mathbf{\Pi} \cdot \hat{\mathbf{e}}|^2 \kappa^2 f \, \delta(\mathbf{k} \cdot \mathbf{v} - \omega) \, d^3\mathbf{v}$$

$$= r_e^2 \int \langle S_i \rangle \, d^3\mathbf{r} \int \left| 1 - \frac{(1 - \hat{\mathbf{s}} \cdot \hat{\mathbf{i}})}{(1 - \beta_i)(1 - \beta_s)} \beta_e^2 \right|^2 \left| \frac{1 - \beta_i}{1 - \beta_s} \right|^2$$

$$\times (1 - \beta^2) f \, \delta(\mathbf{k} \cdot \mathbf{v} - \omega) \, d^3\mathbf{v}. \qquad (7.2.26)$$

The terms inside the velocity space integral have a simple interpretation. The first, which is always less than or equal to 1, is the extent of depolarization of the radiation due to relativistic effects. The second is simply ω_s^2/ω_i^2, the ratio of scattered to incident frequency squared. As such, it can be taken outside the integral, being independent of $\mathbf{v}_\perp$. The third term can be thought of as due to the relativistic mass increase of the electron that decreases its scattering efficiency (remember $r_e \propto 1/m_e$).

As before we can do the $\mathbf{v}$ integral along $\mathbf{k}$ to get

$$\frac{d^2P}{d\omega_s \, d\Omega_s} = r_e^2 \int_V \langle S_i \rangle \, d^3\mathbf{r} \left| \frac{1-\beta_i}{1-\beta_s} \right|^2$$

$$\int \left| 1 - \frac{(1-\hat{\mathbf{s}}\cdot\hat{\mathbf{i}})}{(1-\beta_i)(1-\beta_s)}\beta_e^2 \right|^2 (1-\beta^2)\frac{1}{k}f(\mathbf{v}_\perp, v_k) \, d^2\mathbf{v}_\perp,$$

$$(7.2.27)$$

where $v_k = \omega/k$ and $\boldsymbol{\beta} = (\mathbf{v}_\perp + \mathbf{v}_k)/c$. In this case, without further approximation, we do not obtain simply the one-dimensional distribution because of the extra $\beta_\perp$ dependent terms in the integral. To first order in β these terms are constant and equal to 1, so one easy approximation is to ignore β^2 terms; then

$$\frac{d^2P}{d\omega_s \, d\Omega_s} = r_e^2 \int_V \langle S_i \rangle \, d^3\mathbf{r} \frac{1}{k_i[2(1-\hat{\mathbf{i}}\cdot\hat{\mathbf{s}})]^{1/2}}\left(1 + \frac{3\omega}{2\omega_i}\right)f_k\left(\frac{\omega}{k}\right)$$

$$(7.2.28)$$

(see Exercise 7.3) and we obtain a simply weighted function of the one-dimensional distribution. (Note, we have accounted for the variation of both k and ω_s with ω in this expression.) This mildly relativistic form can be expected to provide reasonable accuracy for electron temperature up to at least 10 keV. The consistent approximation for the thermal distribution for f_k (i.e., ignoring β^2 and higher terms) is the nonrelativistic Maxwellian distribution Eq. (7.2.25). Thus, the temperature is conveniently found by fitting a Gaussian form to a linearly weighted $(1 - 3\omega/2\omega_i)$ multiple of the power spectrum.

For very high temperatures, higher order β terms must be included and the fully relativistic Maxwellian distribution used. The velocity integral of Eq (7.2.27) can then be performed analytically only by making the simplifying approximation of treating the depolarization term as a constant that is independent of velocity. With this approximation one finds that the scattering can be evaluated (Zhuravlev and Petrov 1979) for a relativistic Maxwellian distribution with temperature T, and may be expressed in terms of a differential photon cross section as

$$\frac{\omega_s}{r_e^2}\frac{d^2\sigma_p}{d\omega_s\,d\Omega_s} = q(T)\frac{\left\{2K_2\!\left(m_0c^2/T\right)\right\}^{-1}\omega_r^2}{\sqrt{\left\{1-2\omega_r\hat{\mathbf{i}}\cdot\hat{\mathbf{s}}+\omega_r^2\right\}}}$$

$$\times\exp\left[-\frac{m_0^2c}{T}\sqrt{\left\{1+\frac{\left(\omega_r^2-1\right)^2}{2\omega_r(1-\hat{\mathbf{i}}\cdot\hat{\mathbf{s}})}\right\}}\right], \qquad (7.2.29)$$

where $\omega_r \equiv \omega_s/\omega_i$ and $q(T)$ is the appropriate mean value of the depolarization factor $|1 - \beta_e^2(1 - \hat{\mathbf{i}}\cdot\hat{\mathbf{s}})/(1 - \beta_i)(1 - \beta_s)|^2$, slightly smaller than 1.

Although the modified Bessel function K_2 may be retained in this formula, a more convenient expression is obtained by using the asymptotic approximation

$$[2K_2(x)]^{-1} \approx \left[\frac{x}{2\pi}\right]^{1/2} e^x\left[1+\frac{15}{8x}\right], \qquad (7.2.30)$$

valid for $x = m_0c^2/T \gg 1$. The value of q may be estimated by substituting typical thermal values of β, $\beta_t \sim \sqrt{(T/m_0c^2)}$, into the depolarization factor, giving $q \approx (1 - T/m_0c^2)^2$ for $90°$ scattering. The resulting expression agrees with an exact numerical integration to within negligible error for practical purposes (Selden 1982). The shape of the theoretical spectra is illustrated in Fig. 7.6 for several different electron temperatures.

Fig. 7.6. Spectral shapes for relativistic Thomson scattering.

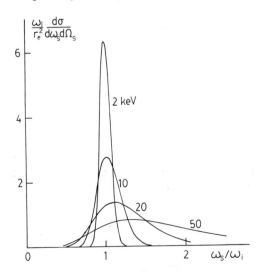

7.2.4 *Experimental considerations*

The order of magnitude of incoherent scattering is determined simply by the total Thomson cross section $\sigma = 8\pi r_e^2/3$, which is a fundamental constant $\sigma = 6.65 \times 10^{-29}$ m^2 ($r_e = 2.82 \times 10^{-15}$ m). Thus if a beam of radiation traverses a length L of plasma of density n_e, a fraction $\sigma n_e L$ of the incident photons will be incoherently scattered. In most laboratory plasmas this fraction is very small; for example, if $n_e = 10^{20}$ m^{-3} and $L = 1$ m, $\sigma n_e L = 6.65 \times 10^{-9}$ and less than 10^{-8} of the photons are scattered. Of these photons, an even much smaller fraction will be detected, since usually one collects scattered radiation only from a short section of the total beam length (perhaps ~ 1 cm) with collection optics that subtend only a rather small solid angle (perhaps 10^{-2} sr). The fraction of scattered photons collected is then $(10^{-2}$ m/1 m$) \times (10^{-2}$ sr/4π sr$) \sim 10^{-5}$. Thus, of the input photons, only perhaps 10^{-13} will be collected.

This fact is the source of most of the practical difficulties involved in performing an incoherent scattering experiment. The first requirement that it forces upon us is that we must have a very intense radiation source in order to provide a detectable signal level. That is why the measurements are almost always performed with energetic pulsed lasers. Actually, of course, the number of scattered photons observed is proportional to the total incident *energy* (regardless of pulse length) for a given frequency. However, the noise from which the signal must be discriminated will generally increase with pulse length. Hence, high incident *power* as well as high energy is required.

A schematic representation of a typical incoherent Thomson scattering configuration is shown in Fig. 7.7. The input laser beam is allowed to pass through the plasma, as far as possible avoiding all material obstructions.

Naturally, it must pass through a vacuum window at its entrance (and possibly exit) to the plasma chamber. At these points unwanted scattering of the laser beam occurs, even from the most perfect windows, whose intensity can far exceed the plasma scattering. This is the second important restriction we face arising from the small magnitude of the Thomson cross section: the need to avoid detection of this "parasitic radiation," usually called stray light. The baffles indicated in the figure are often used to reduce the stray light. Also, removing the windows and other optics far back into the ports is another way to reduce stray light, and the purpose of a viewing dump, when present, is primarily to reduce the effect of multiple scattering of the stray light from vacuum surfaces finally entering into the collection angle. In other words, the viewing dump provides a black background against which to view the scattered light.

Despite these types of precautions the stray light may often still exceed the Thomson scattered light. Fortunately, it can be discriminated against by virtue of the fact that the stray light appears precisely at the input

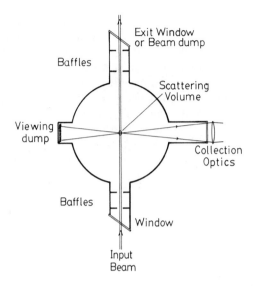

Fig. 7.7. Typical configuration for an incoherent Thomson scattering experiment.

frequency, whereas the Thomson scattered spectrum is broadened out from ω_i by the Doppler effect, which is our main interest. Thus, provided we avoid the frequency ω_i, we can avoid the stray light. Various filters and high rejection spectral techniques exist for this purpose.

Let us suppose, then, that by a combination of these techniques the stray light is eliminated. The other contributions to the noise come either from the detector/electronics used to observe the scattering or from radiation from the plasma. (Strictly, we should add a third possibility: thermal background radiation. This is usually negligible.) The detector noise depends upon the type of detector used; therefore, it is hard to generalize on its contribution. However, in many cases, for example when photomultipliers are used in the visible, it can be ignored.

Therefore, we consider only the plasma radiation as the noise source. Usually the only types of radiation we need to consider are line radiation and bremsstrahlung continuum. Again, a general treatment of line radiation is not feasible since it depends entirely upon the composition of the plasma. In very many cases line radiation is an important or even dominant contribution to the noise signal. However, we shall proceed with a calculation that ignores it. Two points can be made in justifying this approach. First, the bremsstrahlung, which we do include, constitutes the minimum possible plasma radiation that would occur in the absence of any line-emitting impurity atoms. Second, because the lines are narrow it is possible in principle to adopt the same strategy as for avoiding stray light, namely, to discriminate against them by avoiding them in the spectrum. Even when there are many lines in the spectrum this is not as overwhelming a task as it may seem. For example, a "mask" can be constructed to exclude the important lines from being transmitted through a grating spectrometer. Naturally, it is necessary first to perform a spectral survey to discover where these lines are. Because of the extra trouble involved and because the resolution of spectrometers used in scattering experiments is usually insufficient to allow a very fine-wavelength-scale mask to be used, such steps are rarely taken if signal to noise is satisfactory.

Let us therefore calculate the signal to noise expected considering only bremsstrahlung. We suppose the collection optics to subtend a solid angle Ω_s at the scattering volume and to collect light from a region of dimension L along the incident beam and d perpendicular to it. Figure 7.8 illustrates the geometry. We take d to be large enough to see the whole beam, but no larger since that would increase the noise but not the signal. Thus d is essentially also the beam size. The plasma dimension in the direction of the scattered light will generally be much larger than d; call it D. Note that the viewed incident beam length perpendicular to the viewing direction is $L \sin \theta$, where θ is the scattering angle.

The bremsstrahlung emissivity we obtain from Chapter 5, writing it

$$j(\omega_s) = C \frac{n_e^2 Z_{\text{eff}}}{T_e^{1/2}}, \tag{7.2.31}$$

where we take

$$C = \left(\frac{e^2}{4\pi\varepsilon_0}\right)^3 \frac{4}{3\sqrt{3}m^2c^3}\left(\frac{2m}{\pi}\right)^{1/2} \bar{g}_{ff} = 3.3 \times 10^{-64}\bar{g}_{ff} \quad \text{(SI units)} \tag{7.2.32}$$

Fig. 7.8. Geometry for background-light calculation.

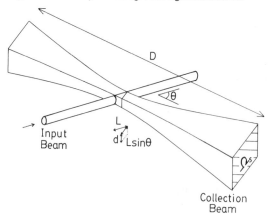

to be constant and to include only the free–free term, which will be a good approximation provided $\hbar\omega_s \ll R_y < T_e$. Remember $R_y = 13.6$ eV so this is satisfied for visible and longer wavelengths. The total number of bremsstrahlung photons per unit frequency collected in a time Δt will then be

$$N_b(\omega) = \Delta t j(\omega)(\Omega_s dL \sin\theta) D/\hbar\omega$$
$$= (\Omega_s dL \sin\theta) D \Delta t \left(Cn_e^2 Z_{\text{eff}}/T_e^{1/2}\hbar\omega_s\right). \qquad (7.2.33)$$

Note here that $\Omega_s dL \sin\theta$ is the étendue, which is therefore constant along the entire collection path. The emissivity j may not be constant along the path because of spatial variation of n_e and T_e. We account for this by taking D to be the *effective* path length of the plasma.

The total number of Thomson scattered photons collected in the nonrelativistic case, when polarization is chosen to make $\hat{s} \wedge (\hat{s} \wedge \hat{e}) = 1$ (E perpendicular to the scattering plane) is

$$N_s = N_i \frac{d\sigma}{d\Omega_s}\Omega_s n_e L = N_i r_e^2 \Omega_s n_e L, \qquad (7.2.34)$$

where N_i is the total number of incident photons. These are spread over a frequency band whose width is approximately

$$\Delta\omega_s \sim k v_t = v_t 2\frac{\omega_s}{c}\sin\frac{\theta}{2}. \qquad (7.2.35)$$

This is the frequency band over which we are also obliged to collect the bremsstrahlung photons, so over the total bandwidth the ratio of signal (scattered) photons to noise (bremsstrahlung) photons is

$$\frac{N_s}{N_b} = \frac{N_i r_e^2 \Omega_s n_e L}{v_t 2(\omega_s/c)\sin(\theta/2)(\Omega_s dL \sin\theta) D \Delta t \left(Cn_e^2 Z_{\text{eff}}/T_e^{1/2}\hbar\omega_s\right)}$$
$$= \frac{N_i \hbar\omega_i}{\Delta t} \frac{1}{2\sin(\theta/2)\sin\theta} \frac{r_e^2}{k_i Dd} \frac{m_e^{1/2}}{n_e Z_{\text{eff}} C}. \qquad (7.2.36)$$

Thus we see that this ratio is proportional to the incident power and inversely proportional to d, D, and n_e. It is independent of T_e and L.

In some cases this ratio is the required signal-to-noise ratio. For example, when Δt (taken equal to the laser pulse length) is of the same order of magnitude as the plasma duration, this is the case. Also, if the plasma has considerable variation (due to turbulence, for example) over the time Δt so that the bremsstrahlung is modulated by a fraction δ, say, the signal to noise is $N_s/\delta N_b$.

If the plasma is essentially constant during the laser pulse, then the true noise level is not the total bremsstrahlung power but the *fluctuations* in the bremsstrahlung. From a practical viewpoint, this may be understood by considering an experiment in which the background average bremsstrahlung emission, determined just before the scattering pulse, is subtracted from the total signal during the pulse. Only the fluctuations in the background remain uncompensated. The minimum level of these fluctuations is due simply to the photon statistics of the bremsstrahlung photons (plus, to be rigorous, the scattered photons). The fluctuation level in the photon number is then just $(N_b + N_s)^{1/2}$ (see Appendix 2). Actually we need to account for the fact that our system does not detect every photon, but has a certain quantum efficiency, Q say, of production of detected photoelectrons. That is, we only detect Q times the number of collected photons. Then the statistical fluctuation level in photoelectrons is $Q^{1/2}(N_b + N_s)^{1/2}$ and the signal-to-noise ratio is $Q^{1/2}N_s/(N_b + N_s)^{1/2}$. Usually the term N_s in the denominator can be ignored, in which case the signal to noise is

$$\frac{S}{N} = \left[\frac{hc}{C 2 \sin(\theta/2)\sin\theta} \right]^{1/2} r_e^2 N_i \left[\frac{\Omega_s L}{dD\Delta t} \right]^{1/2} \frac{T_e^{1/4}}{Z_{\text{eff}}^{1/2}} Q^{1/2}, \quad (7.2.37)$$

which is now independent of density but weakly temperature dependent.

A numerical example discussed by Sheffield (1975) is that of scattering of ruby laser ($\lambda = 694.3$ nm) light from a plasma with parameters $T_e = 500$ eV, $n_e = 2.5 \times 10^{19}$ m^{-3}, $L = 0.7$ cm, $\Omega_s = 2.3 \times 10^{-2}$ sr, $\theta = 90°$, and $Q = 0.025$. The number of scattered photons detected when the laser pulse energy is W joules is $QN_s = 2.7 \times 10^2 W$. For a Q-switched laser pulse of duration 25 ns, the bremsstrahlung estimate of background light suggests that less than one noise photon should be detected. On paper then, the signal to noise with a 6 J pulse looks very good. However, it was found in practice that the background plasma light was about 500 times more intense than expected. The source of this extra light was identified as primarily line radiation, against which no special precautions were taken. This experience serves as a cautionary tale against relying too heavily upon low background light calculations based upon bremsstrahlung alone.

A more modern example illustrating the information to be gained from incoherent scattering is depicted in Fig. 7.9, which shows the so called TV Thomson scattering (TVTS) system originally developed at Princeton University by Bretz et al. (1978). It gains its name from the intensifier tube in the detection system that uses principles similar to a television to receive a two-dimensional image from the detection spectrometer. The optics are arranged so that one axis of this image represents different wavelengths while the other receives scattered light from different spatial positions in the plasma. Thus, a large number of scattered spectra are measured simultaneously, providing good spatially resolved measurements of plasma temperature and density as illustrated in Fig. 7.10. The previous two examples are both based on scattering of ruby laser light.

Fig. 7.9. Schematic illustration of the configuration used for multiple point Thomson scattering (TVTS) [after Bretz et al. (1978)].

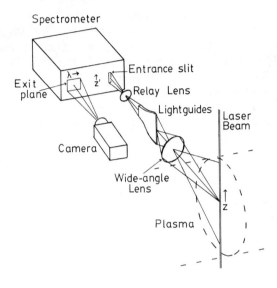

Fig. 7.10. Typical results from the TVTS system on the Princeton large torus.

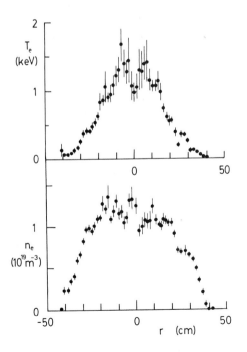

The ruby laser has dominated incoherent plasma scattering experiments since is availability first made them possible in the 1960s. It still offers a well proven way of meeting the requirements of incoherent scattering, primarily high power and energy with good beam quality (low divergence, etc.) at a wavelength (694.3 nm) where sensitive detectors are available. However, recent detector developments have made possible the use of neodymium lasers ($\lambda = 1.06$ μm) with avalanche photodiodes for scattering experiments. The major advantage that such lasers have, particularly when yttrium aluminum garnet (YAG) is used as the solid state laser medium, is the ability to fire repetitively at up to ~ 100 Hz. This can then enable the time evolution of electron temperature and density to be followed in long plasma pulses typical of modern fusion experiments, as has been demonstrated, for example, by Rohr et al. (1982). An incidental advantage also observed is that the plasma light due to impurity line radiation is often less troublesome near 1 μm because there are fewer lines there.

7.3   Coherent scattering
7.3.1   *The scattered field and power*

We must now consider the case in which there is significant correlation between the electrons over the scale length of the inverse scattering k vector, that is $k\lambda_D < 1$. We can no longer invoke the simplification that the total power is the sum of individual electron scattering powers. Instead we must perform a coherent sum of the electric fields from the various electrons. To do this we return to Eq. (7.2.12), which expresses the electric field due to scattering from each electron.

Now any scattering experiment receives scattered power only from a finite volume of plasma. Usually this is defined by the finite extent of the input beam and the finite acceptance angle of the collection optics, although sometimes finite plasma size is important too. Suppose that the position of the jth electron in the vicinity of this volume is $\mathbf{r}_j(t)$; then we can write sums over all electrons conveniently in the form of an integral over a distribution of delta functions

$$\sum_j = \int d^3\mathbf{r}\, d^3\mathbf{v}\, F_e, \tag{7.3.1}$$

where

$$F_e(\mathbf{r},\mathbf{v},t) = \sum_j \delta\big(\mathbf{r}-\mathbf{r}_j(t)\big)\, \delta\big(\mathbf{v}-\mathbf{v}_j(t)\big) \tag{7.3.2}$$

is the Klimontovich point distribution function of the electrons. The total scattered electric field from all the electrons is then

$$\mathbf{E}_s(\omega_s) = \frac{r_e e^{i\mathbf{k}_s \cdot \mathbf{x}}}{2\pi x} \int_{T'} \int_V F'_e \kappa' \mathbf{\Pi}' \cdot \mathbf{E}_i e^{i(\omega_s t' - \mathbf{k}_s \cdot \mathbf{r}')} \, dt' \, d^3 r' \, d^3 v'. \quad (7.3.3)$$

In the dipole approximation (but not in the relativistic case) $\kappa' \mathbf{\Pi}'$ is independent of $\mathbf{v}'$ and so the velocity integral can be done giving

$$\mathbf{E}_s(\omega_s) = \frac{r_e e^{i\mathbf{k}_s \cdot \mathbf{x}}}{2\pi x} \mathbf{\Pi} \cdot \int_{T'} \int_V N'_e \mathbf{E}_i e^{i(\omega_s t' - \mathbf{k}_s \cdot \mathbf{r}')} \, dt' \, d^3 r', \quad (7.3.4)$$

where $N_e \equiv \int F_e \, d^3 v = \sum \delta(\mathbf{r} - \mathbf{r}_j)$ is the (Klimontovich) density and primes remind us that retarded quantities are involved.

These equations (7.3.3) or (7.3.4) are formal solutions when the input wave field $\mathbf{E}_i$ is general. Normally one is interested in scattering of monochromatic incident radiation, so now let us consider, as before, the specific incident wave

$$\mathbf{E}_i(\mathbf{r}, t) = \mathbf{E}_i e^{i(\mathbf{k}_i \cdot \mathbf{r} - \omega_i t)}, \quad (7.3.5)$$

where we shall, for simplicity, take $\mathbf{E}_i$ to be of constant magnitude across the volume V, although the situation may easily be generalized to nonuniform illumination by adopting a weighted integral over space. Bear in mind, too, that we imply in this expression for $\mathbf{E}_i(\mathbf{r}, t)$ that the real part of the right hand side is to be taken.

In the dipole approximation we then have

$$\mathbf{E}_s(\omega_s) = \frac{r_e e^{i\mathbf{k}_s \cdot \mathbf{x}}}{2\pi x} \mathbf{\Pi} \cdot \mathbf{E}_i \int_{T'} \int_V N_e(\mathbf{r}', t') e^{-i(\mathbf{k} \cdot \mathbf{r}' - \omega t')} \, dt' \, d^3 r'. \quad (7.3.6)$$

This particularly simple expression indicates that the scattered field is proportional to the Fourier transform of N_e over finite time and volume. Specifically it is proportional to that component with k vector $\mathbf{k} = \mathbf{k}_s - \mathbf{k}_i$ and frequency $\omega = \omega_s - \omega_i$, the scattering $\mathbf{k}$ and ω. Remember, though, that N_e is the Klimontovich density, including all the "graininess" of the discrete particles, not the smoothed out density n_e that is the ensemble average of N_e.

Now we can obtain the average power during the time T per unit frequency per unit solid angle. The form of Parseval's theorem that we require for this is

$$\int_{-\infty}^{\infty} |\mathcal{R}e(E(t))|^2 \, dt = \pi \int_0^{\infty} |E(\omega) + E^*(-\omega)|^2 \, d\omega, \quad (7.3.7)$$

because we must take real parts before forming products and we cannot distinguish between positive and negative frequencies. Note that $E(\omega) \neq$

$E^*(-\omega)$ when $E(t)$ is not purely real (see Exercise 7.6). The resulting power spectrum is

$$\frac{d^2P}{d\omega_s \, d\Omega_s} = \frac{\pi\varepsilon_0 cx^2}{T} |E_s(\omega_s) + E_s^*(-\omega_s)|^2$$

$$= \frac{r_e^2 \varepsilon_0 c |\mathbf{\Pi}\cdot\mathbf{E}_i|^2}{4\pi T} (2\pi)^8 |N_e(\mathbf{k}_s - \mathbf{k}_i, \omega_s - \omega_i)$$

$$+ N_e^*(-\mathbf{k}_s - \mathbf{k}_i, -\omega_s - \omega_i)|^2,$$

(7.3.8)

where ω_s is positive and the Fourier transform, consistent with our convention, is defined as

$$N_e(\mathbf{k}, \omega) \equiv \frac{1}{(2\pi)^4} \int_{T'} \int_V N_e(\mathbf{r}', t') e^{i(\omega t' - \mathbf{k}\cdot\mathbf{r}')} \, dt' \, d^3\mathbf{r}'.$$

(7.3.9)

If the total incident power across the volume is P_i so that

$$A \tfrac{1}{2}\varepsilon_0 c |E_i|^2 = A\langle S_i \rangle = P_i,$$

(7.3.10)

where A is the incident beam area, that is, the area of V perpendicular to $\mathbf{k}_i$ (A is assumed constant), then the scattered power spectrum can be written

$$\frac{d^2P}{d\omega_s \, d\Omega_s} = \frac{r_e^2 P_i}{2\pi A} |\mathbf{\Pi}\cdot\hat{\mathbf{e}}|^2 n_e V S(\mathbf{k}, \omega),$$

(7.3.11)

where

$$S(\mathbf{k}, \omega) \equiv \frac{(2\pi)^8}{n_e TV} |N_e(\mathbf{k}, \omega) + N_e^*(\mathbf{k} - 2\mathbf{k}_s, \omega - 2\omega_s)|^2.$$

(7.3.12)

$S(\mathbf{k}, \omega)$ is called the scattering form factor.

Several comments should be made concerning this important expression. First, we can obtain an impression of how the two $N_e(\mathbf{k}, \omega)$ terms arise by considering the interference pattern formed between the incident and scattered waves. These are illustrated in Fig. 7.11, taking the waves as plane. The crossing wave fronts form patterns of peaks (lines, say) and troughs (spaces). Where troughs coincide there are enhanced troughs and likewise for peaks. As indicated, there are two diagonal directions that connect the positions of enhancement. These two directions correspond to the wave fronts of the components $N_e(\mathbf{k}, \omega)$ and $N_e(\mathbf{k} - 2\mathbf{k}_s, \omega - 2\omega_s)$ of density perturbation from which the wave can scatter.

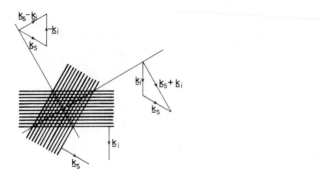

Fig. 7.11. "Beating" between the crossing wavefronts shows where the two terms in $S(\mathbf{k}, \omega)$ come from.

Second, one is usually concerned with radiation frequencies ω_i and ω_s that are much greater than the characteristic frequencies of the plasma (strictly speaking, this is essential for our free space treatment). The second Fourier component $N_e^*(-\mathbf{k}_s - \mathbf{k}_i, -\omega_i - \omega_s)$ is likewise a high frequency component. Moreover, its phase velocity is greater than c. Therefore, there cannot be a normal mode of the plasma with this ω and $\mathbf{k}$. The second term is thus usually negligible compared to the first, whose frequency $\omega_s - \omega_i$ can be much smaller than ω_s. The second term is normally omitted; we shall do so from now on.

Third, we note that the Fourier transform appearing is only over finite time and space. This can be significant; it is generally unnecessary and incorrect to proceed, as many treatments do, to the limit $T, V \rightarrow \infty$. Scattering is done with finite time duration and volume. The effect of taking finite Fourier transforms is to restrict the available $\mathbf{k}$ and ω resolutions to values of the order of $\Delta k \sim 1/V^{1/3}$ and $\Delta \omega \sim 1/T$. If the correlation length of density fluctuations is much smaller than $V^{1/3}$ then no difference is obtained between the finite and infinite spatial transforms. In this case proceeding to the $V \rightarrow \infty$ limit is allowable because in effect the k spectrum is broad and smooth enough that the finite Δk is negligible. In some cases though, for example, scattering from a coherent sound wave, the coherence length exceeds the extent of V. In such a case to proceed to $V \rightarrow \infty$ is incorrect because Δk is not negligible compared to the fluctuation k width (the width is narrow). The same argument applies to T and ω. If we wish to express the result in terms of infinite Fourier transforms we must, in general, note that the finite transform is the convolution of the transform of V or T with the infinite transform (see Appendix 1). This convolution then accounts correctly for the Δk and $\Delta \omega$ resolution.

Fourth, our calculation has considered a single realization of the scattering experiment. Normally T is taken long enough that the averaging process gives a mean power that is equal to the mean over all possible realizations. That is, we can replace $|N_e(\mathbf{k}, \omega)|^2$ by the ensemble average $\langle|N_e(\mathbf{k}, \omega)|^2\rangle$, which is what a statistical treatment of the plasma can calculate.

7.3.2   *Scattering form factor for a uniform plasma*

A particular case of interest for diagnostic purposes occurs when the plasma can be taken as uniform across the scattering volume. We mean by this that the *average* value of the electron density is constant and that all the contributions to the density fluctuation spectrum $N_e(\mathbf{k}, \omega)$ arise from the discreteness of the particles. Provided the number of particles in the shielding cloud of any test plasma particle is large, then particle correlations can be adequately described by regarding this "dressing" as given simply by the dielectric response of the rest of the plasma. In a uniform plasma the plasma response can be expressed in the linearized case by a dielectric constant $\varepsilon = 1 + \chi$, where χ is the susceptibility. We consider only an isotropic case $\mathbf{B} = 0$, although this can be generalized to include an applied magnetic field. The dielectric description can only generally be applied to the *Fourier transformed* Maxwell's equation, so really ε and χ are functions of $\mathbf{k}$ and ω.

Knowledge of $\varepsilon(\mathbf{k}, \omega)$ (or χ) is sufficient to calculate the exact form of the dressing on a test particle of arbitrary velocity: Recall that in a general dielectric medium

$$\mathbf{D} = \varepsilon\varepsilon_0\mathbf{E} = \varepsilon_0\mathbf{E} + \mathbf{P}, \tag{7.3.13}$$

where $\mathbf{P}$ is the electric polarization of the medium

$$\mathbf{P} = \frac{\chi\mathbf{D}}{\varepsilon} = \chi\varepsilon_0\mathbf{E}. \tag{7.3.14}$$

Now we shall actually need to distinguish, in the plasma response, between the polarization (and hence χ) contribution due to electrons and that due to ions, so we write

$$\mathbf{P} = \mathbf{P}_i + \mathbf{P}_e,$$
$$\chi = \chi_i + \chi_e. \tag{7.3.15}$$

The ions contribute negligibly to the scattering so that we shall be interested primarily in the electron part of the polarization $\mathbf{P}_e$ and the charge associated with it, namely,

$$\rho_e = -\nabla \cdot \mathbf{P}_e = -\frac{\chi_e}{\varepsilon}\nabla \cdot \mathbf{D}. \tag{7.3.16}$$

This equation constitutes an expression for the electron charge density dressing our test particle, which is the only particle to be considered in calculating **D**.

Suppose now that the test particle has charge q and velocity **v**, so that the charge density of the test particle is $q\delta(\mathbf{x} - \mathbf{v}t)$. The required Fourier transform of this charge density is then

$$\rho_0 = \frac{1}{(2\pi)^4} \int q\delta(\mathbf{x} - \mathbf{v}t)e^{-i(\mathbf{k}\cdot\mathbf{x} - \omega t)} \, d^3\mathbf{x} \, dt$$

$$= \frac{1}{(2\pi)^3} q\delta(\mathbf{k}\cdot\mathbf{v} - \omega) = \nabla \cdot \mathbf{D}. \tag{7.3.17}$$

Hence the electron cloud dressing this particle is

$$\rho_e(\mathbf{k}, \omega) = \frac{-\chi_e}{\varepsilon} \frac{1}{(2\pi)^3} q\delta(\mathbf{k}\cdot\mathbf{v} - \omega). \tag{7.3.18}$$

In order, now, to obtain the density fluctuation spectrum for the whole plasma particle assembly we consider each plasma particle in turn as being the test particle and add up the contribution to the total $N_e(\mathbf{k}, \omega)$ spectrum from each particle plus shielding cloud. For test electrons we must include the density of the particle itself (ρ_0) as well as the cloud (ρ_e), but not for test ions since they do not scatter significantly; only their shielding electrons are important. The result is

$$N_e(\mathbf{k}, \omega) = \frac{1}{q_e} \sum_{j \text{ electrons}} \left[1 - \frac{\chi_e}{1 + \chi_e + \chi_i} \right] \frac{q_e}{(2\pi)^3} \delta(\mathbf{k}\cdot\mathbf{v}_j - \omega)$$

$$+ \frac{1}{q_e} \sum_{l \text{ ions}} \left[\frac{-\chi_e}{1 + \chi_e + \chi_i} \right] \frac{q_l}{(2\pi)^3} \delta(\mathbf{k}\cdot\mathbf{v}_l - \omega), \tag{7.3.19}$$

where q_e is the electron charge $(-e)$ and q_l the ion charge.

The essence of the dressed particle approach is now to assume that the dressed particles, represented by the two sums in this equation, can be taken as uncorrelated, at least when an ensemble average is taken, all the particle correlations being accounted for by the dressing. This amounts to ignoring higher order correlations of more than two particles. This being so, when we form the quantity $\langle |N_e(\mathbf{k}, \omega)|^2 \rangle$, cross terms vanish and we must simply add up the sum of the individual particle terms squared. We have to note too that we are strictly dealing with finite Fourier transforms when it comes to a practical case of scattering volume V and time T. So the square of a delta function must be interpreted as $\delta^2 = T\delta/2\pi$ (see Appendix 1).

Then we get

$$S(\mathbf{k}, \omega) \equiv \frac{(2\pi)^8}{n_e TV} \langle |N_e(\mathbf{k}, \omega)|^2 \rangle$$

$$= \left\langle \frac{2\pi}{n_e V} \left[\sum_{j \text{ electrons}} \left| 1 - \frac{\chi_e}{1 + \chi_e + \chi_i} \right|^2 \delta(\mathbf{k} \cdot \mathbf{v}_j - \omega) \right. \right.$$

$$\left. \left. + \sum_{l \text{ ions}} \left| \frac{\chi_e}{1 + \chi_e + \chi_i} \right|^2 Z_l^2 \delta(\mathbf{k} \cdot \mathbf{v}_l - \omega) \right] \right\rangle, \quad (7.3.20)$$

where $Z_l \equiv |q_l/q_e|$ is the ion charge number. Now we note that, for example,

$$\left\langle \sum_{j \text{ electrons}} \delta(\mathbf{k} \cdot \mathbf{v}_j - \omega) \right\rangle = \left\langle \int_V d^3x \int d^3v \, F_e(\mathbf{v}) \, \delta(\mathbf{k} \cdot \mathbf{v} - \omega) \right\rangle$$

$$= V \frac{1}{k} \left\langle F_{ek}\left(\frac{\omega}{k}\right) \right\rangle = \frac{V}{k} f_{ek}\left(\frac{\omega}{k}\right), \quad (7.3.21)$$

where the subscript k denotes the one-dimensional distribution function along k. A similar relationship holds for ions. Thus we can write

$$S(\mathbf{k}, \omega) = \frac{2\pi}{k n_e} \left[\left| 1 - \frac{\chi_e}{1 + \chi_e + \chi_i} \right|^2 f_{ek}\left(\frac{\omega}{k}\right) \right.$$

$$\left. + \left| \frac{\chi_e}{1 + \chi_e + \chi_i} \right|^2 \sum_i Z_i^2 f_{ik}\left(\frac{\omega}{k}\right) \right], \quad (7.3.22)$$

where, for completeness, we allow the possibility of different ion species i. Note, too, that we have implicitly assumed that the finite width of $\delta(\mathbf{k} \cdot \mathbf{v} - \omega)$, which arises because of finite V, can be ignored.

All that remains is to obtain appropriate expressions for χ_e and χ_i. We should note that these are related to the electron and ion conductivities by $\chi = -\sigma/i\omega\varepsilon_0$ so that the problem is simply to calculate the conductivity of the plasma, due separately to electrons and ions. The method used in Chapter 4, based on the cold plasma approximation, is clearly not appropriate since we are interested in precisely those values of wave phase velocity ω/k at which $f_k(\omega/k)$ is significant. Thus we cannot ignore the thermal plasma particle velocities, but must treat the response from a kinetic theory viewpoint. To do this we start with the Vlasov equation

$$\frac{\partial f}{\partial t} + \mathbf{v} \cdot \frac{\partial f}{\partial \mathbf{x}} + \frac{q}{m} \mathbf{E} \cdot \frac{\partial f}{\partial \mathbf{v}} = 0, \quad (7.3.23)$$

which we linearize and then Fourier transform to get

$$-i\omega f' + i\mathbf{k}\cdot\mathbf{v}f' + \frac{q}{m}\mathbf{E}\cdot\frac{\partial f}{\partial \mathbf{v}} = 0, \tag{7.3.24}$$

where now f is the zeroth and f' the first order distribution function. Solving this equation for f' and then integrating $q\mathbf{v}f'$ over velocity we get the perturbed current density

$$\mathbf{j} = \frac{q^2}{im}\int\frac{\mathbf{v}(\mathbf{E}\cdot\partial f/\partial \mathbf{v})}{\omega - \mathbf{k}\cdot\mathbf{v}}\,d^3\mathbf{v}, \tag{7.3.25}$$

so that the conductivity tensor is

$$\boldsymbol{\sigma} = \frac{q^2}{im}\int\frac{\mathbf{v}\partial f/\partial \mathbf{v}}{\omega - \mathbf{k}\cdot\mathbf{v}}\,d^3\mathbf{v}. \tag{7.3.26}$$

For our assumed isotropic distribution this can be integrated over the components of $\mathbf{v}$ perpendicular to $\mathbf{k}$ to give

$$\chi = -\frac{\sigma}{i\omega\varepsilon_0} = \frac{q^2}{m\varepsilon_0 k}\int\frac{\partial f_k/\partial v}{\omega - kv}\,dv, \tag{7.3.27}$$

with f_k the one-dimensional distribution along $\mathbf{k}$. In these expressions the improper integrals must be taken along the appropriate Landau contour chosen to satisfy causality. For our sign convention this means a contour in the complex v plane *below* the pole at $v = \omega/k$. Details of this question are discussed in books on plasma kinetic theory; for example, Clemmow and Dougherty (1969) give an insightful account.

Thus the Vlasov treatment provides the required general form of the susceptibility of any species of the plasma. This can be used in our previously calculated form of $S(\mathbf{k}, \omega)$.

For Maxwellian particle distribution,

$$f_{jk}(v) = n_j\left(\frac{m_j}{2\pi T}\right)^{1/2}\exp\left(\frac{-m_j v^2}{2T_j}\right)$$

$$= n_j\left(\frac{1}{2\pi}\right)^{1/2}\frac{1}{v_{tj}}\exp\left(\frac{-v^2}{2v_{tj}^2}\right), \tag{7.3.28}$$

with thermal velocity $v_{tj} = (T_j/m_j)^{1/2}$, the susceptibility for the jth species becomes

$$\chi_j = \frac{\omega_{pj}^2}{k^2 v_{tj}^2}\frac{1}{\sqrt{(2\pi)}}\int\frac{(v/v_{tj})e^{-v^2/2v_{tj}^2}}{\omega/k - v}\,dv. \tag{7.3.29}$$

Recalling that $\lambda_D = v_{te}/\omega_{pe}$, this may be written

$$\chi_j = \frac{1}{(k\lambda_D)^2} \left(\frac{Z_j^2 n_j T_e}{n_e T_j} \right) w(\xi_j), \qquad (7.3.30)$$

where $\xi_j = \omega/kv_{tj}\sqrt{2}$ and

$$w(\xi) = \frac{1}{\sqrt{\pi}} \int \frac{\zeta e^{-\zeta^2} d\zeta}{\xi - \zeta}, \qquad (7.3.31)$$

is equal to minus half the derivative of the plasma dispersion function (Fried and Conte 1961). It can be evaluated for the correct contour to give

$$w(\xi) = 1 - 2\xi e^{-\xi^2} \int_0^\xi e^{\zeta^2} d\zeta + i\pi^{1/2}\xi e^{-\xi^2}. \qquad (7.3.32)$$

The form of $w(\xi)$ is shown in Fig. 7.12.

Rigorously we can now see that when $k\lambda_D \gg 1$, $\chi_j \ll 1$ so the dominant term in $S(k, \omega)$ is just the first term of the electron contribution, that is, the incoherent term, while if $k\lambda_D \lesssim 1$ then $\chi_e \gtrsim 1$ (at least for ξ values that are not very large) and the coherent terms are important. In fact, for $\chi_e \gg 1$ the total electron term will become negligible and, as pointed out before, we are left with the ion feature only.

An understanding of the principles underlying the collective form factor $S(k, \omega)$ may be gained by observing that the denominator in Eq. (7.3.22), $1 + \chi_e + \chi_i$, is simply the dielectric constant ε. Now the dispersion relation for longitudinal waves in the isotropic medium under consideration is

$$\varepsilon = 0. \qquad (7.3.33)$$

Thus, a peak will occur in the collective scattering spectrum [because of a peak in $S(k, \omega)$] for any ω, k corresponding to a longitudinal wave satisfy-

Fig. 7.12. Real and imaginary parts of the plasma dispersion function $w(x)$.

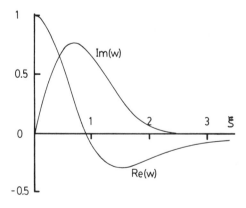

ing this dispersion relation. In general such waves are either electron plasma waves ($\omega \approx \omega_p$) or ion acoustic waves [$\omega/k \approx c_s = (T_e/m_i)^{1/2}$] and these are the points at which we expect special enhancement due to collective effects to occur.

The extent to which this peak is high and narrow in the spectrum depends upon the degree to which these waves suffer damping, primarily Landau damping. If the damping is weak the peak will be very pronounced. However, if the damping is appreciable there will be considerable broadening or perhaps even complete removal of the peak. The reason for this is that damping shows itself in the fact that, for real ω and k, ε has a significant imaginary part that then prevents $|\varepsilon|$ from being exactly zero even when its real part is zero.

Now Landau damping is due to the wave–particle resonance $\omega/k = v$, and occurs when there is an appreciable gradient in the distribution function $\partial f/\partial v$ at velocity ω/k. Recalling the $\lambda_D = v_{te}/\omega_{pe}$ it becomes clear that the condition for collective effects to be dominant ($k\lambda_D \ll 1$) is equivalent to

$$v_{te} \ll \omega_p/k, \tag{7.3.34}$$

that is, the condition for Landau damping of electron plasma waves to be small. In this case the only part of the electron feature that remains is a narrow peak at ω_p.

Ion acoustic waves are always strongly affected by ion Landau damping unless $T_e \gg T_i$. This disparity in temperatures is necessary if the wave phase speed is not to be of the same magnitude as the ion thermal speed. Therefore, the extent to which the ion acoustic resonance shows in the spectrum depends not just on scattering geometry $k\lambda_D$, but also on the plasma electron to ion temperature ratio.

The importance of the wave properties of the plasma in determining the form factor $S(\mathbf{k}, \omega)$ can be thought of as arising because the electrons and ions of the plasma are continuously emitting (and absorbing) waves into the plasma by a Čerenkov (Landau) process. These waves (or more properly fluctuations, since they may be so strongly damped as not to satisfy a normal dispersion relation) are then what a collective scattering process observes.

Returning to a consideration of $S(\mathbf{k}, \omega)$ in Eq. (7.3.22), because of the $f_{jk}(\omega/k)$ factors, the ion and electron terms have very different widths in ω (for a given k), corresponding to velocity ω/k, of the order of the ion or electron thermal speed, respectively. For equal electron and ion densities, in the region $\xi_i = \omega/kv_{ti}\sqrt{2} \sim 1$ this means that $f_i \gg f_e$ (because the integrals of f_j over all velocity space are equal). On the other hand in the region $\xi_e \sim 1$ (near the electron thermal speed) f_i is negligible. Figure 7.13 illustrates this point schematically.

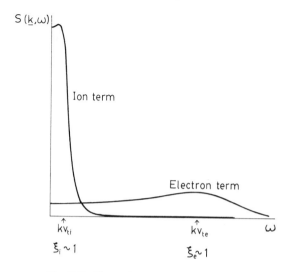

Fig. 7.13. Illustration (not to scale) of the ion and electron contributions to $S(\mathbf{k}, \omega)$ in the collective regime.

As a result of this separation of velocity scales it is possible to express $S(\mathbf{k}, \omega)$ in a convenient approximate form, originally due to Salpeter (1960). This approximation replaces χ_e in the ion term by its value $1/(k\lambda_D)^2$ at $\xi_e \to 0$, and sets χ_i in the electron term equal to zero since $\chi_i \to 0$ as $\xi_i \to \infty$. The resulting Salpeter approximation (see Exercise 7.7) is

$$S(\mathbf{k}, \omega) \approx \frac{(2\pi)^{1/2}}{v_{te}}\Gamma_\alpha(\xi_e) + \frac{(2\pi)^{1/2}}{v_{ti}}Z\left[\frac{1}{(k\lambda_D)^2 + 1}\right]^2\Gamma_\beta(\xi_i),$$

$$(7.3.35)$$

where

$$\alpha^2 \equiv \frac{1}{(k\lambda_D)^2}, \qquad \beta^2 \equiv Z\left[\frac{1}{(k\lambda_D)^2 + 1}\right]\frac{T_e}{T_i}, \qquad (7.3.36)$$

and

$$\Gamma_\beta(\xi) \equiv \frac{\exp(-\xi^2)}{\left|1 + \beta^2 w(\xi)\right|^2} \qquad (7.3.37)$$

is the same shape function for both features, though with different arguments. (Note that this approximation is valid only for a single species of ion.)

In Fig. 7.14 is plotted the shape function Γ_β for various values of β. It shows how the spectral shape of the ion feature varies as a function of the

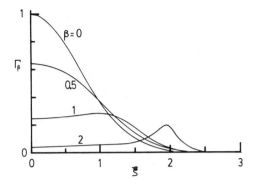

Fig. 7.14. The Salpeter shape function $\Gamma_\beta(x)$ for various values of β.

temperature ratio T_e/T_i, which determines β. For large T_e/T_i (and hence β) we see a sharp peak at the ion sound speed: the ion acoustic resonance. (Care should be exercised here, though, since the Salpeter approximation breaks down for very large T_e/T_i.) As T_e/T_i decreases, the ion acoustic resonance broadens and is absorbed by the thermal spectrum so that at $\beta = 1$, for example, only a vestigial hump remains.

Since $\alpha > 1$ in the collective scattering regimes, if electron and ion temperatures are comparable then $\beta^2 \sim 1$ (assuming $Z \sim 1$). Thus diagnosing the ion distribution using the ion feature will produce not a Gaussian spectrum (which requires $\beta \ll 1$) but one significantly distorted by the dielectric effects of the plasma, the vestigial ion acoustic hump.

7.3.3 *Problems of diagnostics using the ion feature*

There are numerous practical difficulties involved with detecting the ion feature. Many of these are common to incoherent scattering too but some arise specifically because of the requirements of coherent scattering. In particular the requirement $k\lambda_D \lesssim 1$ enforces longer scattering wavelength (smaller k-scattering). This can only be achieved by decreasing the scattering angle or increasing the incident radiation wavelength. The former approach has limits, since the scattered radiation must be separated from the input beam. As a result only rather dense plasmas (typically $n_e \gtrsim 10^{22}$ m^{-3}) are suitable for detection of the ion feature using visible radiation.

For more typical plasmas it is essential to use longer wavelength radiation such as that obtained from a CO_2 laser (10.6 μm) or specially developed far infrared lasers (λ approximately a few hundred micrometers). Particularly for the longer wavelength lasers whole new areas of technology development are required in order to perform the measurement. The cherished objective of a reliable direct measurement of the ion temperature and also of the plasma microstate has been sufficient to motivate extensive development of these techniques [see, e.g., Luhmann (1979)].

Despite the advances in technology, we can see from our theoretical expression that, even if clear measurements of the feature were achieved, it is far from obvious that an accurate unambiguous interpretation of the spectrum obtained would be straightforward.

If we have a pure plasma of (say) hydrogen then, provided the plasma is indeed thermal (i.e., there are no distortions of the ion spectrum due to instabilities, etc.), we might expect to be able to fit appropriate curves to the spectrum so as to deduce T_i. However, impurities (which will be present in any practical laboratory plasma) will tend to confuse the result because if they have similar temperatures their thermal speed will be lower by the square root of the mass ratio. Thus they will tend to enhance the central regions of the ion feature, and unless they are negligible (which requires $Z_{eff} - 1 \ll 1$) or else are carefully compensated for, they will distort the spectrum, giving an incorrect temperature estimate. (One way to try to circumvent this problem is to restrict attention to the wings of the scattering line so as to emphasize the contribution of the lightest ion species.) On the other hand, to look on the positive side, one might possibly be able to deduce detailed information about the impurities from the spectral shape. Figure 7.15 illustrates experimental collective scattering spectra from an investigation of impurity effects in a low-temperature arc plasma (Kasparek and Holtzhauer 1983).

7.3.4   *Scattering from macroscopic density fluctuations*

The formidable difficulties of coherent scattering from the density fluctuations that arise in an otherwise uniform plasma owing to the discreteness of the particles have prevented the routine use of coherent scattering for diagnosing the ions. However, most laboratory plasmas experience density fluctuations·caused by various types of instability within the plasma. These fluctuations generally have wavelengths exceeding the Debye length so we may call them macroscopic. Their frequencies may extend from very low frequency up to the characteristic frequencies of the plasma (ω_p, etc.). Considerable interest focuses upon the ability to diagnose these fluctuations because, in the case of "naturally" occurring fluctuations, they can be responsible for enhanced transport, while in the case of deliberately excited waves, such as those launched for heating purposes, internal detection allows the wave dynamics to be investigated directly.

Generally, the fluctuation levels encountered far exceed the thermal levels calculated in Section 7.3.2, so that by judicious choice of incident frequency the detection problems can be made considerably easier than they are for thermal scattering.

The equations governing this process are again simply those we have had before [Eq. (7.3.6)] except that now we need not consider the Klimontovich density but instead can ignore the discreteness of the particles and use the

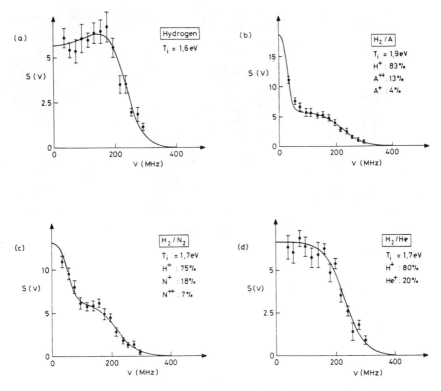

Fig. 7.15. Experimental spectra of composite plasmas. (*a*) Hydrogen alone, (*b*) hydrogen plus argon, (*c*) hydrogen plus nitrogen, and (*d*) hydrogen plus helium [after Kasparek and Holtzhauer (1982)].

smoothed density n_e:

$$\mathbf{E}_s(\omega_s) = \frac{r_e e^{i\mathbf{k}\cdot\mathbf{x}}}{2\pi x} \,\mathbf{\Pi}\cdot\mathbf{E}_i \int_T \int_V n_e e^{-i(\mathbf{k}\cdot\mathbf{r}-\omega t)} \, dt\, d^3\mathbf{r}. \qquad (7.3.38)$$

When, as is often the case, we wish to characterize a density fluctuation spectrum $n_e(\mathbf{k}, \omega)$ that is broad in $\mathbf{k}$ and ω, that is, a rather turbulent spectrum, the frequency spectrum may be obtained by appropriate frequency analysis of the scattered waves. The k spectrum, on the other hand, is most easily obtained by varying the scattering angle so that $k = 2k_i \sin \theta/2$ scans an appropriate domain. Sometimes it is convenient to observe scattering simultaneously at various different scattering angles so as to obtain reasonably complete k information simultaneously. Figure 7.16 shows an example of such a setup and Fig. 7.17 shows some typical k and ω spectra.

Although CO_2 laser radiation has been extensively used for the purposes of density fluctuation measurements, its wavelength (10.6 μm) tends sometimes to be rather smaller than desirable. As a result, very small scattering

Fig. 7.16. An example of a scattering system designed for simultaneous measurement of $S(\mathbf{k})$ at various k values [after Park et al. (1982)].

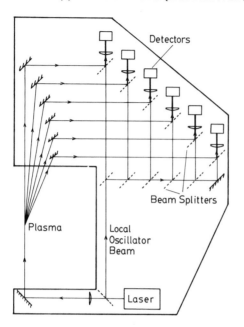

Fig. 7.17. Typical fluctuation spectra obtainable from a collective scattering experiment [after Semet et al. (1980)].

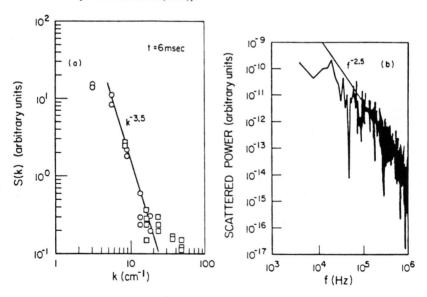

angles are required, which usually prevent one from obtaining spatial resolution along the incident beam. Figure 7.18 illustrates this point. In this respect longer wavelength lasers in the submillimeter spectral region prove more satisfactory, although their technology is less well developed. Microwave sources have also been used extensively. Their main drawback is that the frequency tends to be so low that the beam suffers from considerable refraction by the plasma. Also, diffraction limits the minimum beam size obtainable. From the theoretical point of view, the treatment we have outlined presupposes $\omega_i \gg \omega_p$, which may not be well satisfied for microwaves. Moreover fluctuation wave numbers greater than $2k_i$ are not obtainable so relevant parts of the k spectrum may not be accessible with low k_i microwaves.

More often than not heterodyne (or homodyne) detection techniques are used together with a continuous, rather than pulsed, source. When this is the case the resemblance between a scattering experiment and an interferometer becomes very obvious as illustrated in Fig. 7.19. In fact one can regard interferometry as functionally equivalent to zero angle scattering. This practical similarity reflects a much more fundamental equivalence between the refractive index of any medium and its electromagnetic wave scattering properties. Indeed, we could have approached the problem of scattering from density perturbations from the viewpoint of wave propagation in a medium with refractive-index perturbations, using the dielectric properties of the plasma calculated in Chapter 4. Identical results would

Fig. 7.18. Scattering with shorter wavelength radiation gives poorer spatial resolution along the beam than longer wavelength (for the same **k**).

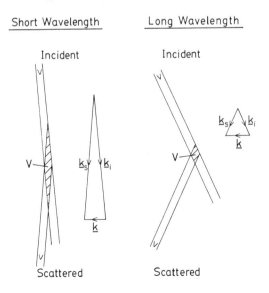

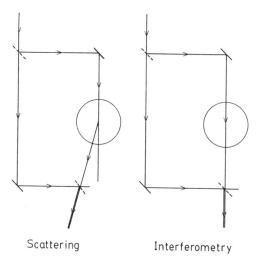

Scattering   Interferometry

Fig. 7.19. The functional similarity between a homodyne scattering experiment and a simple interferometer.

have been obtained for the same assumed plasma. In this respect note that the classical electron radius $r_e \equiv e^2/4\pi\varepsilon_0 mc^2$, which determines the scattering cross section, can be written

$$r_e = \omega_p^2/c^2 4\pi n_e. \qquad (7.3.39)$$

So $n_e r_e$ is proportional to ω_p^2, which is the quantity determining the refractive index $N^2 = 1 - \omega_p^2/\omega^2$ in the field-free case.

7.4   Scattering when a magnetic field is present

When the electrons from which scattering is occurring experience an applied constant magnetic field **B**, considerable additional complexities arise. The most important effect is that the unperturbed electron orbit is now helical instead of being a straight line. The periodicity of the electron motion in the plane perpendicular to the field gives rise to a cyclotron harmonic structure to the scattered spectrum. It turns out, as we shall see, that except in rather special cases this structure is smoothed out by the broadening effects in an integration over the distribution function. The spectrum obtained is then just what we deduced earlier ignoring the magnetic field. If appropriate (rather demanding) precautions are taken so as to observe the "magnetic modulation" (as the cyclotron harmonic structure of the scattered spectrum is often called) then this can in principle be used as a diagnostic of the magnetic field.

7.4.1 *Incoherent scattering from magnetized electrons*

We must return to the treatment of Section 7.1 and generalize the equations to include an applied field. The equation of motion is still solved as

$$\dot{\boldsymbol{\beta}} = \frac{-e}{m_0\gamma}\left\{\frac{\mathbf{E}}{c} - \frac{\boldsymbol{\beta}\cdot\mathbf{E}}{c}\boldsymbol{\beta} + \boldsymbol{\beta}\wedge\mathbf{B}\right\}, \tag{7.4.1}$$

but now **B** has a constant component (**B**$_0$, say) as well as the component arising from the incident wave. We know that the acceleration arising from **B**$_0$ gives rise to radiation. However, that is just the cyclotron radiation that we have already discussed in Chapter 5. It occurs, in any case, at frequencies typically much smaller than the scattering in which we are presently interested. Therefore, we wish to obtain only that part of the acceleration that is due to the incident wave. It is easy to show that this is given, as before, by

$$\dot{\boldsymbol{\beta}} = \frac{-e}{m_0 c\gamma}\left\{\mathbf{E}_i - (\boldsymbol{\beta}\cdot\mathbf{E}_i)\boldsymbol{\beta} + (\boldsymbol{\beta}\cdot\mathbf{E}_i)\hat{\mathbf{i}} - (\boldsymbol{\beta}\cdot\hat{\mathbf{i}})\mathbf{E}_i\right\}, \tag{7.4.2}$$

except that now the $\boldsymbol{\beta}$ appearing on the right hand side of this expression is that appropriate to the *helical* unperturbed orbit. The analysis thus proceeds just as before as far as Eq. (7.2.14):

$$\mathbf{E}_s(\omega_s) = E_i\frac{r_e e^{i\mathbf{k}_s\cdot\mathbf{x}}}{2\pi x}\int_{T'}\kappa'\boldsymbol{\Pi}\cdot\hat{\mathbf{e}}\,e^{i(\omega_s t' - \mathbf{k}_s\cdot\mathbf{r}')}\,dt', \tag{7.4.3}$$

but now we must substitute the quantities [Eq. (5.2.3)]

$$\boldsymbol{\beta} = \beta_\perp(\hat{\mathbf{x}}\cos\omega_c t + \hat{\mathbf{y}}\sin\omega_c t) + \beta_\parallel\hat{\mathbf{z}},$$

$$\frac{\mathbf{r}}{c} = \frac{\beta_\perp}{\omega_c}(\hat{\mathbf{x}}\sin\omega_c t - \hat{\mathbf{y}}\cos\omega_c t) + \beta_\parallel t\hat{\mathbf{z}} \tag{7.4.4}$$

for the helical orbit.

In the general case $\kappa'\boldsymbol{\Pi}'\cdot\hat{\mathbf{e}}$ is a function of $\boldsymbol{\beta}$ and hence is periodic with period $2\pi/\omega_c$ (note $\omega_c = \Omega/\gamma$). Substituting for $\mathbf{r}'$ in the exponential term gives

$$\exp i\left\{-(k_x\sin\omega_c t' - k_y\cos\omega_c t')\frac{\beta_\perp c}{\omega_c} + (\omega - c\mathbf{k}\cdot\boldsymbol{\beta}_\parallel)t'\right\}. \tag{7.4.5}$$

The first term here again gives a period function, while the second term we keep separate. Now the periodic part of the integrand can be expanded as a

Fourier sum,

$$\kappa \, \boldsymbol{\Pi} \cdot \hat{\mathbf{e}} \exp i \left\{ -\left(k_x \sin \omega_c t - k_y \cos \omega_c t \right) \frac{\beta_\perp c}{\omega_c} \right\} = \sum_{n=-\infty}^{\infty} \mathbf{a}_n e^{-in\omega_c t},$$

(7.4.6)

and the integral can then be performed to give

$$\mathbf{E}_s(\omega_s) = E_i \frac{r_e e^{i\mathbf{k}_s \cdot \mathbf{x}}}{x} \sum_{-\infty}^{\infty} \mathbf{a}_n \, \delta \left(\omega - \mathbf{k} \cdot \boldsymbol{\beta}_\| c - n\omega_c \right),$$

(7.4.7)

a sum over discrete cyclotron harmonics much the same as with cyclotron emission, except, of course, that ω here is the difference frequency $\omega_s - \omega_i$.

Writing out the argument of the delta function and setting it equal to zero, the frequency of the nth harmonic ω_{sn} is given by

$$\left(\omega_{sn} - \omega_i \right) - \left(\omega_{sn} \hat{\mathbf{s}} - \omega_i \hat{\mathbf{i}} \right) \cdot \boldsymbol{\beta}_\| - n\omega_c = 0,$$

(7.4.8)

so

$$\omega_{sn} = \frac{\omega_i \left(1 - \hat{\mathbf{i}} \cdot \boldsymbol{\beta}_\| \right) + n\Omega/\gamma}{\left(1 - \hat{\mathbf{s}} \cdot \boldsymbol{\beta}_\| \right)}.$$

(7.4.9)

To understand this result one can regard the electron as a composite entity that moves at the guiding-center velocity $\boldsymbol{\beta}_\|$, but simultaneously gyrates around the guiding center at frequency ω_c. During scattering the electron experiences the incident wave at the frequency $\omega_i(1 - \hat{\mathbf{i}} \cdot \boldsymbol{\beta}_\|)$, Doppler-shifted by the guiding-center motion. It upshifts the frequency by a harmonic of the (relativistically mass shifted) cyclotron frequency $n\omega_c$ and reradiates. The observer sees this frequency Doppler-shifted because of the electron's motion in the direction $\hat{\mathbf{s}}$, that is, divided by $(1 - \hat{\mathbf{s}} \cdot \boldsymbol{\beta}_\|)$.

In the dipole approximation the coefficients $\mathbf{a}_n$ can be calculated straightforwardly because $\kappa \, \boldsymbol{\Pi} \cdot \hat{\mathbf{e}}$ is then independent of β. The result is

$$\mathbf{a}_n = \hat{\mathbf{s}} \wedge (\hat{\mathbf{s}} \wedge \hat{\mathbf{e}}) J_n(k_i v_\perp / \omega_c)$$

(7.4.10)

(see Exercise 7.10). In the fully relativistic case the Fourier coefficients are much more cumbersome to evaluate, although this can be done (Nee et al. 1969). In either case, the most convenient approach is not to attempt to evaluate the appropriate integrals and sums directly, but rather to proceed by reference to the scattering from an unmagnetized plasma, as follows.

Suppose we consider the normal situation where ω_i and ω are much larger than ω_c. If we consider only the scattering that takes place during a time interval $T \ll 1/\omega_c$, then during this time interval an electron rotates by only a small angle about its guiding center. Thus the electron is traveling

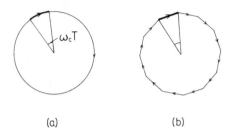

(a)   (b)

Fig. 7.20. The scattering from a very short arc of an electron orbit in a magnetic field is the same as that from a straight line orbit of an unmagnetized plasma when both are averaged over angles.

approximately in a straight line for this duration. The frequency resolution possible for such a short duration is given by $\Delta\omega \sim 1/T$ so that this thought experiment is equivalent to observing the spectrum with a broad resolution $\Delta\omega \gg \omega_c$, in which case all the harmonic structure is averaged out. The angle of the electron in its gyro orbit will be random, of course; therefore, the scattered spectrum from a magnetized electron, when smoothed to remove the harmonic structure, will be just that from an electron traveling in a straight line, but averaged over all orbit angles. Figure 7.20 illustrates the point.

Incoherent scattering from a full velocity distribution of electrons can be treated in just the same way, which demonstrates that smoothing out the harmonic structure will lead to a spectrum precisely that obtained by the unmagnetized plasma treatment (including, of course, the finite resolution $\Delta\omega$ effect). The requirement $\omega \gg \omega_c$ is necessary to allow the harmonic smoothing to be possible without simultaneously smoothing out all the frequency dependence of the unmagnetized spectrum, that is, so that $\Delta\omega$ can be chosen such that $\omega \gg \Delta\omega \gg \omega_c$.

7.4.2   *Presence of the harmonic structure*

Unless rather specific precautions are taken, the incoherent Thomson scattering from a plasma in which $\omega_i \gg \omega_c$ will show no difference from an unmagnetized plasma. In particular most experiments employ a resolution $\Delta\omega$, determined usually by instrumental limitations rather than finite pulse train T, which greatly exceeds ω_c. The question arises, however, if one had sufficiently fine frequency resolution, would the modulation be present? This depends very strongly on the scattering geometry.

When scattering is observed from a full distribution of electrons the frequency of the nth scattering harmonic ω_{sn}, given by Eq. (7.4.9), is different for different electron velocities. This causes the harmonic to have finite width $\Delta\omega_n$. If this width is significantly larger than the spacing of the harmonics, $\omega_{sn+1} - \omega_{sn}$, then the harmonic overlap will cause a smoothing of the spectrum so that the harmonic structure is washed out, regardless of the available spectral resolution $\Delta\omega$.

Just as with cyclotron radiation, there are two broadening mechanisms: Doppler effect and relativistic mass shift. The former is generally stronger since it is of order β, while the mass shift is of order β^2. Unlike the cyclotron radiation case, however, there are now two different frequencies ω_i and $n\Omega$ experiencing different shifts. Because ω_i is usually considerably larger than ω (and hence $n\Omega$), shifts in the ω_i term tend to be more dominant in the broadening, as illustrated by Fig. 7.21.

Let us consider first, then, the condition necessary for the harmonic structure to be visible, taking account only of the shift in the ω_i term. Modulation will occur if

$$\omega_{s0} - \omega_i \lesssim \Omega \qquad (7.4.11)$$

for most electrons in the distribution. Putting β_t for the typical thermal velocity v_t/c, this requires

$$\frac{(1 - \hat{\mathbf{i}} \cdot \hat{\mathbf{z}}\beta_t)}{(1 - \hat{\mathbf{s}} \cdot \hat{\mathbf{z}}\beta_t)} - 1 \lesssim \frac{\Omega}{\omega_i} \qquad (7.4.12)$$

and hence,

$$(\hat{\mathbf{s}} - \hat{\mathbf{i}}) \cdot \hat{\mathbf{z}}\beta_t \lesssim \Omega/\omega_i \qquad (7.4.13)$$

Fig. 7.21. The frequency shift due to the Doppler effect on the incident frequency (*a*) is usually larger than the shift of the cyclotron harmonics (*b*).

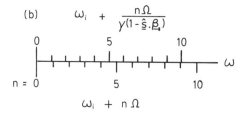

($\hat{z}$ is the direction of the magnetic field). This may be written as

$$k_{\parallel} v_t \lesssim \Omega. \tag{7.4.14}$$

In order to gain an impression of what this criterion involves, consider the specific case of scattering of visible radiation ($\omega_i \sim 3 \times 10^{15}$ s^{-1}) in a magnetic field of ~ 5 T ($\Omega \sim 10^{12}$ s^{-1}) so that $\Omega/\omega_i \sim 3 \times 10^{-4}$ and in a plasma of temperature ~ 1 keV so that $\beta_t \sim 5 \times 10^{-2}$, with angle $2\pi/3$ between $\hat{i}$ and $\hat{s}$ so that $k \approx k_i$. The condition we have derived then requires the angle ϕ between the scattering k vector and the magnetic field to satisfy

$$\left(\frac{k}{k_i}\right)\cos\phi = (\hat{s} - \hat{i}) \cdot \hat{z} \leq \frac{1}{\beta_t}\frac{\Omega}{\omega_i}. \tag{7.4.15}$$

Thus $|\phi - \pi/2| < 6 \times 10^{-3}$. We require **k** to be perpendicular to $\mathbf{B}_0$ to within $\sim 0.3°$.

Provided the scattering geometry satisfies this criterion, the Doppler width due to parallel electron motion will be sufficiently small as to allow the harmonic modulation to be present. Near the center of the scattering spectrum (i.e., where n is small) the magnetic modulation will appear provided the distribution is not highly relativistic. However, for the modulation not to be smoothed out at large n requires, in addition to this criterion, that the spread of the second, harmonic, term in Eq. (7.4.9) be small enough. The tendency of this term to cause harmonic overlap is greatest at greatest n and hence greatest ω. There will, therefore, always be some n for which overlap occurs. However, the bulk of the scattered spectrum will be modulated provided overlap is avoided out to $\omega \sim k v_t$, that is, for $n < k v_t/\Omega = (k/k_i)\beta_t\omega_i/\Omega$. Substituting, therefore, the typical thermal values, we find that overlap is avoided if

$$\Omega\left[\frac{k\beta_t\omega_i}{k_i\Omega}\right]\left[\frac{1}{\gamma_t(1 - \hat{s} \cdot \hat{z}\beta_t)} - 1\right] < \Omega. \tag{7.4.16}$$

The Doppler-broadening effect on the cyclotron harmonic frequency $(1 - \hat{s} \cdot \hat{z}\beta_t)^{-1}$ will be sufficient to cause overlap unless

$$(k/k_i)\hat{s} \cdot \hat{z}\beta_t^2(\omega_i/\Omega) < 1, \tag{7.4.17}$$

thus indicating that $\hat{s}$ (as well as $\hat{s} - \hat{i}$) must be nearly perpendicular to **B**. The angular constraint is less severe though: In our preceding example $\hat{s}$ must be within about 6° of perpendicular.

The broadening effect due to mass shift is unaffected by the scattering geometry and so represents the irreducible minimum harmonic broadening from a given distribution. It will be small enough to allow the modulation to appear (when the other criteria are met) only if

$$(k/k_i)\beta_t^3(\omega_i/\Omega) < 1, \qquad (7.4.18)$$

which may also be written

$$\frac{T}{mc^2} < \left(\frac{\Omega}{\omega_i}\frac{k_i}{k}\right)^{2/3}. \qquad (7.4.19)$$

In the case of visible radiation and at 5 T magnetic field this becomes $T < 2.5(k_i/k)^{2/3}$ keV, which is satisfied in our example but would not be for very much higher electron temperature.

A final effect that must be considered is the variation of B_0 within the scattering volume. In inhomogeneous magnetic fields the variation ΔB_0 across the scattering volume must satisfy

$$\frac{\Delta B_0}{B_0} < \frac{1}{n} \approx \frac{k_i}{k}\frac{\Omega}{\omega_i}\frac{1}{\beta_t}. \qquad (7.4.20)$$

This places an upper bound on the dimensions of the scattering volume in the direction of field inhomogeneity if modulation is to be observed.

7.4.3   *Magnetic field measurement*

In view of the discussion of the previous section it is clear that the mere observation of the modulation provides extremely precise information on the angle of the magnetic field, namely, it is perpendicular to the scattering k direction $(\hat{s} - \hat{i})$. This seems to be the most promising application.

Early proof of principle experiments (Carolan and Evans 1972) demonstrated the presence of modulation and, several individual peaks were able to be resolved. Figure 7.22 gives an example. These were in low-temperature high-density plasmas. For hot plasma applications when a large number of harmonics are present (~ 150 in our example), the scattered intensity in each harmonic is too low to be detectable, often less than one photon per harmonic. Therefore, a multiplexing technique is necessary. In an elegant experiment, based in part on an idea of Sheffield (1972), Forrest et al. (1978) used a Fabry–Perot interferometer whose free spectral range was equal to the cyclotron frequency (known because $|B|$ was accurately known), which then allowed all the harmonics to pass simultaneously, causing a brightening of the scattered image at the scattering angle that satisfied the criteria previously discussed. This thus gave the direction of the magnetic field from which the small poloidal component, of great importance in tokamaks, could be deduced.

The practical difficulties of such experiments can hardly be overstated. Perhaps this explains why the preceding experiment remains as yet the only

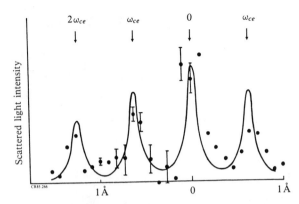

Fig. 7.22. Experimental observation of magnetic modulation [after Carolan and Evans (1972)].

example of the use of the magnetic modulation of Thomson scattering in fusion research.

7.4.4 Coherent scattering in a magnetic field

As we have seen, the effect of magnetization of the electrons is to introduce cyclotron modulation into the spectrum. When one is interested in the ion feature of coherent scattering, naturally the helical motion of the ions introduces an ion cyclotron harmonic structure into the spectrum, peaks spaced by $\Omega_i \equiv Z_i eB/m_i$. This modulation is generally in practice even harder to see than the electron cyclotron modulation. In addition the presence of the magnetic field greatly complicates the calculation of the plasma susceptibility χ. In fact, it becomes an anisotropic tensor. Thus, additional effects can occur due to the change in plasma dielectric properties caused by the magnetic field.

In view of the uncertain applicability of observations of the ion feature and the considerable mathematical complexities of the theoretical treatment, we shall not here venture into further discussion of the effects of a magnetic field on coherent scattering. The interested reader may refer to the specialist literature.

Further reading

Many of the details of the theory and practice of Thomson scattering are addressed in a monograph devoted to the subject:

Sheffield, J. (1975). *Plasma Scattering of Electromagnetic Radiation*. New York: Academic.

In addition to articles on Thomson scattering in the review compendia mentioned previously, the theoretical aspects are reviewed in:

Bernstein, I., Trehan, S. K. and Weenink, M. P. (1964). *Nucl. Fusion* 4:61.

Both theory and experiment are discussed in the excellent review:

Evans, D. E. and Katzenstein, J. (1969). *Rep. Prog. Phys.* 32:207.

Of the basic plasma physics texts, a substantial discussion of various aspects of scattering is included in:

Krall, N. A. and Trivelpiece, A. W. (1973). *Principles of Plasma Physics.* New York: McGraw-Hill.

Bekefi, G. (1966). *Radiation Processes in Plasmas.* New York: Wiley.

Exercises

7.1 Obtain Eq. (7.1.8) from the formulas in the text.

7.2 It is possible in principle to measure the electric current density $\mathbf{j}$ in a plasma by using incoherent scattering to determine the appropriate moment of the distribution function. Show how this might be done, and in particular show how to choose the scattering geometry in order to measure a specific geometrical component of the current density. If the order of magnitude of the current is $j \sim 0.1n_e ev_{te}$ and we may ignore noise arising from background plasma light, how many scattered photons would have to be detected in order to measure j to an accuracy of 10%?

7.3 Obtain Eq. (7.2.28).

7.4 Suppose we measure electron temperature T_e by observing only the short wavelength side of an incoherent scattering spectrum and do not account for relativistic effects but simply fit a straight line to a plot of $\log[dP/d\omega]$ versus ω^2, for $0 < \omega < 2kv_{te}$, and obtain T_e from its slope. By what approximate percentage shall we be in error if the true temperature is (a) 100 eV; (b) 1 keV; (c) 5 keV?

7.5 In a certain incoherent Thomson scattering experiment the stray light arises primarily from scattering from the vacuum windows. Suppose these are such that a fraction F of the light passing through them is scattered isotropically in all directions, the rest of the beam passing through unaffected. The windows are flush with a spherical vacuum vessel wall. They are directly visible, one from another, and are arranged so that the laser beam crosses a diameter before reaching the exit window. The scattering angle is 90° and the laser input beam polarization is perpendicular to the scattering plane. Show that the ratio of stray light to Thomson scattered light (in the whole scattering spectrum) is equal to at least

$$\frac{1}{(4\pi)^2\sqrt{2}}\frac{F^2 d}{r_e^2 R^2 n_e},$$

where d is the laser beam diameter in the scattering volume, R is the radius of the sphere, n_e is the plasma density, and r_e is the classical electron radius. Evaluate this ratio for $n_e = 10^{19}$ m^{-3}, $R = 0.2$ m, and $d = 10^{-3}$ m $F = 10^{-2}$, and comment on the viability of this scattering experiment.

7.6 Prove Eq. (7.3.7).

7.7 Obtain Eq. (7.3.35).

7.8 Show that the total scattered power P_s from a coherent density fluctuation $\tilde{n}_e \cos(kx - \omega t)$ propagating perpendicular to an incident laser beam of power P_i and k vector k_i such that $k_i \gg k$ is

$$P_s = P_i \tfrac{1}{4} r_e^2 \tilde{n}_e^2 \lambda_i^2 L^2,$$

where r_e is the classical electron radius, $\lambda_i = 2\pi/k_i$, and L is the length of the scattering region along k_i. How will this result change if $k \sim k_i$?

7.9 Coherent scattering is to be used to measure a broad turbulent density fluctuation spectrum. The k resolution required of the $\tilde{n}_e(k, \omega)$ measurement is fixed as Δk. The plasma size (and hence scale length) is D such that $\Delta k D \gg 1$ and the anticipated width of the $\tilde{n}_e(k)$ k spectrum is k_w ($\gg \Delta k$). Calculate the spatial resolution (along the beam) achievable when measuring fluctuations $\tilde{n}_e(k_w)$ as a function of the incident radiation k vector k_i.

7.10 Obtain Eq. (7.4.10).

8   Ion processes

Although they may involve the ions in the physical process taking place, the diagnostic principles we have discussed so far, with the exception of plasma particle flux, do not depend upon detection of the ions themselves. Instead, they rely upon detection of electromagnetic fields, photons, and so forth. In this chapter we turn to discussion of various diagnostics based on the direct use of heavy particles: ions, atoms, and neutrons. As we shall see, these provide valuable information on the ion parameters of the plasma and, less obviously, on certain other plasma properties such as electric potential.

8.1   Neutral particle analysis

Although most hot plasmas are almost completely ionized, there are, nevertheless, neutral atoms that are continually being formed within the plasma. Because these travel straight across any confining magnetic field, significant numbers can escape from the plasma without suffering a collision. These atoms then carry information out of the plasma about the state of the inner regions. They are called fast neutrals to distinguish them from the more numerous neutrals that tend to surround even a relatively hot plasma and that are edge particles, providing no information about the interior.

8.1.1   Collision processes

The proportion of fast neutrals that can reach the plasma edge (and hence be detected) without suffering a collision depends upon the collision cross sections for the various possible collisions. The most significant types of collisions are generally:

1. Ionizing collisions with electrons.
2. Ionizing collisions with ions.
3. Charge-exchange collisions.

The relative importance of these depends on the velocity of the neutral. For definiteness and because it is the most frequent case, let us consider a neutral hydrogen atom in a proton–electron plasma.

Electron-impact ionization has already been considered in Chapter 6. For a stationary ion the cross section (σ_e) is zero for electron energy less than the ionization potential R_y, rises rapidly to a value of the order of πa_0^2 as the electron energy rises above R_y, and then falls off slowly at much higher energy as illustrated in Fig. 8.1.

Because electrons move generally more rapidly than ions the motion of the atom will not affect this type of collision until the translational energy of the atom exceeds that of the electrons by a factor (m_i/m_e) (~ 1800 for H). One can thus usually approximate the ionization rate as independent of atom velocity. This leads to an effective cross section (per electron) for an atom moving through a distribution of electrons

$$\sigma_{\text{eff}} = \frac{\langle \sigma_e v_e \rangle}{v_a}, \tag{8.1.1}$$

where $\langle \sigma_e v_e \rangle$ is the electron ionization rate coefficient with v_e the electron velocity and v_a the atom velocity. (Note the change of notation from that in Chapter 6: σ_e instead of σ_i to avoid confusion with ionic processes.)

Ion-impact ionization requires the ejection of the atomic electron by the colliding ion. Consider, again, a stationary atom. If we think of the ion as simply a point charge moving past the atom at velocity v_i then it will have approximately the same ionizing effect as an electron moving past at the same velocity. Of course, when this imagined electron is at an energy near the threshold (R_y) it does not simply move in a straight line, so the effect will be different from that of our ion, but for reasonably high velocities the

Fig. 8.1. Cross sections for electron loss. Electron-impact ionization σ_e; ion-impact ionization σ_p; charge exchange σ_c for hydrogen atoms in an electron–proton plasma. The electron-impact ionization can be regarded as for a stationary electron and moving atom (bottom scale) or a stationary atom and moving electron (top scale).

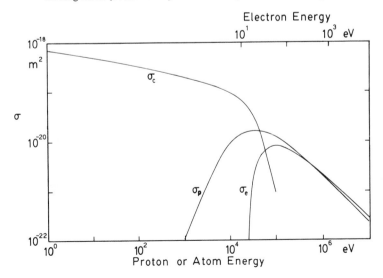

comparison is a good one. Thus the proton-impact ionization cross section is approximately equal to the electron-impact cross section except that the proton must have energy m_i/m_e higher. Figure 8.1 shows a fit to experimental results from a very useful summary of relevant cross section data by Freeman and Jones (1974). The scaling argument works quite well for proton energy above about 50 keV. Below that energy, the cross section falls, though not as abruptly as for the electron since there is still sufficient energy for ionization. (The experimental data at lower energies is rather sparse.)

Charge-exchange refers to collisions in which the captive electron is transferred from the atom to an ion. The two nuclei largely retain the energy they had prior to the collision, but the nucleus of the old atom is now trapped by the magnetic field whereas the new atom, which departs from the scene in a direction virtually random compared to the initial atom's direction, carries away its "memory" of life as an ion in the vicinity of the collision.

Charge-exchange collisions, such as the one we are considering, in which the nuclei involved are identical are termed symmetric. Because the atom formed is of the same species as the incident atom, the process need involve no change in the energy of the bound electron other than the translational energy change associated with the different atomic velocities. Since the ions move more slowly than electrons by the ratio m_e/m_i, this translation will be a small effect for collision energies less than about $R_y(m_i/m_e)$. The importance of the transfer being possible between atomic levels (usually the ground state) with the *same* energy is that the charge-exchange process is then *resonant*.

The resonance phenomenon can only be properly understood quantum mechanically. One may think of the wave functions of the electron attached to one or the other of the protons as being oscillators with frequency E/h. Because of the equality of energies, these two oscillators have the same frequency: They are in resonance. As a result, the oscillator of the initial atom, which initially has unity amplitude corresponding to 100% probability of the electron being attached to it, can gradually transfer oscillation amplitude to the second oscillator (which initially has none). This resonant transfer can take place even if the coupling is extremely weak, such as will be the case if the nuclei are considerably further apart than a_0, the Bohr radius. Calculation of the transfer probability can be performed quite accurately based on this coupling approach [see Exercise 8.1 and, e.g., Gurney and Magee (1957)] and shows that the transfer can occur with large probability in slow encounters even for distant collisions with impact parameter up to perhaps 5 to 10 times a_0. Moreover, the cross section increases logarithmically as the collision energy decreases because the nuclei are close for a longer time, allowing greater transfer. Thus, the resonant charge-exchange cross section can be as much as $50\pi a_0^2$, much larger than the ionization cross sections. Figure 8.1 shows the experimental form of the cross section. Note the dramatic drop in σ_c as the velocity exceeds the electron orbital velocity (at ~ 20 keV energy, $R_y m_i/m_e$) and the translational effect breaks the resonance.

For ion-impact ionization and charge exchange, the relevant parameter is the relative velocity of ion and atom. Therefore, if we are dealing with deuterium rather than hydrogen, the cross sections are essentially the same at similar velocities. The energy scale must then be interpreted as energy *per nucleon*.

8.1.2   *Neutral transport*

We now have in hand the important processes affecting the passage of a neutral out of the plasma. If the mean number of collisions per unit path length is α, then the probability of an atom surviving from a point A in the plasma interior to a point B at the edge, say, without suffering a collision is

$$P_{AB} = \exp\left[-\int_A^B \alpha(l)\,dl\right]$$ (8.1.2)

(just as for the problem of radiation transport; see Section 5.2.4).

In evaluating α, account must be taken, in general, of the velocity of the background plasma particles as well as the atom. The rate at which collisions occur between a species j and a stationary atom is given by the usual expression

$$n_j\langle \sigma v\rangle_0 = \int \sigma(v)\,vf_j(\mathbf{v})\,d^3\mathbf{v}.$$ (8.1.3)

Now if the atom is moving at velocity v_a the cross section is a function only of the relative velocity of atom and ions so (ignoring relativistic effects) the collision rate becomes

$$n_j\langle \sigma v\rangle_{v_a} = \int \sigma(|\mathbf{v} - \mathbf{v}_a|)|\mathbf{v} - \mathbf{v}_a|f(\mathbf{v})\,d^3\mathbf{v}$$ (8.1.4)

or, writing $\mathbf{u} = \mathbf{v} - \mathbf{v}_a$,

$$\langle \sigma v\rangle_{v_a} = (1/n_j)\int \sigma(u)\,uf_j(\mathbf{u} + \mathbf{v}_a)\,d^3\mathbf{u}.$$ (8.1.5)

Thus, the rate coefficient we require to obtain the collision rate for a specific process is that corresponding to a shifted distribution of particles. For example, if f is Maxwellian, we require the rate coefficient for a shifted

Maxwellian distribution, not simply the normal Maxwellian rate coefficient. This is illustrated in Fig. 8.2.

Two limits are particularly simple to deal with. First, if v_a is much smaller than the typical particle thermal velocity, it may be ignored and we recover the unshifted (Maxwellian) rate coefficient. Second, if v_a is much larger than the thermal speed, then the thermal motion may be ignored and the rate coefficient becomes simply $\sigma(v_a)v_a$.

The general form for the attenuation coefficient α due to the three processes we are considering is

$$\alpha = \frac{1}{v_a}\Big[\langle\sigma_e v_e\rangle_{v_a} n_e + \big(\langle\sigma_p v_i\rangle_{v_a} + \langle\sigma_c v_i\rangle_{v_a}\big)n_i\Big], \tag{8.1.6}$$

where σ_e, σ_p, and σ_c are, respectively, the electron ionization, proton (ion) ionization, and charge-exchange cross sections. The electron collisions are almost always in the first limit $v_a \ll v_{te}$. The ion processes may require the general shifted rate coefficient to be calculated, but in the case of most diagnostic interest v_a is significantly greater than v_{ti} and it is often

Fig. 8.2. The shifted distribution over which integration must be performed for shifted rate coefficients. (*a*) The shift is small so one obtains approximately the unshifted coefficient. (*b*) The shift is large so one obtains approximately $\sigma(v_a)v_a$.

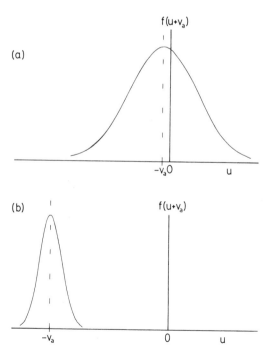

reasonable to approximate α by

$$\alpha \approx \frac{1}{v_a} \langle \sigma_e v_e \rangle_0 n_e + (\sigma_p + \sigma_c) n_i \qquad (8.1.7)$$

for fast neutrals.

Because of the $1/v_a$ dependence, the electron term is important only at low atom velocities. It is less than the charge-exchange term when $\sigma_c(v_a) v_a > \langle \sigma_e v_e \rangle$ (assuming quasineutrality $n_i = n_e$). The Maxwellian rate coefficient $\langle \sigma_e v_e \rangle$ has a broad maximum near $T_e = 100$ eV with peak value about 3×10^{-14} m^3 s^{-1}; this value is exceeded by the charge-exchange rate for atom energy greater than ~ 20 eV. Figure 8.3 shows the relevant rates for the three processes, but note that the graph is versus electron temperature not atom energy for the electron ionization process.

For most of the relevant atom energy range, therefore, the electron ionization is smaller than the other effects (although not by a very large factor) so that $\alpha \approx n_i (\sigma_p + \sigma_c)$ and the transmission becomes

$$P_{AB} \approx \exp \left[-(\sigma_p + \sigma_c) \int n_i \, dl \right]. \qquad (8.1.8)$$

Inspection of the cross section magnitudes plotted in Fig. 8.1 reveals the fact that in many cases of interest the absorption depth $\int \alpha \, dl$ is large and hence P_{AB} rather small. For example, at 10 keV energy and 10^{20} m^{-3} density the attenuation length is about 0.1 m, so that only shallow depths of this order into such a plasma can be diagnosed. In short, the applicability of fast neutral analysis is limited to plasmas for which $\int n_i \, dl \le 10^{19}$ m^{-2} or a few times larger.

Fig. 8.3. The electron loss rates appropriate when $v_{ti} \ll v_a \ll v_{te}$. The electron rate coefficient is plotted versus electron temperature.

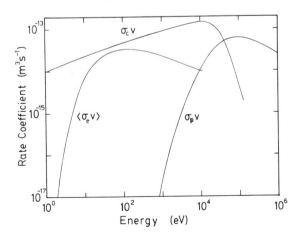

8.1.3   *The fast neutral spectrum*

In order to calculate the fast neutral spectrum to be observed we must consider the number and sources of neutrals in the plasma interior. Clearly, one source is the radiative recombination of electrons with the plasma ions. We have already obtained an estimate of this rate [Eq. (6.3.7)] based on our knowledge of radiative recombination. In the absence of other neutrals due to transport, the neutral density would be obtained by setting the recombination and ionization rates equal:

$$n_e n_i \langle \sigma_r v_e \rangle = n_e n_a \langle \sigma_e v_e \rangle. \tag{8.1.9}$$

We ignore proton collisional ionization for this estimate, and charge exchange, of course, does not create or remove additional neutrals; it involves only a "change of identity." If we substitute typical values for these rate coefficients in the expression for the neutral/ion density ratio $\langle \sigma_r v_e \rangle / \langle \sigma_e v_e \rangle$, we obtain extremely small numbers for virtually all electron temperatures significantly above R_y. For example, $n_a/n_i \sim 5 \times 10^{-8}$ at $T_e \sim 1$ keV. This is because radiative recombination is much slower than ionization (by a factor of the order of the fine structure constant α cubed; see Exercise 8.3).

Because this equilibrium neutral density due to recombination is so small, in most cases where the plasma attenuation depth is small enough to allow observable numbers of neutrals to escape out to the plasma edge, the transport of neutrals inward from the edge to the center is the most important neutral source there. Naturally, a neutral arriving at the center from the edge carries information about its existence as an ion at the edge. If it subsequently passes on and emerges without a collision at the opposite side of the plasma it will not provide the information we seek about the interior. If, on the other hand, it suffers a charge exchange with an ion in the interior, the neutral formed will be a fast neutral appropriate for diagnosis of interior ion parameters. Because charge exchange is usually the most important source of fast neutrals, this type of diagnostic is sometimes called charge-exchange analysis.

Consider then a typical diagnostic experiment. Figure 8.4 shows schematically the typical configuration. The neutrals emerging from the plasma along a specific collimated path are allowed to pass through a stripping cell (outside the magnetic field of the plasma confinement region) where, by

Fig. 8.4. Schematic configuration of a typical fast neutral analysis experiment.

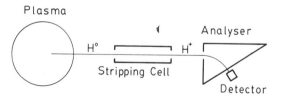

collisions with an appropriate stripping gas, they are ionized. The ions may then be analyzed, using either a magnetic or electric field to cause their orbits to depend on their energy, thus allowing specific energy to be selected. Often both electric and magnetic analysis are used, enabling the mass as well as the energy of the ions to be resolved.

A typical fast neutral energy spectrum is shown in Fig. 8.5. Qualitatively, we may understand its shape as follows. For $E \gg T_i$ the variation is dominated by the inverse exponential dependence of the number of ions (and hence fast neutrals) in a thermal distribution. At these high energies it is the high temperature parts of the plasma in the line of sight that tend to give the greatest contribution. The logarithmic slope is thus roughly characteristic of the highest temperature visible. At lower energies $E \lesssim T_i$ the spectrum rises above a straight line fitted to the higher-energy part. The reason is that the lower temperature parts of the plasma are contributing a significant number of particles to the spectrum so that the total flux at this lower energy is correspondingly greater.

To interpret quantitatively such a spectrum, suppose that the area and solid angle viewed are A and Ω_s so that the étendue is $A\Omega_s$. Then the particle flux to the detector per second in the energy interval dE is

$$F(E)\,dE = v^2\,dv\,A\Omega_s \int_{-a}^{a} \exp\left[-\int_x^a \alpha(l)\,dl\right] S(x,\mathbf{v})\,dx. \qquad (8.1.10)$$

Here the plasma is supposed to extend from $-a$ to a and $S(x,\mathbf{v})\,d^3\mathbf{v}$ is the total birth rate of neutrals of velocity $\mathbf{v}$ in the velocity element $d^3\mathbf{v}$ at a position x. That is, S includes, in general, all processes by which ions are converted into neutrals. We normally assume that charge exchange dominates S (although recombination could be included if significant). So

Fig. 8.5. Typical fast neutral spectrum from a tokamak plasma (courtesy C. Fiore).

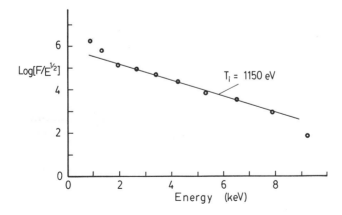

let us write S as

$$S(\mathbf{v}_i) = f_i(\mathbf{v}_i) \int \sigma_c(|\mathbf{v}_a - \mathbf{v}_i|) |\mathbf{v}_a - \mathbf{v}_i| f_a(\mathbf{v}_a) \, d^3 \mathbf{v}_a$$

$$= f_i(\mathbf{v}_i) \langle \sigma_c v_a \rangle_{v_i} n_a, \tag{8.1.11}$$

where $\langle \sigma_c v_a \rangle_{v_i}$ is the rate coefficient over the neutral atom distribution (v_a) shifted by the ion velocity $\mathbf{v}_i$. Naturally $v_i = (2E/m)^{1/2}$ in the direction along the viewing chord toward the detector: The subscript i refers to the ion that becomes the neutral detected; a refers to the atom from which it gains the electron.

The velocity distribution of neutrals $f_a(\mathbf{v}_a)$ is usually difficult to calculate since it involves a random walk of a rather small number of steps whereby the bound electron, coming in from the edge is charge-exchanged, possibly several times, before arriving at the position x. Some treatments calculate f_a by computer modeling with assumed plasma profiles using a Monte Carlo technique [e.g., Hughes and Post (1978)] or, for plasmas of greater attenuation depth, a diffusion approach. However, let us obtain a useful approximation by considering escaping neutrals that started life as ions with energy much larger than the typical neutrals inside the plasma. In other words, we may take the atom and ion velocities prior to the charge exchange to be such that $|\mathbf{v}_a - \mathbf{v}_i| \approx v_i$. Then the neutral velocity distribution is unimportant and

$$S(\mathbf{v}_i) \approx f_i(\mathbf{v}_i) \sigma_c v_i n_a, \tag{8.1.12}$$

where $\sigma_c v_i$ is to be taken at energy E. Hence,

$$F(E) \, dE \approx v_i^2 \, dv_i \, A\Omega_s \big[\sigma_c(E) v_i(E) \big]$$

$$\times \int_{-a}^{a} \exp\left[-\int_x^a \alpha(l, E) \, dl \right] n_a(x) f_i \, dx. \tag{8.1.13}$$

This equation is naturally insufficient to determine $f_i(v_i, x)$ without many further additional assumptions (or knowledge) about $n_i(x)$, $n_a(x)$, and so on.

Suppose, first, that we assume f_i to be thermal, that is,

$$f_i(\mathbf{v}_i, x) = n_i(x) \left(\frac{m_i}{2\pi T(x)} \right)^{3/2} \exp\left(-\frac{E}{T(x)} \right), \tag{8.1.14}$$

and ask: At what position does the maximum contribution to the integral of Eq. (8.1.13) occur? To discover this we must find the maximum of the integrand. Differentiating the integrand with respect to x and setting the result equal to zero we find that the maximum occurs when

$$\alpha + \frac{n_a'}{n_a} + \frac{n_i'}{n_i} - \frac{3T'}{2T} + \frac{ET'}{T^2} = 0, \tag{8.1.15}$$

where primes denote x differentiation. The first term here comes from the attenuation factor and the last from the thermal exponential $\exp(-E/T)$.

If now we choose to observe at very high energy E ($\gg T$), then the final term will dominate all the others and the condition will become $T' = 0$. That is, the maximum contribution to the integral will arise from the position where T is maximum, as noted qualitatively before. Moreover, the width of this maximum in the integrand may be estimated by noting that since the exponential term is dominant, the integrand will have decreased by e^{-1} when E/T has increased by 1, that is, when

$$\frac{E}{T} - \frac{E}{T_{\max}} = 1, \tag{8.1.16}$$

which leads to

$$T \approx T_{\max}(1 - T_{\max}/E) \tag{8.1.17}$$

($T_{\max}/E \ll 1$ by supposition). Thus, virtually all the contribution to the integral occurs from points where T is within a small fraction $T_{\max}/E$ of its maximum value $T_{\max}$.

This fact allows us to obtain a simple approximate relationship between the slope of the flux spectrum and this maximum temperature; for, if we differentiate the expression for the flux [Eq. (8.1.13)] with respect to E (noting $v_i \propto E^{1/2}$) we find (Exercise 8.4) that

$$\frac{d}{dE}\ln\left|\frac{F}{\sigma_c E}\right| \approx -\frac{1}{T_{\max}} - \left[\int_{x_m}^{a}\alpha\,dl\right]\frac{d}{dE}\ln|\alpha|. \tag{8.1.18}$$

This expression depends on the approximation that n_i, n_a, T, and x can be taken as equal to their values at the position x_m where the integrand is maximum (i.e., the hottest part of the plasma). Recalling that $\alpha \approx n_i(\sigma_p + \sigma_c)$ is a rather weak function of E in the energy region below about 20 keV, if the attenuation depth is small we may ignore the last term. Then the temperature may be obtained as the reciprocal of the slope of a plot of $\ln|F/(\sigma_c E)|$ versus E. Often, additional approximations to σ_c are made, such as taking $\sigma_c v$ ($\propto \sigma_c \sqrt{E}$) to be constant. The appropriate combination to be plotted is then $\ln|F/\sqrt{E}|$. The fractional error in temperature deduced, caused by this approximation, will be at most of the order of $T_{\max}/2E$, for $E \leq 20$ keV, and hence small. This approximation is adopted in Fig. 8.5. Thus, for cases where absorption is modest, a rather simple analysis yields reasonably accurate estimates of the peak ion temperature in the line of sight.

8.1.4 *Dense plasma cases*

Now let us consider what happens when the absorption depth is greater than 1. Returning to the question of where the maximum contribution to the flux at energy E comes from, if the absorption depth is significant then the first two terms in Eq. (8.1.15) may not be negligible. Generally, these two terms will be of the same order of magnitude because they arise from the attenuation of neutrals on their way out of the plasma (first term) and on their way in (second term). Their effect will be to shift the position of maximum contribution away from the point $T = T_{\max}$ (where $T' = 0$) to the point where

$$\frac{ET'}{T^2} + \alpha + \frac{n'_a}{n_a} = 0. \tag{8.1.19}$$

Expanding the temperature profile about its maximum as the first two terms of a Taylor series $T \approx T_{\max}(1 - x^2/b^2)$ and denoting

$$\alpha + n'_a/n_a \equiv 1/\lambda \tag{8.1.20}$$

so that λ will be of the order of (half) the atom mean free path, one may readily show (Exercise 8.5) that the maximum contribution to the flux occurs from a point where

$$T \approx T_{\max}\left[1 - \left(\frac{T_{\max}}{E}\frac{b}{2\lambda}\right)^2\right]. \tag{8.1.21}$$

Therefore, determining the temperature from the logarithmic slope as before will reflect not the maximum temperature but a somewhat smaller temperature, as given by this equation, corresponding to a point closer to the observation position.

The second term on the right hand side of Eq. (8.1.18) may also be significant and needs to be accounted for. Fortunately, its effect is to give greater flux at higher energy because the cross section for attenuation is decreasing with energy; therefore, it increases the apparent temperature deduced simply from the logarithmic slope. This fractional increase is approximately (Exercise 8.6)

$$\left[E\frac{d}{dE}\ln|\sigma_p + \sigma|_c\right]\left[(\sigma_p + \sigma_c)\int_{x_m}^{a} n_i\, dl\right]\frac{T}{E} \tag{8.1.22}$$

and in many cases substantially cancels the correction due to the position of maximum contribution, so that the simple analysis, ignoring attenuation depth, fortuitously gives quite accurate results.

In sum, for thermal plasmas, by looking at energies considerably greater than the thermal energy one can deduce the maximum temperature along the line of sight from the logarithmic slope of the flux spectrum in cases where the attenuation depth is small. When the depth is considerable, corrections to the simple analysis should be applied either in the fashion we have discussed or using more elaborate specific numerical modeling of the various effects. Fortunately, these corrections tend to cancel so that the simple analysis is often still effective.

8.1.5 *Nonthermal plasmas*

Since the preceding approach is based on observations of energies rather greater than the thermal energy, the results will be very sensitive to any deviations of the tail of the distribution function from Maxwellian. From the viewpoint of determining bulk temperature this is a serious handicap. However, in situations where the high-energy nonthermal part of the distribution is of direct interest, fast neutral measurements allow considerable information to be gained.

As an example, in plasmas with additional heating by energetic neutral beams, the capture and slowing down of the beam particles can be diagnosed. This is done by observing the fast neutrals arising from nuclei that enter the plasma in the energetic neutral beam, are ionized (usually by charge exchange with thermal ions) and hence trapped by the magnetic field, and then undergo charge exchange a second time to exit the plasma. Figure 8.6 shows the fast neutral spectrum during neutral-beam injection heating of the TFR tokamak. The region below about 33 keV shows a flat spectrum representing the classical slowing down of the fast ions in the plasma. At half the nominal injection energy, there is a jump corresponding to the substantial proportion of the heating beams that have half the accelerating energy (because they are accelerated as D_2^+ ions). These too

Fig. 8.6. Fast neutral spectrum from a nonthermal neutral-beam heated plasma showing the slowing down spectrum of the ions [after Equipe TFR (1977)].

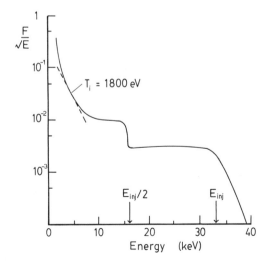

have a flat slowing down spectrum until the bulk thermal part is reached. Obviously, in such a case there is only a rather narrow energy range in which the thermal temperature is reflected by the slope. Thus, the ion "temperature" measurement is somewhat less certain.

Another case where nonthermal ion diagnosis is important is when radio frequency heating of ions is used. This also can often create ion tails.

In either case it is sometimes possible to obtain a more reliable bulk temperature measurement by observing a mass species not affected by the nonthermal effects. For example, deuteron neutral injection into hydrogen plasmas allows the bulk temperature to be deduced from hydrogen charge exchange. Alternatively, ion cyclotron heating of hydrogen minority in deuterium majority plasmas can be diagnosed to give bulk temperature from deuterium charge exchange [see, e.g., Hosea et al. (1980)].

8.1.6   *Neutral density measurement*

So far we have discussed deducing the ion temperature from what is essentially the relative shape of the fast neutral spectrum. However, measurements of the absolute magnitude of the fast neutral flux provide additional information. Referring to Eq. (8.1.13), we see that the flux through a specified étendue ($A\Omega_s$) at energy E is proportional to a weighted integral along the line of sight of the product of neutral atom density and ion density at energy E; that is, $n_a f_i$. Usually the ion total density n_i, is rather well known via measurements of n_e and the principle of quasineutrality. If, in addition, the ion temperature (profile) is known, then from this knowledge the factor f_i can be calculated. What a measurement of the absolute fast neutral flux then, in principle, tells us is the value of n_a.

In a uniform plasma the deduction of n_a from the flux measurement is more or less trivial. However, practical plasma measurements require spatial variation to be accounted for. The difficulties arise in trying to deconvolve the spatial profiles of n_a from flux measurements that give an integral over the line of sight. In some cases these difficulties can be overcome and reasonably reliable n_a measurements obtained.

The case when the plasma is thin (i.e., $\int \alpha \, dl \ll 1$) is easiest to deal with because then the flux is proportional to

$$P(E) = \int n_a(x) f_i(E, x) \, dx, \qquad (8.1.23)$$

where we ignore for the moment the space-independent factors v, σ_c, and so forth. If the ions are Maxwellian with density and temperature $n_i(x)$ and $T_i(x)$, then this becomes

$$P = \int n_a(x) n_i(x) \left(\frac{m_i}{2\pi T} \right)^{3/2} \exp\left(-\frac{E}{T} \right) dx. \qquad (8.1.24)$$

Now suppose we consider an energy E substantially larger than T. As we saw in Section 8.1.3, the dominant variation in the integrand becomes the exponential factor; therefore, we may approximate the other factors as being constants equal to their values at the place where T is maximum and take them outside the integral

$$P \approx n_a n_i \left(\frac{m_i}{2\pi T_{max}}\right)^{3/2} \int \exp\left[\frac{-E}{T(x)}\right] dx. \tag{8.1.25}$$

To evaluate the integral, take $1/T$ to be given by the first two terms of its Taylor series expansion about its minimum $1/T(x) \approx (1/T_{max})(1 + x^2/b^2)$. (The b here is, to relevant order, essentially the same as that in Section 8.1.4.) With this substitution we obtain a simple Gaussian integral that gives

$$\int \exp\left[\frac{-E}{T_{max}}\left(1 - \frac{x^2}{b^2}\right)\right] dx = b\left(\frac{\pi T_{max}}{E}\right)^{1/2} \exp\left(\frac{-E}{T_{max}}\right), \tag{8.1.26}$$

so the fast neutral flux is

$$F(E)\, dE = A\Omega_s \sigma_c(E) \left[\frac{2E}{m_i}\right]^{1/2} \frac{b\, dE}{2\pi T_{max}} \exp\left(\frac{-E}{T_{max}}\right) n_i n_a, \tag{8.1.27}$$

where we have restored the energy-dependent factors and written v_i in terms of E. Thus we have an explicit equation that allows us to deduce the neutral density at the place where $T = T_{max}$ from $F(E)$. For this we need to know T_{max}, the ion density n_i, and the width of the temperature maximum b.

The dense plasma case can be analyzed similarly including the additional factor $\exp[-\int \alpha\, dx]$. The discussion of Section 8.1.4 has already indicated that for $E > T$ the position of maximum contribution (x_m) is shifted outward from the peak temperature position. In the neighborhood of x_m we make the approximation

$$n_a \exp\left[-\int \alpha\, dx\right] = K \exp[(x - x_m)/\lambda] \tag{8.1.28}$$

[equivalent to Eq. (8.1.20)] and account only for the variation of this quantity and the temperature exponential. That is, we take

$$P \approx K n_i \left(\frac{m_i}{2\pi T}\right)^{3/2} \int \exp\left[\frac{-E}{T_{max}}\left(1 + \frac{x^2}{b^2}\right) + (x - x_m)\frac{1}{\lambda}\right] dx. \tag{8.1.29}$$

Completing the square in the exponent we get a simple Gaussian integral again, in which the exponent is

$$-\frac{E}{T_{\max}}\frac{(x-x_m)^2}{b^2}-\frac{E}{T_{\max}}-\frac{b^2 T_{\max}}{4E\lambda^2} \qquad (8.1.30)$$

and we have used the condition for x_m to be truly the point at which the exponent is maximum:

$$x_m = \frac{b^2 T_{\max}}{2E\lambda}. \qquad (8.1.31)$$

Performing the integral and substituting as before we get the flux

$$F(E)\,dE = A\Omega_s\sigma_c \left[\frac{2E}{m_i}\right]^{1/2}\frac{b\,dE}{2\pi}\frac{T_{\max}^{1/2}}{T(x_m)^{3/2}}n_i(x_m)n_a(x_m)$$

$$\times\exp\left[-\int_{x_m}^{a}\alpha\,dl\right]\exp\left[\frac{-E}{T(x_m)}\right]. \qquad (8.1.32)$$

This equation confirms that the slope of F is characteristic of the temperature $T(x_m)$ at the point of maximum contribution. It shows also that we obtain an equation for the magnitude of the flux that replaces n_i and n_a by their values at x_m and accounts for the attenuation by the factor corresponding to this position, $\exp(-\int_{x_m}^{a}\alpha\,dl)$.

Information on the spatial profile of n_a is, in principle, obtainable from the shape of the $F(E)$ spectrum at lower energies [assuming the shape of $T_i(x)$ to be known, which it often is not]. This is indicated in an approximate way by Eq. (8.1.32). The contribution to the flux comes from progressively further out toward the colder edge of the plasma as E is lowered. More often, however, a numerical model is used that includes a priori theoretical ideas of the way in which the neutral density is liable to diffuse into the plasma. The parameters of this elaborate model may then be adjusted to get consistency with the low-energy shape of the fast neutral spectrum. Nevertheless there remain considerable ambiguities in the neutral profile deduced.

8.2   Active probing with neutral particles

As with other types of particles, it can be fruitful to use active neutral-beam probing techniques, as well as merely passive detection of plasma emission, as a diagnostic tool. In the case of neutrals one obvious possibility is to measure the attenuation of a beam of neutral particles transmitted through the plasma. The hope here (not entirely fulfilled) is that one can thereby obtain information on the ion density. This is of interest in plasmas with unknown composition (because of impurities, for example) since then there is no simple relationship between ion density and the electron density measured by, for example, an interferometer.

In addition to straightforward attenuation, other more sophisticated diagnostic possibilities exist.

8.2.1 *Neutral-beam attenuation*

The attenuation of a neutral beam by the plasma is given by our earlier equation (8.1.2),

$$I(B) = I(A)\exp\left[-\int_A^B \alpha \, dl\right],$$ (8.2.1)

where $\alpha(\ell)$ is the attenuation coefficient and I the intensity. Now we are interested in situations with more than one relevant ion species, so we write

$$v_a \alpha = \langle \sigma_e v_e \rangle_{v_a} n_e + \sum_{j \text{ ions}} \left[\langle \sigma_{\ell j} v_j \rangle_{v_a} n_j \right],$$ (8.2.2)

where we abbreviate $\sigma_{\ell j} = \sigma_p + \sigma_c$ as the total electron-loss cross section for the jth species. Of course, this measurement is a chordal one, so we shall only gain information about chord averages unless a detailed Abel inversion can be carried out using many chords. However, we shall not worry about that complication here, but simply take equations appropriate for a uniform plasma. Evidently a measurement of the attenuation then gives us directly the value of α. Incidentally, one advantage of the neutral transmission technique is that, unlike an interferometer, it does not have to keep track of any phase ambiguity, so it has no need of information about past (or future) plasma history. This advantage is rarely of overwhelming importance, though.

For a single ion species plasma, for which the relevant cross sections are known, knowledge of α allows us immediately to deduce n_i (and n_e). Suppose, now, that the plasma has two ion species and that an independent measurement of n_e is available. Can we deduce the two ion densities n_1, n_2 from α? Yes we can, in principle, provided we know the relevant rate coefficients. We can obtain two simultaneous equations for the densities:

$$Z_1 n_1 + Z_2 n_2 = n_e,$$
$$\langle \sigma_{\ell 1} v_1 \rangle_{v_a} n_1 + \langle \sigma_{\ell 2} v_2 \rangle_{v_a} n_2 = \alpha v_a - \langle \sigma_e v_e \rangle_{v_a} n_e.$$ (8.2.3)

The first of these comes from the quasineutrality condition of course. These equations can generally be solved to give n_1 and n_2. The exception to this statement is when the determinant formed from the coefficients on the left hand side of the equations is zero. When this is the case no unique solution is possible because the equations are linearly dependent. In practice, even if the determinant is not exactly zero but it is small, an attempt at solution will give a result extremely sensitive to small errors in the coefficients (or n_e or α). In such a case the equations are said to be ill conditioned and solutions will be unreliable. The conditioning of the Eqs. (8.2.3) is determined by the quantity

$$C_j = \langle \sigma_{\ell j} v_j \rangle_{v_a} / Z_j. \tag{8.2.4}$$

Provided C_j is different for the different species, a reasonable solution will be obtained. If the C_j are nearly equal then the equations are ill conditioned.

Suppose we do not have just two species of ions but more, N say. Clearly, we do not have enough information in just two equations to deduce all n_j. However, the rate coefficients appearing in the equations are dependent on particle velocity v_a (or, equivalently, energy), so the obvious thing to do is to measure the attenuation α for different neutral-beam energies. If we choose a total of $N-1$ energies we shall have N equations for N unknowns. (Actually we could regard n_e as unknown and then use N energies giving $N+1$ equations with $N+1$ unknowns.) Again the conditioning problem arises, only this time rather more stringently in that all the equations must be linearly independent. This requires not only that the C_j's be different (for some energy), but also that they vary differently with energy.

It is evident that in order to make any real progress, we must have some knowledge of the relevant rate coefficients. Reasonably reliable values of these coefficients have become available only during the past few years. One useful approximate rule, for ions with $Z \gtrsim 5$, that emerges from these calculations [e.g., Olsen et al. (1978)] is

$$\sigma_{\ell j} \approx 4.6 \times 10^{-20} Z_j \left\{ Q_j (1 - e^{-1/Q_j}) \right\} \text{ m}^2, \tag{8.2.5}$$

where $Q_j = 32 Z_j / E$, E being the neutral-beam energy (per atomic mass) in keV. (Strictly, this formula is numerically accurate only for energies greater than 50 keV. However, the Z dependence is appropriate even to lower energies.) In the case where $v_a \gg v_{tj}$, so that the (velocity shifted) rate coefficient is proportional to $\sigma_{\ell j}$, this leads to

$$C_j \propto Q_j (1 - e^{1/Q_j}). \tag{8.2.6}$$

For practical experiments where the beam energy $E < 100$ keV, one finds $Q \gg 1$ for all but the very lightest ions [for which, in any case, Eq. (8.2.5) is not accurate]. This means that C_j is approximately constant, independent of ion species. The result is disastrous for our hopes of being able to deduce n_j because the system is ill conditioned for all n_j.

Physically, one can summarize the situation by saying that if one replaces an ion of charge Z_j with Z_j/Z_k ions of charge Z_k (hence keeping n_e

constant) it makes very little difference to the attenuation. Despite the fact that there is some deviation in detail from the approximation Eq. (8.2.5), one concludes that it is extremely difficult to get reliable n_j measurements from neutral-beam attenuation. There is a silver lining, though. The proton–hydrogen charge-exchange cross section is substantially larger than Eq. (8.2.5) primarily because the process is symmetric, and hence *resonant* at energies $\lesssim 30$ keV. Therefore, the attenuation will be mostly caused by protons despite any impurities present. One can therefore take the attenuation as indicative of the proton density, or, more properly, return to the two species equations (8.2.3) and lump all impurity effects into the second species n_2, taking its cross section to be given by Eq. (8.2.5) or some other appropriate form. Then one can solve for n_1 the proton density.

To conclude, then, neutral hydrogen beam attenuation gives a reasonable measurement of proton density. Unfortunately, the difference between the proton density and the electron density is a less sensitive measurement of impurities than Z_{eff}. For a single impurity of charge Z, the proton density is given by $n_p = n_e(Z - Z_{\text{eff}})/(Z - 1)$ (see Exercise 8.7). This relative insensitivity to low levels of impurity, the availability of interferometry for electron density, and the complexity of neutral-beam systems largely explain the fact that neutral-beam attenuation is very infrequently used as a diagnostic.

8.2.2 Active charge exchange

As we have seen, one of the problems with neutral particle analysis of emitted charge-exchange neutrals is that when the plasma is dense, the energy spectrum obtained reflects the temperature along the line of sight in a complicated way. Thus uncertainty and ambiguity arise in the interpretation. An effective way to alleviate this problem is to use an active neutral beam to enhance the neutral particle emission locally. Using a view across the active (doping) beam, one can then achieve much better localization and a less ambiguous spectrum.

In Fig. 8.7 are shown energy spectra with and without an active beam. The characteristically curved spectrum without doping becomes essentially straight with the active beam on because then only a single temperature is represented.

A complication with this technique is that the enhancement of emission occurs not merely by a single charge exchange between a plasma ion and the probing beam but also by several charge-exchange events. The point is that in a thick plasma the beam is surrounded by a diffusive "halo" of neutrals formed by charge exchange from the beam. The size of the halo will be of the order of the mean free path of the neutrals and the enhanced emission may well be dominated by the halo rather than by direct exchange with the beam. Thus, the localization perpendicular to the beam is not quite

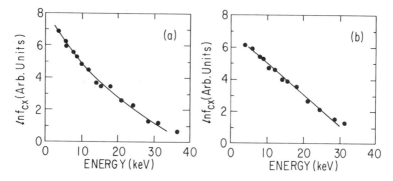

Fig. 8.7. Fast neutral spectra (*a*) without and (*b*) with an active doping beam to remove the ambiguity of the interpretation [from tokamak PLT, Goldston (1982)].

as good as might first be thought. Nevertheless, active charge exchange is used to advantage in various experiments, despite the obvious added difficulty (relative to passive charge-exchange analysis) of operating the doping beam.

8.2.3   *Fast ion diagnostics*

The ability of an energetic neutral beam to inject fast ions in a controlled way into a plasma enables one to utilize the properties of these ions for diagnostic purposes. When combined with detection of the fast ions via reemitted charge-exchange neutrals, a number of possibilities arise.

One example mentioned earlier is that the slowing down spectrum of the ions, including their distribution in pitch angle, can be obtained from neutral particle measurements. This allows detection of anomalies in the ion scattering due, for example, to impurities or to collective plasma instabilities. Such information can be extremely useful.

On the other hand, the trajectories of the fast ions are determined by the internal magnetic fields and so may be used as a diagnostic of these fields. Perhaps the most successful application of this approach is the determination by Goldston (1978) of the central current density in a tokamak.

In this application advantage is taken of the fact that the orbits of fast ions in a tokamak are shifted from the magnetic surfaces by a distance $\Delta = q_s v_i / \Omega_i$ (see Section 8.4.2), where q_s is the magnetic field "safety factor" ($q_s = r B_\phi / R B_\theta$) and near the plasma center is inversely proportional to the toroidal current density j_ϕ. By injecting neutrals tangentially along the field outside the magnetic axis and observing reemitted neutrals tangentially inside the magnetic axis, the position of the orbit of the fast ions can be determined, and thence the q_s-value and current density. Figure 8.8 shows an example of the measured evolution of q during the early phases of a discharge, when the current profile shape is of particular interest. These

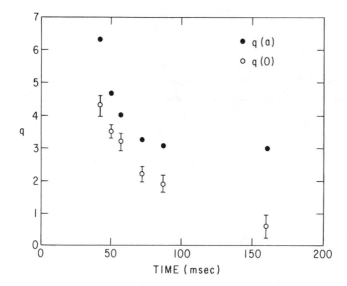

Fig. 8.8. The evolution of the central q value (open points) measured by the orbit displacement of fast ions from a tangentially injected neutral beam [from tokamak PDX, Goldston (1982)].

results are notable in the diagnostic context because they constitute a rare example of a nonperturbing measurement of current density with sufficient accuracy to compare with estimates from Ohm's law.

8.2.4 *Doping species, lithium beams*

There are some situations in which it can be extremely useful to use neutral beams consisting of atoms other than those typical of the plasma as a whole. One advantage that these have is that they are relatively easily distinguished from the plasma species, usually spectroscopically, and thus they can be used as tracers in the plasma. This is the principle behind the laser blow-off technique. A suitable species of atoms is deposited on a glass slide, which is then placed facing the plasma. At an appropriate moment a powerful pulsed laser is fired onto the slide, causing ablation of the tracer atoms that are sprayed into the plasma. They can be tracked by their line radiation and provide valuable information about the ion transport properties of the plasma.

Another advantage to be gained with doping beams is that the atoms may be chosen so as to make possible some spectroscopic technique that the majority species does not allow. An atom of particular importance in this respect is lithium. The difficulty with many atoms, especially hydrogen, is that the transitions to and from their ground state are sufficiently energetic as to involve ultraviolet rather than visible radiation. Because of the greater difficulties of ultraviolet techniques, some applications are presently impossible with these transitions. Lithium, though, has a very convenient line to the ground state ($\lambda = 670.8$ nm), which lies in the visible.

As an illustration of the potential and also some of the difficulties of using a lithium doping beam we consider the attempts to measure the internal magnetic field direction using the Zeeman effect on a lithium beam. In its latest form (West 1986) this experiment involves a configuration such as that illustrated in Fig. 8.9. A high-energy lithium ion beam is steered so as to be collinear with a CW dye laser. It is then neutralized and both lithium and laser beams are sent into the plasma. Inside the magnetic field of the plasma, the lithium line of interest is split by the Zeeman effect (see Section 6.6.2) into symmetrically shifted (σ) and unshifted (π) lines whose polarization, for perpendicular propagation, is perpendicular and parallel to the magnetic field, respectively.

The energy of the neutral lithium beam (typically up to ~ 100 keV) must be very carefully controlled so that the broadening of the split lines due to the Doppler effect does not cause them to overlap. (Here already is an advantage with a doping beam: Thermalized atoms in a hot plasma generally experience so much Doppler broadening that the Zeeman structure is very hard to see.) The dye laser is tuned to have a frequency that coincides with one of the Zeeman components (e.g., the π line) but not the

Fig. 8.9. Schematic diagram of the lithium beam Zeeman effect measurement of magnetic field direction.

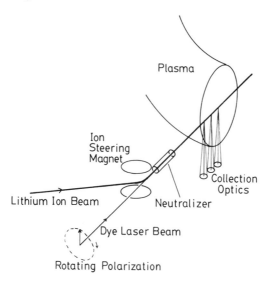

others. It thus causes fluorescence by populating (via absorption) the upper level, which then emits a photon in decaying back to the ground level. The photon can be detected by a spectrometer viewing the beam line from the side. This then provides good spatial localization of the measurement (another advantage of doping beams).

The input polarization of the laser beam is caused to rotate rapidly. When it points parallel to the local field, excitation of the π line can occur. However, when it is perpendicular no excitation occurs because of the polarization properties of the Zeeman components. The signal detected by the spectrometer is therefore modulated at the frequency of the polarization rotation. The *phase* of this modulation provides the information we seek on the direction of the magnetic field.

The main difficulties with this type of experiment are mostly associated with obtaining sufficient fluorescence signal and hence giving adequate signal to noise. The intensity available from CW dye lasers is not sufficient to pump the transition into saturation; therefore, the signal is approximately proportional to laser power. Normally, it is easier to control the laser beam diameter than the lithium beam diameter, so we may assume that the laser beam is entirely within the lithium beam. Therefore, the fluorescence from any specified length of beam is proportional to the lithium density and the laser power. Naturally, the lithium density is proportional to beam current (at fixed energy), so a high current beam is required. However, the density is also inversely proportional to the beam cross sectional area. Therefore, the beam quality is also of great importance. A low-angular-divergence beam is required. A very great deal of the effort involved in attempting this type of experiment has been devoted to obtaining high power lithium beams of sufficiently low divergence.

A final major difficulty should be mentioned, since it appears to be a fundamental restriction on the plasmas that may be diagnosed by this technique: that is, beam attenuation by the plasma. Just as with fast neutrals or hydrogenic beams, the lithium beam is attenuated by ionization and charge-exchange processes. There is a limit, therefore, on the plasma depth that can be probed. Estimates of the attenuation coefficient may be made using known values of the various cross sections. Ionization by hydrogen ions has a cross section of about 10^{-19} m^2 at typical (100 keV) energy, and electron ionization may contribute at most about the same, the exact value depending on electron temperature. Charge exchange with hydrogen contributes rather more to the total electron-loss cross section. Experiments on proton beams in lithium vapor (Gruebler et al. 1970) indicate the charge-exchange cross section is about 6×10^{-19} m^2 at 5 keV proton energy, falling to 1.5×10^{-19} m^2 at 15 keV (the error bars are large, nearly a factor of 2). These results correspond to lithium beam energies of

35 and 105 keV, respectively. Hence, the total electron-loss cross section (σ_ℓ) will be typically between $\sim 8 \times 10^{-19}$ and $\sim 3 \times 10^{-19}$ m^2, depending on the lithium beam energy.

The characteristic depth $\int n_e \, dl$ of plasma to which the beam penetrates is then $1/\sigma_\ell \sim 1$ to 3×10^{18} m^{-2}. For example, at the center of a plasma of 0.25 m radius, using 90 keV beam energy, this estimate predicts e^{-1} attenuation when the density is $\bar{n}_e = 1/\sigma_\ell r \sim 10^{19}$ m^{-3}. In practice, however, experiments have shown rather greater attenuation (by a factor up to ~ 2) than this estimate indicates. The reasons for this discrepancy are not fully understood. One possibility is that ionization takes place much more readily by a stepwise process of collisional excitation of the atom followed by ionization of the excited state. At any rate, the consequences of the enhanced attenuation are very serious for practical diagnosis of the larger denser plasmas of interest in fusion. The limitation to small plasma depths (less than $\sim 4 \times 10^{18}$) for acceptable beam attenuation considerably detracts from the usefulness of the technique. Unless this limitation can be overcome, the restriction to small plasmas, combined with the very considerable technical complexities of the diagnostic (requiring a commitment of resources typically available only on larger experiments) may prevent the measurement from having any widespread usefulness.

8.3   Neutron diagnostics

8.3.1   *Reactions and cross sections*

Nuclear reactions (the primary objective of fusion research) occurring within the plasma can be used as a convenient diagnostic for the ions. For this purpose the reaction product of most immediate interest is the neutron because, being uncharged, it is able to escape immediately from the plasma and hence be detected.

In mixed deuterium–tritium (D–T) plasmas the neutron-producing reactions are

$$D + T \rightarrow (^4He + 3.5 \text{ MeV}) + (n + 14.1 \text{ MeV}), \tag{8.3.1}$$

$$D + D \rightarrow (^3He + 0.82 \text{ MeV}) + (n + 2.45 \text{ MeV}), \tag{8.3.2}$$

$$T + T \rightarrow (^4He + 3.8 \text{ MeV}) + (2n + 7.6 \text{ MeV}). \tag{8.3.3}$$

In future devices the tritium reactions may (one hopes!) be important, but to date the D–D reaction has been the primary diagnostic tool. In some experiments deuterium has been used as the plasma constituent in preference to hydrogen specifically for diagnostic purposes. We shall concentrate on this reaction, though much of what is said will be applicable, mutatis mutandis, to the other reactions.

The cross section for the neutron D–D reaction has been measured with considerable accuracy down to about 12 keV energy. At the lower end of the energy range these results are reasonably well fitted with the expression

$$\sigma_n = \frac{71}{E_D} \exp\left(-\frac{44}{E_D^{1/2}}\right) \text{b,} \tag{8.3.4}$$

where E_D is the incident energy (in the laboratory frame) of a deuteron in keV colliding with a stationary target deuteron [a barn (b) is 10^{-28} m^2]. The form of this expression is that obtained from an approximate theoretical treatment (Gamow and Critchfield 1949; see Exercise 8.8), but with the coefficients modified slightly to fit experiment. Using this form, which has good theoretical justification, is expected to provide reasonable accuracy even extrapolated to lower energies than the experimental results. Figure 8.10 shows how rapidly the cross section increases with energy up to around 50 keV.

For a Maxwellian ion distribution the rate coefficient for this reaction may be evaluated by integrating the cross section over the velocity distribution. Again, it is usually convenient to use an analytic approximation guided by theory, but with the coefficients adjusted to give a more accurate

Fig. 8.10. Neutron-producing fusion reaction cross sections (stationary target).

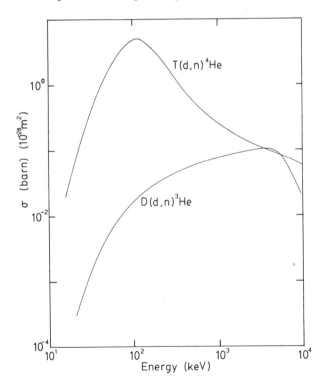

fit of the numerically integrated results. One convenient form (see Exercise 8.12) is

$$\langle \sigma_n v \rangle = \frac{3.5 \times 10^{-20}}{T_D^{2/3}} \exp\left(-\frac{20.1}{T_D^{1/3}} \right) \text{ m}^3 \text{ s}^{-1} \qquad (8.3.5)$$

(Hively 1977), which is accurate to about 10% for deuteron temperatures T_D from 1 to 80, expressed in keV. Again, for temperature below about 20 keV this rate coefficient is an extremely strong function of temperature, as illustrated in Fig. 8.11.

The rate at which neutrons are produced per unit volume by this reaction is then

$$S = \tfrac{1}{2} n_D^2 \langle \sigma_n v \rangle \text{ m}^{-3} \text{ s}^{-1}. \qquad (8.3.6)$$

(The factor $\tfrac{1}{2}$ is required in order to count all possible reactions only once for like particles.)

On the one hand, the strong dependence of $\langle \sigma_n v \rangle$ on T_D is helpful in providing an estimate of temperature that is very insensitive to uncertainties in either the cross section or the deuteron density. On the other hand, however, we must note that this strong dependence is a reflection of the fact that the neutrons are produced mostly by reactions of ions on the tail of the Maxwellian distribution. For $T_D \leq 50$ keV we shall show in Section 8.3.3

Fig. 8.11. Fusion reaction rate in (equal temperature) thermal plasma.

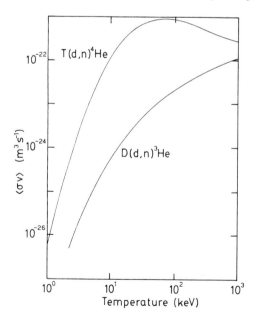

that the mean energy of the reacting particles (in the lab frame) is approximately

$$\bar{E} = \tfrac{7}{6}T + 3.1T^{2/3} \quad (\text{keV}). \tag{8.3.7}$$

Thus, for example, at 1 keV temperature the neutron production is from ions with mean energy 4.3 keV.

As a result, the neutron production rate is sensitive to deviations of the ion distribution from Maxwellian. Any enhanced tail on the distribution can lead to an overestimate of T_D. Conversely, tail depletion (for example, due to enhanced loss of high-energy ions) can lead to an underestimate. Note, however, that this dependence of the measurement on high-energy particles is actually less pronounced than for fast neutral analysis, in which it is not uncommon to depend upon particle energies 5 to 10 times the temperature.

8.3.2 Complicating factors

One source of uncertainty is deducing deuteron temperature from neutron production rate arises because of the need to know n_D, which enters into the rate as a squared power. The unavoidable impurities present in hot plasmas dilute the number of deuterons. This effect can be estimated, as we have seen before, from the Z_{eff} of the plasma. If Z_{eff} is significantly greater than 1 (values of 2–5 are not uncommon in fusion experiments) then that indicates that n_D may be different from n_e, the density whose value is usually best known. Although n_D is often quite uncertain (by up to perhaps a factor of 2), fortunately the strong dependence of neutron rate on temperature reduces the consequent T_D uncertainty to manageable proportions. For example, our previous equations show that at 1 keV the percentage error in T_D caused by n_D uncertainty is 0.3 times the percentage error in n_D (see Exercise 8.13).

Nonfusion neutrons are another source of possible error in the temperature estimates. The primary cause of processes that produce such neutrons is high-energy electrons. These electrons can be created in toroidal devices by electron runaway along the magnetic field. Runaway is a process in which the decreasing electron collision cross section at higher energy allows sufficiently energetic electrons to accelerate continuously in an applied electric field up to very high energy. Radio frequency heating schemes (for example, electron cyclotron resonance heating) can also create very energetic electrons.

Nonfusion neutrons can arise from these high-energy electrons in several ways. They can cause disintegration of deuterons either directly, by the reaction $D(e, e'n)H$ or by producing energetic photons (gammas) by bremsstrahlung, which then cause photodisintegration by the reaction $D(\gamma, n)H$. [Nuclear reactions are conventionally written $A(x, y)B$ where A and x are the reactants and B and *y* are the products.] The former process occurs throughout the plasma volume, but the latter is concentrated primarily in those locations where the energetic electrons can collide with solid structures at the plasma edge, since the photons are produced most copiously by thick-target bremsstrahlung in solids. Both of these deuteron disintegration processes have threshold energy (below which no reactions will occur) equal to the binding energy of the deuteron (2.2 MeV).

There are other processes of neutron production that are usually more important. They involve nuclear reactions not of deuterium but of the material from which the edge structures are made (with which the electrons can collide). These are frequently refractory metals such as molybdenum or tungsten for which reactions such as $^{97}Mo(e, e'n)^{96}Mo$ or $^{97}Mo(\gamma, n)^{96}Mo$ are possible. Subsequent observation of activation of the structures due to the radioisotopes produced can confirm these processes.

In order to distinguish the nonfusion neutrons from the component that is to be used for T_D diagnosis one requires, in general, an energy spectrum of the neutrons to be measured. It is sometimes possible to discriminate against locally produced nonfusion neutrons (due to solid interactions) by virtue of their localization. However, for volumetric processes such as the *e*-D direct disintegration, one must appeal to the distinctive energy spectrum of D–D neutrons (namely peaked at 2.46 MeV), compared to the endoenergetic disintegration reactions, which produce spectra usually falling monotonically with energy. Figure 8.12 illustrates an interpretation of an observed spectrum in which the D–D fusion peak is only a relatively small fraction of the total neutron yield.

Of course, if it can be established by these and other cross checks that the neutrons observed *are* dominantly D–D, then subsequent spectral measurements are less important and an observation of simply the total neutron rate may be sufficient to determine the ion temperature.

Assuming then that the fusion neutrons can be unambiguously distinguished, to obtain a local ion temperature requires the calculation of the local neutron production rate in the plasma from measurements of a detection rate at some point(s) outside the plasma. It is rather difficult to obtain collimated neutron measurements along a specified viewing path. Instead, more often, an uncollimated detector, sensitive to the total neutron production rate, is used. To deduce the absolute neutron rate then requires the details of the machine and plasma geometry to be taken into account. This is usually done by in situ calibration using appropriate calibrated neutron sources inside the actual machine itself.

Evidently the temperature and density profile effects are also rather important and need to be modeled before a temperature can be deduced. Fortunately, the strong dependence of the neutron rate on temperature helps in two ways. First, it means that one is sensitive mostly to the peak

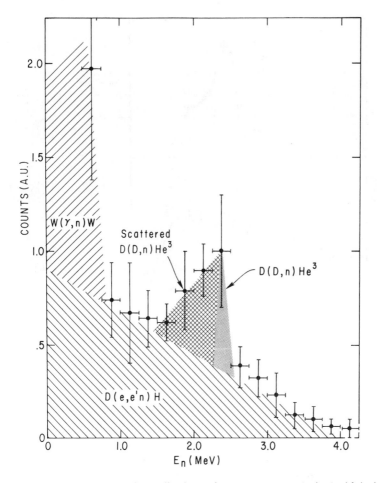

Fig. 8.12. An experimentally observed neutron spectrum (points) with its interpretation in terms of various completing production processes in a tokamak plasma with appreciable high-energy electron component [after Strachan and Jassby (1977)].

temperature (normally at the plasma center). Second, the temperature deduced is not very sensitive to the (assumed) temperature and density profile widths. As a result, a quite reliable continuous measurement of ion temperature is often possible. Figure 8.13 shows an example.

8.3.3 *Thermal neutron spectrum*

Apart from the important role of distinguishing fusion from nonfusion neutrons, measurements of the energy spectrum can, in principle, provide a more direct measurement of the (thermal) deuteron energy. The reason for this is that the exact neutron kinetic energy depends on the

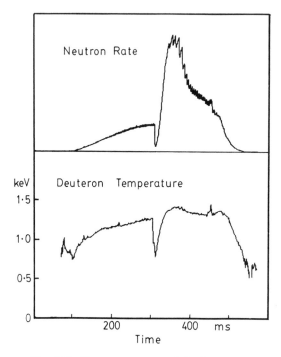

Fig. 8.13. The neutron rate and the ion temperature derived from it (using other data also) for a case with negligible nonfusion neutron production (Alcator C tokamak, courtesy R. Granetz). The sharp decrease at about 300 ms is due to a pellet of solid deuterium injected into the plasma for fueling purposes. It cools the plasma.

kinetic energy of the reacting deuterons in a way that can be determined by analyzing the particle kinetics of the reaction. We shall continue to speak of the D–D neutron-producing reaction in the following analysis but it should be noted that the results are completely general for a reaction that produces only two product particles. Thus D–T reactions and proton-producing reactions are also covered by this treatment. In the center-of-mass frame of two reacting deuterons the conservation of momentum requires that for the neutron (subscript n) and the ^{3}He (subscript α) emerging from the reaction the total momentum is zero,

$$m_n \mathbf{u}_n + m_\alpha \mathbf{u}_\alpha = 0, \qquad (8.3.8)$$

where $\mathbf{u}$ is particle velocity. Conservation of energy then gives

$$Q + K = \frac{1}{2} m_n u_n^2 + \frac{1}{2} m_\alpha u_\alpha^2 = \frac{1}{2} m_n u_n^2 \left(\frac{m_n + m_\alpha}{m_\alpha} \right), \qquad (8.3.9)$$

where Q is the total energy released in the reaction and K is the relative kinetic energy of the reacting deuterons, and we have used Eq. (8.3.8). Now we consider the center of mass to have velocity $\mathbf{V}$ in the lab frame and so the neutron velocity in the lab frame is

$$\mathbf{v}_n = \mathbf{V} + \mathbf{u}_n \tag{8.3.10}$$

and its kinetic energy is

$$E_n = \tfrac{1}{2} m_n v_n^2 = \tfrac{1}{2} m_n \left(V^2 + u_n^2 + 2\mathbf{V} \cdot \mathbf{u}_n \right). \tag{8.3.11}$$

Substituting for u_n we get

$$E_n = \frac{1}{2} m_n V^2 + \frac{m_\alpha}{m_n + m_\alpha}(Q + K)$$
$$+ V \left[\frac{2 m_n m_\alpha}{m_n + m_\alpha}(Q + K) \right]^{1/2} \cos\theta, \tag{8.3.12}$$

where θ is the angle between $\mathbf{V}$ and $\mathbf{u}_n$, and is hence random. When we take the mean value of this expression over all possible reaction configurations the $\cos\theta$ term averages to zero provided the deuterium distribution is isotropic:

$$\bar{E}_n = \frac{1}{2} m_n \overline{V^2} + \frac{m_\alpha}{m_n + m_\alpha}(Q + \bar{K}). \tag{8.3.13}$$

The term in Q represents the usual fraction of the fusion yield that goes to the neutron (2.45 MeV for D–D), while the $\overline{V^2}$ and $\bar{K}$ terms represent a small shift in mean energy.

The displacement in energy of any specific observed neutron from the mean neutron energy is

$$E_n - \bar{E}_n = \frac{1}{2} m_n \left(V^2 - \overline{V^2} \right) + \frac{m_\alpha}{m_\alpha + m_n}(K - \bar{K})$$
$$+ \left[\frac{2 m_\alpha m_n}{m_\alpha + m_n} \right](Q + K)^{1/2} V \cos\theta. \tag{8.3.14}$$

Now provided the temperature is much less than Q, which will almost always be true since Q is so large (3.3 MeV for D–D), this equation will be completely dominated by the final term and may be written

$$E_n - \bar{E}_n \approx \left[\frac{2 m_\alpha m_n Q}{m_\alpha + m_n} \right]^{1/2} V \cos\theta. \tag{8.3.15}$$

Since $V\cos\theta$ represents simply the component of $\mathbf{V}$ in the direction of $\mathbf{u}_n$, the distribution of $E_n - \bar{E}_n$ directly reflects the distribution of $\mathbf{V}$ in one dimension (viz., along $\mathbf{u}_n$, which is approximately along $\mathbf{v}_n$). However, it is important to realize that the number of reactions that occur with a given neutron energy will naturally be weighted by the neutron-producing reaction rate σv, where v is the relative velocity. All the averages denoted by an overbar in this section must be interpreted as integrals of the quantity in question times σv. Thus, the distribution of V reflected in the energy spectrum is weighted by σv toward those deuterons that contribute most to the reaction rate.

Now we must explore the averaging process over the velocity distributions of the reactants in order to determine the distribution of V and hence of E_n. First we note that the total reaction rate coefficient for two species $1, 2$ (which we treat as distinct now to maintain generality) may be written

$$n_1 n_2 \langle \sigma v \rangle = \int\int \sigma(|\mathbf{v}_1 - \mathbf{v}_2|)|\mathbf{v}_1 - \mathbf{v}_2| f_1(\mathbf{v}_1) f_2(\mathbf{v}_2)\, d^3\mathbf{v}_1\, d^3\mathbf{v}_2. \quad (8.3.16)$$

We may change the variables in this integration to express it in terms of the c.m. velocity $\mathbf{V}$ and the relative velocity $\mathbf{v} = \mathbf{v}_1 - \mathbf{v}_2$. The Jacobian of the transformation is unity (see Exercise 8.9) so the result is

$$n_1 n_2 \langle \sigma v \rangle = \int\int \sigma(|\mathbf{v}|)|\mathbf{v}| f_1\left(\mathbf{V} + \frac{m_2}{m_1 + m_2}\mathbf{v}\right) f_2\left(\mathbf{V} - \frac{m_1}{m_1 + m_2}\mathbf{v}\right) d^3\mathbf{v}\, d^3\mathbf{V}.$$

$$(8.3.17)$$

Clearly, then, the contribution to the reaction rate from collisions with c.m. velocity $\mathbf{V}$ in the velocity element $d^3\mathbf{V}$ is

$$d^3\mathbf{V} \int \sigma v f_1\left(\mathbf{V} + \frac{m_2}{m_1 + m_2}\mathbf{v}\right) f_2\left(\mathbf{V} - \frac{m_1}{m_1 + m_2}\mathbf{v}\right) d^3\mathbf{v}. \quad (8.3.18)$$

To the extent that Eq. (8.3.15) is a good approximation, it shows that neutrons of a specific energy arise from collisions in which the component of $\mathbf{V}$ in the neutron emission direction (z say), $V \cos \theta$, has a specific value. The number of neutrons with this energy is obtained by a partial integration over the components of $\mathbf{V}$ perpendicular to this direction, as well as the integration over $\mathbf{v}$. This integration can be done (numerically) for any general forms for f_1 and f_2, so that the deuteron energy spectrum provides some (limited) information on the distribution shape.

When f_1 and f_2 are equal temperature Maxwellians the $\mathbf{V}$ and $\mathbf{v}$ dependences separate, because then

$$f_1 f_2 = n_1 n_2 \left(\frac{m_1}{2\pi T}\right)^{3/2} \left(\frac{m_2}{2\pi T}\right)^{3/2} \exp\left(-\frac{MV^2}{2T} - \frac{mv^2}{2T}\right), \quad (8.3.19)$$

where $M = m_1 + m_2$ is the total mass and $m = m_1 m_2/(m_1 + m_2)$ is the reduced mass (see Exercise 8.10), so that

$$\langle \sigma v \rangle = \int \left(\frac{M}{2\pi T}\right)^{3/2} \exp\left(-\frac{MV^2}{2T}\right) d^3\mathbf{V} \int \left(\frac{m}{2\pi T}\right)^{3/2} \sigma v \exp\left(-\frac{mv^2}{2T}\right) d^3\mathbf{v}.$$

$$(8.3.20)$$

Since $dE_n \propto dV_z$ (taking $\mathbf{v}_n$ to be in the z direction) we can therefore identify immediately the neutron-production energy distribution $[P(E_n)]$ as also having a Gaussian shape:

$$P(E_n)\, dE_n \propto \exp\left(-\frac{MV_z^2}{2T}\right) dE_n$$

$$\propto dE_n \exp\left[-M(E_n - \bar{E}_n)^2 \Big/ \left\{\frac{4m_\alpha m_n}{m_\alpha + m_n} QT\right\}\right]. \quad (8.3.21)$$

The potential advantage of basing a temperature on this energy spectrum width rather than an absolute measurement of neutron intensity is that it removes the necessity for absolute calibration. Also uncertainties arising from deuteron density are removed. On the other hand, to obtain a neutron spectrum with sufficient resolution to determine a width of the order of the geometric mean of Q and T at energy Q is rather challenging, especially since neutrons scattered from the various material structures surrounding the plasma must be excluded. (Their energy is degraded by the scattering.) Thus measurements based on the spectral width are only just beginning to prove feasible on the latest machines with large neutron rates [e.g., Fisher et al. (1984) or Jarvis et al. (1986)]. Moreover, it should be clear, from the comments above, that using the energy spectral width does not avoid the difficulties with non-Maxwellian distributions that arise because neutron production comes mostly from the tail of the distribution. Instead the spectral width will reflect roughly the tail temperature.

In one example of measurements based on the details of the neutron spectrum, Strachan et al. (1979) observed the energy shift of the neutron peak due to directed motion of injected energetic deuterons in a neutral-beam heated tokamak. The absence of shift when only protons were injected indicated that no directed motion of the deuterium background plasma then occurred.

To complete our discussion of the neutron spectrum we demonstrate our previous formula for the mean energy of reacting particles. From Eq. (8.3.20) integrated over all $\mathbf{V}$ we have

$$\langle \sigma v \rangle = \left(\frac{m}{2\pi T}\right)^{3/2} \int \sigma v \exp\left(-\frac{mv^2}{2T}\right) d^3\mathbf{v}$$

$$= \left(\frac{1}{\pi m}\right)^{1/2}\left(\frac{2}{T}\right)^{3/2} \int_0^\infty \sigma \exp\left(-\frac{K}{T}\right) K\, dK, \quad (8.3.22)$$

where, as before, K is the relative energy $\tfrac{1}{2}mv^2$. As a convenient shortcut to obtaining the mean energy (Brysk 1973) we note also that

$$\bar{K} \equiv \frac{\langle \sigma v K \rangle}{\langle \sigma v \rangle} = T^2 \frac{d}{dT}\left(\ln|T^{3/2}\langle \sigma v \rangle|\right). \quad (8.3.23)$$

This may be verified by direct differentiation of the previous equation (see Exercise 8.11). Now we take a rate coefficient in the form of Eq. (8.3.5),

$$\langle \sigma v \rangle \propto T^{-2/3} \exp(-C/T^{1/3}),\tag{8.3.24}$$

and it follows that

$$\bar{K} = T^2 \frac{d}{dT}[\ln T^{5/6} - CT^{-1/3}] = \frac{5}{6}T + \frac{C}{3}T^{2/3}.\tag{8.3.25}$$

Now the mean energy attributable to center-of-mass motion is

$$\overline{\tfrac{1}{2}MV^2} = \tfrac{3}{2}T,\tag{8.3.26}$$

which must be added to K to give the total mean kinetic energy of the reacting particles. This energy is shared between the two particles, so the mean per particle is

$$\bar{E} = \frac{1}{2}\left[\left(\frac{3}{2}+\frac{5}{6}\right)T + \frac{C}{3}T^{2/3}\right] = \frac{7}{6}T + \frac{C}{6}T^{2/3},\tag{8.3.27}$$

as previously stated [Eq. (8.3.7)].

8.4   Charged particle diagnostics
8.4.1   *Charged reaction products*
Thermal-energy ions in magnetically confined plasmas are, by design, unable to escape readily from the plasma. However, sufficiently energetic ions can have Larmor radii comparable to or greater than the plasma size, or, in electrostatic confinement schemes, sufficient energy to overcome the potential barrier. In that case it is possible to perform diagnostics by observing these particles outside the plasma.

The Larmor radius is, of course, $m_i v_i/ZeB$, which may be written in terms of (perpendicular) particle energy W (in MeV), mass μ (in units of the proton mass), and magnetic field B (in tesla) as

$$\rho_i = 0.144\,(\mu W)^{1/2}/ZB \text{ m.}\tag{8.4.1}$$

For a 14 MeV proton ($\mu = 1$, $Z = 1$) in 1 T field the radius is about 0.5 m, which can be comparable to plasma dimensions. Thus, it is possible under some circumstances to use charged nuclear reaction products for diagnosis.

The reactions of interest, in addition to those mentioned in Section 8.3, which of course produce charged products such as ^{4}He in addition to neutrons, include

$$D + D \rightarrow (T + 1 \text{ MeV}) + (p + 3 \text{ MeV})\tag{8.4.2}$$

and particularly

$$D + {}^3He \rightarrow ({}^4He + 3.6 \text{ MeV}) + (p + 14.7 \text{ MeV}),\tag{8.4.3}$$

as well as other reactions involving tritium, etcetera.

The D–D proton-producing reaction has a very similar cross section and energy release to the neutron reaction. It was used in one of the earliest studies of (controlled) fusion product energy spectra. In this work (Nagle et al. 1960) the plasma configuration was a "θ-pinch," which has open field lines along which the charged particles can easily escape. Detection and energy resolution is much easier for charged particles than it is for neutrons because the charged particles have short stopping lengths in, for example, emulsions, semiconductor detectors, or ionization chambers. In the θ-pinch studies, consistent ion temperatures were deduced from the energy spectral widths. Later work on plasma configurations, where the charged particles do not easily escape, has tended to concentrate on the neutron branch of the reaction, although continued use has been made of the charged particles in open field configurations [e.g., Foote (1979)].

In plasmas with ^{3}He as a major ion component the option of diagnosing this component with neutrons is not open because ^{3}He has no neutron-producing reactions (except with T and that has a much smaller cross section). Moreover, the cross section for the production of a neutral ^{3}He atom by charge exchange is small since the ^{3}He nucleus needs to acquire two electrons. Double charge exchange is much less probable than single. Therefore, charge-exchange fast neutral analysis of ^{3}He is very difficult. Because of these difficulties there has been considerable motivation to employ the D–^{3}He reaction proton as a diagnostic of the ^{3}He population. Some measure of success in this direction has been achieved by using particle detectors at the plasma edge (Chrien and Strachan 1983) appropriately collimated to intercept the large ion orbits. Energy spectra have been obtained, although they are not yet sufficiently free of scattering and other spurious effects to provide an independent measurement of thermal temperatures.

Future interest in fusion research is likely to focus on diagnosis of the alpha particles produced by the D–T reaction in proposed ignition experiments. The ultimate success of such experiments hinges on the behavior and particularly the confinement of these alphas, so naturally their diagnosis assumes considerable importance.

As we have noted, in open field configurations, and even more obviously, in inertially confined plasmas, the charged particles can escape immediately. These prompt escaping particles carry out the information about their birth, which, by analysis identical to that in Section 8.3.3 for neutrons, can give ion-temperature information. On the other hand, the charged particles that do not escape immediately, gradually slow down by collisions on the bulk plasma. During this slowing down process, some fraction of them may escape and be able to be detected at the plasma edge. Therefore, a measurement of the energy spectrum of the escaping particles should give information about the slowing down process and the confinement of particles during slowing down.

Whether a fast particle undergoes prompt loss or is confined depends on its orbit, which in a practical machine is usually very complicated; however, a brief elementary introduction to this subject is given in the next section.

8.4.2   *Orbits of energetic charged particles*

We began our discussion of charged particle diagnosis by a consideration of the size of the Larmor radius ρ_i. To lowest order, the particle orbit consists of a gyration about the magnetic field lines, with this radius, plus a motion along the field given by the parallel velocity $v_{\parallel}$. However, in nonuniform fields it is well known that there are, in addition, slower drifts of the center of the gyroorbit across the field lines. These drifts are fast enough that unless they lead to gyrocenter motion of only limited extent they will constitute a loss mechanism that is essentially prompt. Thus a fast particle will be confined only if both the gyroradius and also the size of the drift orbit motion of the gyrocenter are smaller than the plasma. Because the gyrocenter drift orbit is usually larger than the gyroradius, the drift orbit is usually the dominant effect.

Drift orbit analysis depends strongly on the magnetic field geometry. We consider here the case of a tokamak, both as a conveniently simple illustration of some of the principles and as the configuration of greatest current interest as far as controlled fusion ignition is concerned.

Fast particle orbits can be taken to be given by an approximate adiabatic analysis, ignoring collisions. [The reader not familiar with particle drift analysis may find a discussion in any elementary plasma text, e.g., Chen (1984).] Because of the helical form of the magnetic field lines and the toroidicity, the magnetic field strength varies along the field line; B is approximately inversely proportional to the major radius R. As a result of this variation and the conservation of magnetic moment $\mu = \frac{1}{2}mv_{\perp}^2/B$, some of the particles are mirror trapped in the outer lower-field regions of the plasma. The others, which have a small enough pitch angle between their velocity and the field, can stream continuously in one direction along the field. They are sometimes called passing particles. All the particles experience a cross-field drift due to the bending of the field lines. Since the total magnetic field in a tokamak is mostly toroidal and only weakly influenced by the plasma currents, the drift is quite well approximated by

$$v_d = \frac{\left(\frac{1}{2}mv_{\perp}^2 + mv_{\parallel}^2\right)}{qB_{\phi}R} = \frac{\left(\frac{1}{2}v_{\perp}^2 + v_{\parallel}^2\right)}{\Omega R} \tag{8.4.4}$$

for a particle with charge q, mass m, and parallel velocity $v_{\parallel}$. The direction of v_d is vertical (i.e., in the $\hat{\mathbf{z}}$ direction).

The drift orbit is determined by a superposition of the drift motion and the parallel motion along the field lines. The field lines orbit around the flux surfaces $1/q_s$ times the short way (poloidally) for each revolution the long way (toroidally), where q_s is the safety factor. Therefore, the $v_\parallel$ motion produces an effective rotational velocity in the poloidal cross section, $\mathbf{v}_p \approx (B_p/B_\phi)v_\parallel$, which is added to $\mathbf{v}_d$. For simplicity we shall consider only a circular cross section so that $B_p = B_\theta$ and $q_s = rB_\phi/RB_\theta$, although the principles remain similar for shaped cross sections.

In the poloidal cross section, then, the equation of motion of the gyrocenter is

$$\mathbf{v} = \mathbf{v}_d + \mathbf{v}_p = v_d\hat{\mathbf{z}} + v_p\hat{\boldsymbol{\theta}}. \tag{8.4.5}$$

This may be written in terms of minor radius r as

$$\frac{1}{r}\frac{dr}{d\theta} = \frac{v_d\sin\theta}{v_d\cos\theta + v_p}. \tag{8.4.6}$$

For passing particles it is reasonable to approximate v_d and v_p as constant to lowest order. We can then solve this orbit equation to get

$$r = r_0\left[1 + \frac{v_d}{v_p}\cos\theta\right]^{-1}, \tag{8.4.7}$$

where r_0 is the drift orbit minor radius when $\theta = \pi/2$. This orbit is approximately a circle shifted in the major radius direction by a distance

$$\Delta = r_0\frac{v_d}{v_p} = r_0\frac{\left(\frac{1}{2}v_\perp^2 + v_\parallel^2\right)}{\Omega R}\bigg/\frac{r_0 v_\parallel}{Rq_s}$$

$$= q_s\frac{\left(\frac{1}{2}v_\perp^2 + v_\parallel^2\right)}{\Omega v_\parallel}. \tag{8.4.8}$$

The direction of this shift depends on the relative signs of v_d and v_p, and hence on the direction of $v_\parallel$ with respect to the plasma current. A positive ion, moving in the same direction as the current, shifts outward. In the opposite direction the shift is inward. Notice that Eq. (8.4.8) confirms and generalizes the expression given in Section 8.2.3 for the orbit shift of fast ions. There we had implicitly assumed that $v_\perp$ was negligible. The $v_\perp = 0$ case gives the smallest shift for particles of given total energy. Then $\Delta = q_s\rho_i$, which will generally be substantially larger than ρ_i since tokamaks operate with q_s near the plasma edge typically ~ 3. Thus even for passing particles the drift orbit size substantially exceeds the Larmor radius (which we regard in this section as defined using the total particle energy: $\rho_i = m|\mathbf{v}|/ZeB$).

Mirror trapped particles have even bigger drift orbits and, more particularly, orbits with a very different shape. As illustrated in Fig. 8.14, unlike the approximately circular orbits of passing particles, mirror trapped particles have orbits that are banana shaped, as they bounce back and forth between their mirror reflection points. To calculate the orbits analytically from the drifts is very difficult for trapped particles. However, an appeal to the constants of the motion enables one to write down quite quickly the equations governing the orbit: They are the conservation of magnetic moment μ, the conservation of total energy $\frac{1}{2}mv_{\parallel}^2 + \mu B$, and finally a new conservation equation, [which arises because of the ignorability of the toroidal (ϕ) coordinate], the conservation of the toroidal canonical momentum

$$R\left(mv_{\parallel} + qA_{\phi}\right) = \text{constant.} \qquad (8.4.9)$$

The toroidal component of the vector potential A_{ϕ} is directly related to the poloidal flux (see Section 2.2) by $2\pi RA_{\phi} = \psi$.

Knowing A_{ϕ} the conservation equations provide an equation for the orbit, given initial values of $v_{\parallel}$ and $v_{\perp}$. These are still difficult to solve in general but by considering just the points where the orbit crosses the midplane we can rapidly estimate the drift orbit dimensions. Denote quantities by the suffixes 1 and 2 at these points. Then the ϕ momentum conservation equation may be written

$$-q\left(R_1 A_1 - R_2 A_2\right) = m\left(R_1 v_1 - R_2 v_2\right). \qquad (8.4.10)$$

Fig. 8.14. The projection in the poloidal plane of orbits of energetic particles in a tokamak. Passing particles have shifted approximately circular orbits. Trapped particles have larger banana orbits.

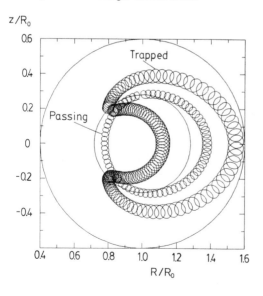

Now for trapped particles, the two points are both on the outside of the machine so we write

$$R_1 A_1 - R_2 A_2 = \int_2^1 \frac{\partial}{\partial R} (R A_\phi) \, dR = \int_2^1 R B_z \, dR \equiv \overline{R B_\theta} \Delta R,$$

(8.4.11)

where z is the vertical component, which is also the poloidal component on the midplane. Also the signs of $R_1 v_1$ and $R_2 v_2$ are opposite (since the $v_\parallel$ is opposite), so denoting their mean magnitude by $R v_\parallel$ we get the width of the banana orbit as

$$|\Delta R| = \frac{m}{q \overline{R B_\theta}} 2 R v_\parallel \approx 2 \frac{m v_\parallel}{q B_\theta},$$

(8.4.12)

where the averages are now left implicit. The trapping condition shows that a typical value of $v_\parallel / v$ is $\sim (r/R)^{1/2}$. If we substitute this value and express B_θ in terms of the safety factor q_s, we get

$$\Delta R \sim 2 q_s (R/r)^{1/2} \rho_i.$$

(8.4.13)

Incidentally, one can calculate the orbit shift Δ of passing particles by the same technique. In this case the quantity $R_1 A_1 - R_2 A_2$ must be written as $\approx R B_\theta 2 \Delta$, recognizing that the midplane points are now on opposite sides of the axis. Also v_1 and v_2 have the same sign, so

$$\Delta = \frac{m}{2 q R B_\theta} (R_1 v_1 - R_2 v_2).$$

(8.4.14)

If the perpendicular velocity $v_\perp$ is negligible, then $v_1 = v_2$ and so, since $R_1 - R_2 = 2r$, we get

$$\Delta = \frac{m v_\parallel}{q B_\theta} \frac{r}{R} = q_s \rho_i,$$

(8.4.15)

as before. The more general expression for nonzero $v_\perp$ [Eq. (8.4.8)], may be obtained by accounting for the $v_\parallel$ variation (see Exercise 8.14). Physically, the drift integration and momentum conservation are entirely equivalent.

Summarizing, then, the drift orbits are substantially larger than the Larmor radius, by a factor that for trapped particles may be typically $2 q_s (R/r)^{1/2} \sim 8$. This means that there will be substantial prompt loss of charged particle fusion products in all the present generation of magnetic confinement experiments. Moreover, if the pitch angle of the escaping particle can be measured, then one can calculate the trajectory of the drift orbit by numerical integration of the drift (or adiabatic conservation) equations. This then would give information on the birth position of the detected particle. Studies and development of these possibilities are currently proceeding.

8.4.3   *Ion probing beams*

In order to use active charged particle beams injected into the plasma rather than passively observing the particles produced in nuclear reactions, it proves convenient to employ much heavier species. This allows a sufficiently large Larmor radius to be obtained without having to enter the MeV energy range. For example, beams of singly charged thallium ions (atomic mass 203 or 205) can be produced with conventional ion accelerators with sufficient energy (100–200 keV) to penetrate tokamak plasmas. Other elements sometimes used include cesium and rubidium.

Two plasma parameters are of primary interest in diagnostic measurements with such heavy ion beam probes. The first, rather obviously, is the magnetic field itself, which may be deduced from the trajectory of the ion beam. Naturally, the emerging direction of the beam is determined by an integral over the beam path of the magnetic field, so that a complicated integral deconvolution is normally required to determine the spatial variation.

The most sensitive way to do this deconvolution is generally to observe, in addition to the primary beam, the second beam (or rather array of beams) generated by ionization in the plasma of the injected ions. For example, in the tokamak (where it is the small poloidal field that is of greatest interest, since the toroidal field is scarcely perturbed by the plasma) the birth position of a Tl^{++} ion produced from a Tl^{+} primary beam can be deduced from the poloidal position of the orbit (see Fig. 8.15). In principle, the poloidal field may then be deduced from the toroidal orbit displace-

Fig. 8.15. The orbits for ion beam probing are determined by the magnetic fields. In this illustration B is mostly into the page.

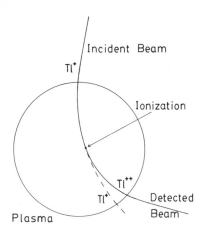

ment. In practice, although consistency with other estimates has been obtained [e.g., Hosea et al. (1973)], the uncertainties accumulating from various field perturbations prevent this from being an independently successful poloidal field measurement (as yet).

However, greater success attaches to measurements of the second plasma parameter of interest, namely, the plasma potential. This is measured by virtue of the energy change in the secondary beam (equal to eV_p) due to the

Fig. 8.16. The potential inside the EBT plasma as measured by an energetic ion beam experiment [after Colestock et al. (1978)].

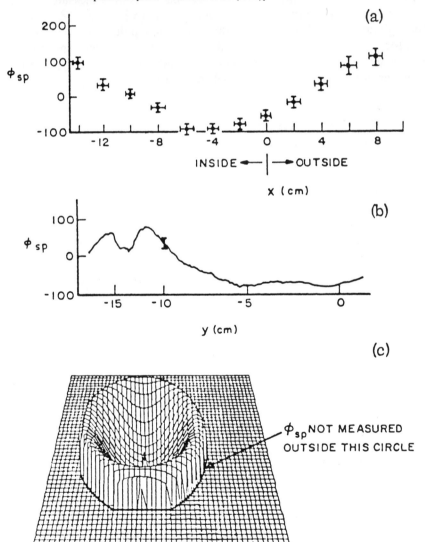

additional electrostatic potential energy gain of the ion. The location of the birth point is determined again by the orbit analysis and the potential by subsequent energy analysis of the secondary beam. This is naturally quite challenging since one is interested in energy changes of the order of typically a few hundred electron volts or less at an energy of about 100 keV. Nevertheless, essentially complete radial profiles of V_p have been obtained that no other noncontact diagnostic has been able to provide. These measurements are of particular interest for mirror-confined plasmas where the electrostatic potential plays a fundamental role in particle confinement.

In Fig. 8.16 an example is shown of the space potential measured in an Elmo bumpy torus plasma using a rubidium beam with energy 8–30 keV. The significance of these results is that the hollow potential profile forms only for certain types of discharge with more favorable confinement properties. Many of the operational details of such an ion beam potential measurement are discussed by Bienisek et al. (1980).

Further reading
The subject of electronic and ionic collisions is a vast and complicated one. Refer to such encyclopedic works as:

Massey, H. S. W., Burhop, E. H. S., and Gilbody, H. B. (1974). *Electronic and Ionic Impact Phenomena*. 2nd ed. London: Oxford.

More recent reviews in the fusion context may be found in:

McDowell, M. R. C. and Ferendeci, A. M., eds. (1980). *Atomic and Molecular Processes in Controlled Thermonuclear Fusion*. New York: Plenum.

Fusion reactions are discussed in most books on controlled fusion, for example:

Glasstone, S. and Lovberg, R. H. (1960). *Controlled Thermonuclear Reactions*. New York: Van Nostrand.

Valuable reviews of ionic process diagnostics can be found in:

Stott, P. E., et al., eds. (1982). *Diagnostics for Fusion Reactor Conditions*, Proc. Int. School Plasma Phys., Varenna. Brussels: Commission of E.E.C.
Sindoni, E. and Wharton, C. (1978). *Diagnostics for Fusion Experiments*, Proc. Int. School Plasma Phys., Varenna. London: Pergamon.

Exercises
8.1 The purpose of this exercise and the next is to obtain an estimate of the resonant charge-exchange cross section. The perturbation theory of quantum mechanics shows that the probability of finding an electron in state 2 (attached to ion 2) at a time t after it was initially in state 1 (attached to ion 1) in the situation in which ions are treated as stationary, is

$$P_{12} = \left| \langle 2|V|1 \rangle \frac{t}{\hbar} \right|^2 = \frac{t^2}{\hbar^2} \left| \int \psi_2^* V \psi_1 \, d^3 r \right|^2 ,$$

where ψ_1, ψ_2 are the wave functions and V the perturbation Hamiltonian. In our case of interest where we have two protons, if we take the origin at proton 1 and let the position vectors of the electron and proton 2 be **r** and **R**, respectively, then

$$\psi_1 = \frac{1}{\pi^{1/2}a_0^{3/2}}\exp\left(\frac{-r}{a_0}\right),$$

$$\psi_2 = \frac{1}{\pi^{1/2}a_0^{3/2}}\exp\left(\frac{-|\mathbf{r}-\mathbf{R}|}{a_0}\right),$$

$$V = \frac{e^2}{4\pi\varepsilon_0 r};$$

ψ_1 and ψ_2 are the ground state wave functions for hydrogen. By evaluating the overlap integral using these formulas show that the transition probability is

$$P_{12} = \frac{t^2}{\hbar^2}\left|\frac{e^2}{4\pi\varepsilon_0}\frac{1}{a_0}\left(\frac{R}{a_0}-1\right)e^{-R/a_0}\right|^2.$$

8.2 Now we consider actual collisions with impact parameter b and velocity v. We can calculate what happens in such cases by approximating the collision as an encounter between the two ions of time duration $t = b/v$ at constant distance b. If the collision is distant ($b \gg a_0$), P_{12} will be small. As b becomes smaller, P_{12} increases. Obtain an equation for the critical impact parameter (b_c) at which $P_{12} = \frac{1}{4}$. For impact parameters less than b_c the electron may, with roughly equal probability, emerge attached to either ion, so write the cross section as approximately

$$\sigma = \tfrac{1}{2}\pi b_c^2.$$

From your b_c equation evaluate (by some simple numerical solution) σ for proton–hydrogen charge exchange at 100 eV collision energy and compare it with Fig. 8.2.

8.3 Using the following identities for the atomic parameters,

$$\alpha \equiv \frac{e^2}{4\pi\varepsilon_0 \hbar c}, \qquad R_y \equiv \left(\frac{e^2}{4\pi\varepsilon_0}\right)^2\frac{m_e}{2\hbar^2}, \qquad a_0 \equiv \frac{4\pi\varepsilon_0}{e^2}\frac{\hbar^2}{m_e},$$

$$2R_y a_0 = \frac{e^2}{4\pi\varepsilon_0}, \qquad \left(\frac{2R_y}{m}\right)^{1/2} = \alpha c, \qquad \alpha = \frac{2R_y a_0}{\hbar c},$$

show that if the ionization coefficient is taken as $\langle \sigma_e v_e \rangle \approx \pi a_0^2 2\alpha c$ then the ratio of recombination to ionization rates is approximately

$$\frac{\langle \sigma_r v_e \rangle}{\langle \sigma_e v_e \rangle} \approx 5\sqrt{2}\alpha^3 \left(\frac{R_y}{T_e} \right)^{3/2}$$

8.4 Obtain Eq. (8.1.18).

8.5 Show that the temperature at which maximum contribution to the fast neutral flux occurs is given by Eq. (8.1.21).

8.6 Prove that the increase in apparent temperature due to attenuation is given by Eq. (8.1.22).

8.7 Consider a plasma with two ion species: protons and a single type of impurity with charge Z. From the definition of Z_{eff} show that the proton density is given by

$$n_p = \left[\frac{Z - Z_{eff}}{Z - 1} \right] n_e.$$

8.8 The Gamow cross section [Eq. (8.3.4)] is based on a calculation of the probability that an incident nucleus can tunnel quantum mechanically through the repulsive Coulomb barrier of another and so allow a nuclear reaction to occur. A simple approximate way to do the calculation is to use the one-dimensional Schrodinger equation

$$\left[\frac{\hbar}{2m} \frac{\partial^2}{\partial x^2} + (E - V(x)) \right] \psi = 0,$$

where E is the total energy and the potential energy is taken as

$$V(x) = \frac{e^2}{4\pi\varepsilon_0 x} \qquad \text{for } x > 0$$

$$= 0 \qquad \text{for } x < 0.$$

Show by a WKBJ solution that the probability of a particle, incident from $x = +\infty$, tunneling through this barrier is proportional to

$$\exp - \left[\frac{e^2}{4\pi\varepsilon_0} \frac{\sqrt{2m}}{\hbar} \frac{\pi}{2E^{1/2}} \right].$$

Calculate the numerical value of the exponent for an energy of 1 keV. How does it compare with Eq. (8.3.4)? (*Note*: the $1/E$ term in σ_n arises because the real collision is three dimensional and angular momentum must be accounted for.)

8.9 Write the reaction rate as an integral over two velocity distributions in the lab frame. Transform to velocity coordinates $\mathbf{V}$ and $\mathbf{v}$ (the c.m. and relative velocities, respectively) to obtain Eq. (8.3.16); in particular show that the Jacobian of the transformation is unity.

8.10 Consider the general problem of collisions of two particles m_1, m_2 whose (lab frame) velocities are $\mathbf{v}_1, \mathbf{v}_2$. Show that the total kinetic energy of the system can be written

$$E = \tfrac{1}{2}m_1 v_1^2 + \tfrac{1}{2}m_1 v_2^2 = \tfrac{1}{2}MV^2 + \tfrac{1}{2}mv^2,$$

where $M = m_1 + m_2$ and $\mathbf{V} = (m_1\mathbf{v}_1 + m_2\mathbf{v}_2)/(m_1 + m_2)$ are the center-of-mass mass and velocity, and $m = m_1 m_2/(m_1 + m_2)$ and $\mathbf{v} = \mathbf{v}_1 - \mathbf{v}_2$ are the reduced mass and the relative velocity, respectively. If the distributions of particles 1 and 2 are Maxwellian with equal temperature T show that the reaction rate (per particle) is

$$\langle \sigma v \rangle = \left(\frac{1}{\pi m} \right)^{1/2} \left(\frac{T}{2} \right)^{-3/2} \int_0^\infty \sigma(K) \exp\left(\frac{-K}{T} \right) K \, dK.$$

Here $K = \tfrac{1}{2}mv^2$ (σ is independent of V, of course).

8.11 To calculate the mean energy E, weighted by the reaction rate, show that the mean c.m. energy $(\tfrac{1}{2}MV^2)$ is $\tfrac{3}{2}T$ and that the mean relative energy (K) is given by

$$\overline{K} \equiv \frac{\langle \sigma v K \rangle}{\langle \sigma v \rangle} = T^2 \frac{d}{dT} \left(\ln\left[T^{3/2} \langle \sigma v \rangle \right] \right),$$

regardless of the form of σ.

8.12 Substitute a Gamow form of the cross section

$$\sigma(K) = \frac{A}{K} \exp\left(\frac{-B}{K^{1/2}} \right)$$

into the expression (8.3.22) for $\langle \sigma v \rangle$. Evaluate the integral approximately by expressing the total exponent in the integrand approximately as a Taylor expansion up to second order about its maximum value to obtain

$$\langle \sigma v \rangle \approx \frac{4}{\sqrt{m}} \left(\frac{2}{3} \right)^{1/2} \left(\frac{B}{2} \right)^{1/3} A \frac{1}{T^{2/3}} \exp\left[-3 \left(\frac{B}{2} \right)^{2/3} \frac{1}{T^{1/3}} \right].$$

By using the values in Eq. (8.3.4) appropriately converted to SI units and center-of-mass energy K (rather than stationary target energy E_D), obtain values for the coefficients in this expression and compare them to Eq. (8.3.5).

8.13 Show from the expression (8.3.5) for the reaction rate that the error ΔT in deduced temperature caused by an error Δn_D in deuteron density n_D

is given by

$$\frac{\Delta T}{T} = \frac{2}{\left(6.7/T^{1/3} - 2/3\right)} \frac{\Delta n_D}{n_D} \approx \frac{T^{1/3}}{3.3} \frac{\Delta n_D}{n_D}.$$

8.14 Show that for a passing particle the conservation of energy and magnetic moment leads to a difference in $v_\parallel$ between the inner and outer orbit midplane positions of

$$\Delta v_\parallel \approx \frac{v_\perp^2}{v_\parallel} \frac{r}{R}.$$

Hence approximate Eq. (8.4.14) on the basis $\Delta v_\parallel \ll v_\parallel$ and $\Delta R \ll R$ to obtain Eq. (8.4.8).

Appendix 1   Fourier analysis

For a function $f(t)$, defined for $-\infty \leq t \leq \infty$, we define the *Fourier transform* as

$$F(\omega) = \frac{1}{2\pi} \int_{-\infty}^{\infty} e^{i\omega t} f(t)\, dt. \tag{A1.1}$$

We shall presume this integral to exist (in some sense) for all the functions in which we are interested. The *inverse transform* is

$$f(t) = \int_{-\infty}^{\infty} e^{-i\omega t} F(\omega)\, d\omega. \tag{A1.2}$$

The rationale for the particular sign convention we are adopting is that we can regard $F(\omega)$ as the amplitude of the particular component of $f(t)$ that varies as $\exp(-i\omega t)$. We shall mostly be concerned with waves, which it is convenient to regard as given by a variation $\exp i(\mathbf{k} \cdot \mathbf{x} - \omega t)$, so that the wave propagates in the direction $+\mathbf{k}$ for positive ω. That is why we take the ω component of f to be proportional to $\exp(-i\omega t)$ (not $+i\omega t$). For spatial transforms we shall reverse the sign. Associating the factor $1/2\pi$ with the forward transform rather than the inverse is also appropriate if we wish to regard $F(\omega)$ [rather than $F(\omega)/2\pi$] as the mode amplitude.

The convolution of two functions F and G is defined as

$$F \mathbf{x} G(\omega) \equiv \int_{-\infty}^{\infty} F(\omega') G(\omega - \omega')\, d\omega'. \tag{A1.3}$$

The *convolution theorem* of Fourier transforms states that the transform of the convolution of two functions is the product of the transforms. That is

$$\int e^{-i\omega t} F \mathbf{x} G\, d\omega = \int e^{-i\omega t} \int F(\omega') G(\omega - \omega')\, d\omega'\, d\omega$$

$$= \int\int e^{-i(\omega - \omega')t} e^{-i\omega' t} F(\omega') G(\omega - \omega')\, d\omega'\, d\omega$$

$$= \int e^{-i\omega' t} F(\omega')\, d\omega' \int e^{-i\omega'' t} G(\omega'')\, d\omega''$$

$$= f(t) g(t). \tag{A1.4}$$

A particular example of recurring interest is when one has a sample of the function $f(t)$ over only a finite time duration T. One can consider this as defining a new function $f'(t)$ that is equal to $f(t)$ for $-T/2 \leq t \leq +T/2$, but zero otherwise. Clearly

$$f'(t) = f(t)g(t),$$ (A1.5)

where

$$g(t) = 1, \qquad -T/2 \le t \le +T/2,$$
$$= 0, \qquad \text{otherwise.}$$ (A1.6)

The Fourier transform of f', say F', is then immediately given by the convolution theorem as

$$F'(\omega) = F \mathbf{x} G \equiv \int_{-\infty}^{\infty} F(\omega')G(\omega - \omega')\, d\omega'.$$ (A1.7)

Now the Fourier transform of g may be readily evaluated:

$$G(\omega) = \frac{1}{2\pi} \int_{-\infty}^{\infty} e^{i\omega t} g(t)\, dt = \frac{1}{2\pi} \int_{-T/2}^{T/2} e^{i\omega t}\, dt = \frac{\sin(\omega T/2)}{\pi\omega},$$
(A1.8)

whose shape is illustrated in Fig. A1.1. The central lobe of $G(\omega)$ has a full width at half maximum approximately equal to the distance to the first zero, viz. $\omega \sim 2\pi/T$. The effect of the convolution upon the spectrum is that features in $F(\omega)$ are smoothed out over a range of the order of the width of $G(\omega)$. Thus, a very narrow feature on $F(\omega)$, say a single line, acquires a finite width, its shape being that of G but centered on the line center. This is the basis for the general principle that a wave train of finite duration has a finite frequency width of order $1/T$. The principle is equally valid for any shape of $g(t)$ of some finite width, though the detailed shape of $G(\omega)$ will obviously be different.

Fig. A1.1. The function $\sin(\omega T/2)/\pi\omega$, which is the Fourier transform of the box function. It acts as the convolution shape when dealing with Fourier transforms of finite duration T.

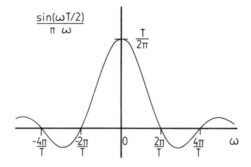

Since $F \mathbf{x} G$ and fg are transform pairs we may write down the transform in the form

$$\frac{1}{2\pi} \int e^{i\omega t} f(t) g(t) \, dt = \int F(\omega') G(\omega - \omega') \, d\omega'. \qquad (A1.9)$$

As a special case of this relationship we can take $\omega = 0$, in which case we obtain a form of *Parseval's theorem*,

$$\int f(t) g(t) \, dt = 2\pi \int F(\omega) G(-\omega) \, d\omega. \qquad (A1.10)$$

The more usual form of this relationship is obtained by noting that

$$G(-\omega) = \frac{1}{2\pi} \int e^{-i\omega t} g(t) \, dt = \left[\frac{1}{2\pi} \int e^{i\omega t} g^*(t) \, dt \right]^*, \qquad (A1.11)$$

where the asterisk denotes complex conjugate. Therefore, if we replace the function $g(t)$ by $g^*(t)$ (and vice versa) in both sides of the expression (A1.10) we get

$$\int f(t) g^*(t) \, dt = 2\pi \int F(\omega) G^*(\omega) \, d\omega \qquad (A1.12)$$

and in particular, putting $g = f$,

$$\int_{-\infty}^{\infty} |f(t)|^2 \, dt = 2\pi \int_{-\infty}^{\infty} |F(\omega)| \, d\omega^2. \qquad (A1.13)$$

The importance of this result arises when f is some physical quantity such as electric field, for which $|f|^2$ corresponds to an energy or power (density). If f is a physical quantity it is real, and one may readily demonstrate then that $F(-\omega) = F^*(\omega)$ and hence that $|F(-\omega)|^2 = |F(\omega)|^2$. We then can write Parseval's theorem in the form

$$\int_{-\infty}^{\infty} |f(t)|^2 \, dt = 4\pi \int_{0}^{\infty} |F(\omega)|^2 \, d\omega \qquad (A1.14)$$

by combining together the positive and negative frequencies. In fact, positive and negative frequencies in a physical quantity are never distinguishable. For, suppose they were – suppose for example, that we could construct a "super spectrometer" that filtered out all but the positive frequency ω; then the output of this spectrometer would be $F(\omega)\exp(-i\omega t)$, a complex quantity. But physical quantities are real. Therefore, this is impossible. (One should not confuse the common practice of regarding a physical quantity as being represented by a complex number as implying that the quantity itself is complex. This representation is shorthand for the quantity being the *real part* of the complex number. In our present discussion $f(t)$ is the quantity itself, so if a complex representation is being used, the real part must be taken before applying Parseval's theorem in its final form.)

Naturally, one can construct spectrometers or filters that allow only a passband $d\omega$ at frequency $\pm\omega$ to pass (ω a positive number). The output of such a filter may be denoted $f'(t)$, that is,

$$f'(t) = \int_{d\omega} \left[F(\omega')e^{i\omega't} + F(-\omega')e^{-i\omega't} \right] d\omega'. \tag{A1.15}$$

Then applying Parseval's theorem to the filtered signal, we find that it has a total "energy"

$$dW = \int_{-\infty}^{\infty} |f'(t)|^2 \, dt = 4\pi |F(\omega)|^2 \, d\omega. \tag{A1.16}$$

Thus the energy dW contained in $f(t)$ in the frequency element $d\omega$ is $4\pi |F(\omega)|^2 \, d\omega$. That is, the spectral density of the energy is

$$\frac{dW}{d\omega} = 4\pi |F(\omega)|^2, \qquad \omega \geq 0. \tag{A1.17}$$

This then demonstrates that once having written down Parseval's theorem, one may identify the energy at a given frequency with the frequency integrand in the manner indicated.

When Fourier transforms in space as well as time are taken, the requirement that the quantity be real imposes still only one symmetry relation on the transform, viz.

$$F(-\mathbf{k}, -\omega) = F^*(\mathbf{k}, \omega). \tag{A1.18}$$

When accounting for this in Parseval's theorem we may again combine positive and negative frequencies. However, $\mathbf{k}$ may then range over all possible directions. This is a way of saying we *can* distinguish the direction of propagation of a wave even though we cannot distinguish positive and negative frequencies.

The topic of *generalized functions*, sometimes called distributions, arises frequently in using Fourier transforms. They are most easily understood as a singular limit of a sequence of (ordinary) functions. As the most important example, consider the *Dirac delta function*, whose defining property can be considered to be that its convolution with any function gives just the function itself: $F \mathbf{x} \, \delta = F$. The delta function arises naturally from our previous discussions of the box function $g(t)$ defined in Eq. (A1.6). The Fourier transform of the box is the sinc function $G(\omega) = \sin(\omega T/2)/\pi\omega$. If we now let $T \to \infty$ the width of the sinc function $\to 0$ but its height simultaneously $\to \infty$ in an area-preserving manner. Denote the limit of G

as $T \to \infty$ by G_∞. Then, since clearly in the limit eq. (A1.7) becomes

$$F'(\omega) = F \mathbf{x} G \to F(\omega), \tag{A1.19}$$

we have

$$F(\omega) = \int_{-\infty}^{\infty} F(\omega') G_\infty(\omega - \omega') \, d\omega'. \tag{A1.20}$$

But this is just the property required of the delta function. Thus,

$$\delta(\omega) = G_\infty(\omega) = \lim_{T \to \infty} \frac{1}{2\pi} \int_{-T/2}^{T/2} e^{i\omega t} \, dt. \tag{A1.21}$$

Frequently, the notation is simplified by putting $\pm \infty$ as the limits of integration, and we say that the Fourier transform of a constant is a delta function.

Sometimes, when computing power, we need the "square of a delta function." This must be understood in terms of these limiting processes. Consider

$$\lim_{T \to \infty} \int_{-\infty}^{\infty} |G|^2 h(\omega) \, d\omega = \int_{-\infty}^{\infty} \delta^2(\omega) h(\omega) \, d\omega. \tag{A1.22}$$

Since G becomes more and more localized at $\omega = 0$ as $T \to \infty$ it is clear that if $h(\omega)$ is any function sufficiently well behaved at $\omega = 0$ this becomes

$$
\begin{aligned}
\int \delta^2(\omega) h(\omega) \, d\omega &= \lim_{T \to \infty} h(0) \int_{-\infty}^{\infty} |G|^2 \, d\omega \\
&= \lim_{T \to \infty} h(0) \int_{-\infty}^{\infty} \frac{1}{2\pi} |g(t)|^2 \, dt \\
&= \lim_{T \to \infty} \left[h(0) \frac{T}{2\pi} \right] \\
&= \int_{-\infty}^{\infty} \lim_{T \to \infty} \frac{T}{2\pi} \delta(\omega) h(\omega) \, d\omega,
\end{aligned}
\tag{A1.23}
$$

where we have used Parseval's theorem. Thus we must identify

$$\delta^2(\omega) = \lim_{T \to \infty} \left[\frac{T}{2\pi} \delta(\omega) \right]. \tag{A1.24}$$

Naturally, the nonconvergence of this limit is associated with the fact that integrating a constant power over an infinite time leads to infinite energy. More often than not one is interested in the mean energy per unit time. This will involve $\delta^2(\omega)/T$, which naturally has a perfectly good (generalized function) limit $\delta/2\pi$.

When dealing with periodic functions having period τ, say, the continuous Fourier transform gives rise to a series of delta functions at frequencies

$\omega_n = 2\pi n/\tau$, where n is an integer. It is then often more convenient to work in terms of *Fourier series* (rather than being involved in the mathematical complexities of the generalized functions). These are

$$F_n = \frac{1}{\tau} \int_0^\tau e^{i\omega_n t} f(t)\, dt,$$

$$f(t) = \sum_{n=-\infty}^{\infty} e^{-i\omega_n t} F_n. \tag{A1.25}$$

Sometimes it is preferable to use sine and cosine series rather than complex series. These are

$$C_n = \frac{2}{\tau} \int \cos(\omega_n t) f(t)\, dt,$$

$$S_n = \frac{2}{\tau} \int \sin(\omega_n t) f(t)\, dt, \tag{A1.26}$$

$$f(t) = \frac{C_0}{2} + \sum_{n=1}^{\infty} C_n \cos \omega_n t + S_n \sin \omega_n t.$$

Appendix 2 Errors, fluctuations, and statistics

The errors that can enter into any kind of measurement, and so limit its accuracy, may be classified into two main types.

First *systematic* errors arise from inaccuracies in calibration or the general performance of experimental instruments. The errors are systematic when they are consistent and reproducible. For example, in measuring the length of an object using a ruler whose own length markings are, say 1% too close together, a consistent overestimate by 1% will be obtained. The measurement may be repeated many times using the same ruler, but will give the same error. Of course, in the complicated electronic and mechanical systems used in sophisticated diagnostics, many more complicated possibilities exist for systematic errors to arise. Nevertheless, the principle remains that these errors cannot be revealed by repeated measurement with the same instrument. There is very little to be said in the way of general analysis of systematic errors except that the best way to reveal them is to compare measurements of the same quantity using different instruments (or even different techniques). In the absence of such a check, the experimenter must attempt to estimate the systematic uncertainties from a fundamental knowledge of how an instrument works and what potential flaws there are, or else from his own experience.

The second type of error is random or *statistical*. There are many possible sources of such errors. In our length measurement example they may arise from slight misreadings of the scale due to misalignment or parallax, for example. The governing principle here is that statistical errors *are* revealed by repeated measurements with the same instrument. Therefore, it is possible, in principle to overcome them in part by repeated measurement. Also considerable insight into their effect is possible through a general statistical analysis, some key features of which will be summarized in a moment.

It should be noted that a third possible category of error, all too familiar to most of us, is a plain blunder or mistake, such as writing down the wrong number or doing arithmetic incorrectly. We assume hereafter that care or self-correction can exclude such mistakes.

A major source of statistical uncertainty is the *fluctuation* of physical quantities involved in the measurement. These may be present in the quantity to be measured (for example the temperature), in the physical property used for the measurement (for example the radiation intensity), or in the measuring instruments (for example the detector or amplifying electronics). In the present discussion we shall lump these all together, even though when the fluctuations are genuinely in the quantity to be measured, we may be "correct" when we get different results from succeeding measurement. (In other words the temperature, for example, really did change.)

The expression "accuracy" is used technically to mean the overall reliability of a measurement value, accounting for both systematic and random errors. "Precision," on the other hand, expresses how reproducible a value is; that is, to what extent random errors have been minimized, regardless of the effect of systematic errors. Thus a precise measurement may not necessarily be particularly accurate.

Suppose we make a series of measurements of some statistically varying quantity x. Let the number of measurements be N and the result of the ith measurement be x_i. We call this set of measurements a sample, and then the mean value of the sample is

$$\mu_N = \frac{1}{N} \sum_{i=1}^{N} x_i \tag{A2.1}$$

and the standard deviation of the sample is

$$\sigma_N = \left[\frac{1}{(N-1)} \sum_{i=1}^{N} (x_i - \mu_N)^2 \right]^{1/2}. \tag{A2.2}$$

The standard deviation obviously represents the spread of the different measurement values about the mean ($N \geq 2$).

Intuitively we can see that as N becomes very large the values of μ_N and σ_N will tend to constants. (We make no pretence of rigor here.) It is sometimes helpful to think of the measurements as being random samples drawn from a large population of possible results. The limiting values of μ and σ are then the population mean and population standard deviation, respectively. From the viewpoint of attempting to obtain reliable measurements of some quantity subject to random errors, the usual presumption is that the required value is the population mean μ. Then the statistical uncertainty to be attributed to a *single* measurement x_k is approximately the population standard deviation σ.

Of course, if we have N measurements and take their mean μ_N, we actually have considerably greater confidence that this is close to μ than we would have in a single measurement. In other words, we may regard μ_N as a random variable, different for different measurement sets, itself distributed with a certain mean and standard deviation, and the standard deviation of μ_N is smaller than σ. How much smaller is determined by the central limit theorem of statistics. It states that for large enough N the quantity μ_N is a random variable distributed with a Gaussian probability distribution having a standard deviation $\sigma/\sqrt{N}$. The standard deviation of the sample means is usually called the *standard error* of the measurement, since it is an estimate of the statistical uncertainty in attributing the correct value (μ) to the sample mean (μ_N). We normally don't know what the population standard deviation σ is. The obvious estimate to use instead is σ_N, which we do know. Using this estimate it is usual to write the standard error as

$$s_N = \frac{\sigma_N}{\sqrt{N}} = \left[\frac{1}{N(N-1)} \sum_{i=1}^{N} (x_i - \mu_N)^2 \right]^{1/2}. \tag{A2.3}$$

To summarize, if we attempt to estimate the mean value of a population (of measurements) using the mean of a sample of N measurements, the standard error in the estimate decreases as the square root of N. Eventually, of course, if time allows us to increase N sufficiently, the statistical error will become small enough that systematic errors will dominate. At that point no further accuracy is gained even though the estimate becomes more precise.

The probability distribution $p(x)$ of a population gives the probability that the result of a single measurement will lie in the range dx at x as $p(x)\,dx$. Two specific forms of probability distribution are of dominant importance in error and fluctuation analysis.

The first is the Gaussian distribution

$$p(x) = \frac{1}{\sigma\sqrt{(2\pi)}} \exp\left(\frac{-x^2}{2\sigma^2} \right), \tag{A2.4}$$

where σ is the population standard deviation. It gains its importance in part from the central limit theorem, as mentioned earlier, which shows that any quantity that is the mean of a large enough sample of independent measurements will be distributed with Gaussian probability distribution. The ubiquity of this form is recognized in the practice by statisticians of referring to the Gaussian as the normal distribution.

The second very important distribution is the Poisson distribution. It is a discrete distribution for counting; the variable can take integral values k with probability

$$p_k = \frac{e^{-\mu}\mu^k}{k!}. \tag{A2.5}$$

The Poisson distribution governs situations in which events of some type occur at a constant rate, but independent of one another. For example, the number of radioactive decays observed in a given time interval in a large sample of radioactive material is governed by a Poisson probability distribution, or the number of photons arriving in a certain time from a source of constant intensity. As implied by the notation already, the mean of a Poisson population is μ, the value of the exponent. Another property, which may be verified by elementary summation, is that the standard deviation of the Poisson distribution is

$$\sigma = \sqrt{\mu}. \tag{A2.6}$$

For our purposes, the most important application of the Poisson distribution is in discussing the fluctuations in detected signal attributable to the statistical arrival of electromagnetic radiation quanta, that is, photon statistics. Suppose we have a source of constant intensity that we wish to measure. We observe for a time T and (in effect) count the number of photons detected, k, say. If the mean photon arrival rate per unit time is ν, then clearly k has a Poisson distribution with mean $\mu = \nu T$. The standard deviation in k will give us the uncertainty in our estimate of ν from a single measurement. It will be $\sigma = \sqrt{\mu}$. Notice that when we know that the photons are Poisson-distributed, one measurement allows us to estimate both μ and σ. (This may be thought of as arising from the fact that photon counting is adding up lots of little time duration elements into the duration T; so, in a way, we have already done many measurements.) The relative uncertainty in our estimate of the intensity (in other words, of ν) is, therefore,

$$\frac{\Delta \nu}{\nu} \approx \frac{\sigma}{\mu} = \frac{1}{\sqrt{\mu}} \approx \frac{1}{\sqrt{k}}. \tag{A2.7}$$

The relative error is inversely proportional to the square root of the total number of photons counted. The longer the time duration T, the larger μ, and so the more accurate the estimate of ν.

When, as is often the case, errors arise in different ways at different stages in the measurement process one usually assumes that they are independent. That being so, the total estimate of uncertainty is obtained as the square root of the sum of the squares of the error in the final value arising from each individual error process. Specifically, if the quantity required, (x) is a sum of other quantities (x_j),

$$x = \sum_j x_j, \tag{A2.8}$$

then

$$\sigma_x = \left[\sum_j \sigma_j^2 \right]^{1/2}. \tag{A2.9}$$

Whereas, if x is a product

$$x = \prod_j x_j, \tag{A2.10}$$

then we express each x_j as having a difference Δ_j from its mean μ_j, $x_j = \mu_j + \Delta_j$, and then retain only the lowest order terms in the Δ_j, so that

$$x \approx \left[\prod_j \mu_j \right] \left[1 + \sum_j \frac{\Delta_j}{\mu_j} \right] \approx \mu_x \left[1 + \sum_j \frac{\Delta_j}{\mu_j} \right]. \tag{A2.11}$$

We can, therefore, regard x as approximately a sum of random variables $\mu_x \Delta_j / \mu_j$ and, hence,

$$\frac{\sigma_x}{\mu_x} \approx \left[\sum_j \frac{\sigma_j^2}{\mu_j^2} \right]^{1/2}. \tag{A2.12}$$

Summarizing, the error in a *sum* of independent variables is the square root of the sum of the squares of the individual errors. The fractional error in a *product* is approximately the square root of the sum of the squares of the individual fractional errors.

Appendix 3   Survey of radiation technology

The purpose of this appendix is to provide a brief overview of the currently available technology of radiation detection and generation. Through this material an impression may be gained and guidance obtained on what is technically feasible in terms of plasma experiments. The tables of detectors and sources are far from complete and give only representative examples of what is available. Naturally, complete design of experiments requires much more detailed information, which is usually available only from manufacturers of the equipment. In the Further reading section some guidance is given on how to go about obtaining the details one needs as well as references to more general texts and reference works on radiation technology.

Detectors

The performance of detectors of electromagnetic radiation may be characterized by a variety of parameters. These parameters are not all independent and are not all appropriate for describing any specific detector, but the more important of them are described in the following.

The most immediate question concerning a detector's operation is what *frequency* of electromagnetic radiation it is sensitive to. This may, of course, also be specified in terms of wavelength λ or photon energy, the latter being more appropriate for radiation, such as x-rays, in which the photon energy is large.

A second parameter is the *speed of response* to a change in the radiation intensity. This may be expressed as a response time or as a (video) bandwidth. The (video) bandwidth is usually *not* the same as the width of the radiation frequency spectrum of the detector sensitivity (although it cannot be greater).

The *responsivity* describes the detector signal output (volts) per radiation power input (watts). This is important in optimizing the connection to preamplifiers and other signal conditioners. It sometimes determines the signal-to-noise performance although most detectors are packaged with amplifying electronics so that the responsivity is of less fundamental significance for system design.

The *noise equivalent power* (NEP) is a fundamental parameter of detector performance, defined as that radiation power that would produce a signal equal to the root mean square (r.m.s.) noise level of the detector. The noise usually consists of a broad spectrum across the video bandwidth of the detector with the result that the mean square noise is equal to the integral of the mean square noise power spectrum over the video passband of the conditioning electronics. A wider passband thus leads to higher noise level, so the NEP must be defined in terms of the mean square voltage per unit frequency. Since for most video detectors the signal voltage (not voltage squared) is proportional to radiation power, this means that the NEP is equal to the square root of the mean square noise voltage spectrum divided by the responsivity. Thus it is expressed in watts per $\sqrt{\text{Hz}}$. An exception to this is when heterodyne radiation detection is used, because then the signal (for fixed local oscillator power) is proportional to the square root of the radiation power. Therefore, the NEP for a heterodyne system is in the form of watts per hertz. Note however, that the heterodyne NEP depends on local oscillator power so that it is a property of an entire receiver system not just of the detector alone. In heterodyne detection there is a direct relationship between the bandwidths of the radiation and of the intermediate frequency (IF) at least prior to any subsequent second detector and smoothing. Therefore, the signal to noise at the first IF is proportional to the radiation's spectral power density (watts per hertz). A heterodyne system generally responds to a single electromagnetic mode, and as a result has an étendue equal to λ^2. Blackbody radiation at a temperature T [see Eq. (5.3.36)] then would give rise to a power accepted by the system of $T\,d\nu$ (watts) in a frequency bandwidth $d\nu$ ($= d\omega/2\pi$). Therefore, the noise equivalent power (in watts per hertz) can be expressed as a *noise temperature* T_n, where T_n (J) = NEP (W Hz^{-1}). (If T_n is expressed instead in degrees, it must be multiplied by Boltzmann's constant in this formula.) The noise temperature is the temperature a blackbody must have if its radiation is to give unity signal to noise.

Naturally, for improving signal to noise, the lower the NEP the better. In order to have a parameter for which "bigger is better" one defines the detectivity, D as the inverse of the NEP. Its units are therefore hertz$^{1/2}$ watts^{-1}. (Detectivity is rarely used for heterodyne systems.) For many types of detector, the detectivity is inversely proportional to the square root of the area of the detector. The basic reason for this is that the effective mean square noise voltage produced by the detector is proportional to its area while the responsivity is effectively constant. These scalings are only approximate and only appropriate when the noise is dominated by the detector element (or background radiation); nevertheless it is useful to define a parameter D^*, equal to the product of D times the square root of the area, which is called the *specific detectivity*. When comparing performance of different detectors, D^* is often used. Its units are usually centimeters$^{1/2}$ hertz$^{1/2}$ watts^{-1} (SI not withstanding).

Detectors of more energetic photons for the visible or shorter wavelengths tend increasingly to be limited not by noise arising from thermal fluctuations in the detector or electronics, but by photon statistics and related fluctuations. For these detectors the most convenient figure of merit describing their performance is the *quantum efficiency* (QE). In such detectors, many of which work by electron multiplication, a single photon, if detected, causes a macroscopically measurable signal. The quantum efficiency is simply the proportion of photons striking the detector that are detected. For high-energy photons this can be made close to unity. In most plasma applications the noise levels in such photon detectors are dominated by the photon statistics of the plasma "light." However, in very low-level detection situations, the limiting noise is the fluctuations in the *dark current*, which is the current signal in the absence of any photons.

Many of the photon-counting detectors for higher photon energy work well also for the detection of charged elementary particles (such as protons) with similar characteristics to photon detection. More specialized detectors are necessary for detection of neutrons because their interactions are much weaker. In either case the detection efficiency is the key parameter, like the photon efficiency for photon detection.

Table A1 summarizes representative examples of the types of detectors that have been used in plasma diagnostic applications. The list is by no means exhaustive or complete. It merely attempts to give an indication of the typical wavelength/photon energy range, NEP/quantum efficiency, and response speed of available detectors. These parameters are the most crucial in determining the appropriateness of a detector in a specific application.

Coherent sources of electromagnetic radiation

In almost all plasma diagnostics that use externally generated radiation to probe the plasma, the source used is of a coherent type. The reason for this is that more often than not the radiation is required to be both intense and narrow band. Thermal broad band sources, such as lamps, rarely meet these requirements satisfactorily.

Broadly speaking, one may divide the sources available into continuous wave (CW) and pulsed sources. Table A2 summarizes some illustrative examples of CW sources from millimeter wavelength down to the shortest wavelength readily available. The different types of sources are briefly as follows.

Electron tube types, notably klystrons, backward wave oscillators (carcinotrons), and gyrotrons use free-electron beams in vacuum resonators for the wave generation. They tend to be cumbersome and expensive but give high power and, in some cases, good tunability and stability.

Solid state microwave type generators based on high frequency oscillations in semiconductors include Gunn diodes and Impatt diodes. They are more compact and robust. Gunn diodes are not easily tunable. Impatt

Table A1. *Radiation detectors.*

Type	Wavelength (μm)	NEP (W Hz$^{-1/2}$)	Speed	Comments
Rectifiers				
Point contact	≥ 300	10^{-9}	Fast	Fragile
Schottky	≥ 100	10^{-10} 10^4 K (heterodyne)	Depends on IF and matching,	Most used rectifier
Josephson 4.2 K	≥ 300	to 10^{-12}	to ~ 10 GHz	Fragile
Thermal bolometers				
Thermopile	Broad band	To 10^{-9}	3 Hz typical	
Pyroelectric	$\leq 10^3$	10^{-9} 10^{-7}	@ 100 Hz @ 1 MHz	
Bismuth microbolometer	$\leq 10^3$	10^{-9}	1 MHz	Arrays possible
Thermistor bolometer	Broad band	10^{-9}	@ 30 Hz	
Composite 4.2 K bolometer	$\leq 3 \times 10^3$	10^{-13}	@ 100 Hz	
InSb hot-electron 4.2 K	$100-3.10^3$	10^{-13}	200 kHz	
Photoconductors				
Germanium 4.2 K various dopings	2–160	10^{-13}	1 MHz	To GHz with special matching
HgCdTe 77 K	2–20	$10^{-13}-10^{-11}$	1 MHz to 1 GHz	Various alloys for different λ
Silicon photodiode (various configurations)	x-ray To ~ 1	10^{-12} (50% QE)	@ 1 MHz to 500 MHz	Arrays possible; also particle detection
Silicon avalanche	0.4–1	10^{-13} (40% QE)	300 MHz	
Si(Li)	1 keV–15 keV Res. ~ 0.5 keV @ 4 keV	100% QE	≤ 1 MHz count	Pulse height spectrometer

Type	Photon energy	Quantum efficiency	Speed	Comments
Ionization/electron multipliers/scintillators				
Phototubes	Visible	$\leq 10\%$ QE	To 1 GHz	Various photocathodes
Microchannel/ plate CCD		Characteristics similar to phototubes but provides imaging capabilities at high speed and sensitivity		
Proportional counter	Res.: 0.7 keV($\leq 10\%$ QE) @5 keV		Pulse-height ≤ 1 MHz rate	Also particles and noncounting mode
Multiwire proportional counter		For high spatial resolution imaging, etc.		
Scintillators, e.g., NaI(Tl)	Res: 4 keV @50 keV 50 keV @1000 keV	100% QE	Pulse-height ≤ 1 MHz	Also particles and noncounting mode
Emulsions		Various types and uses		
Neutron detectors				
BF$_3$/long counter		0.1% typical		Counters with moderator for total flux measurement
He^3 + moderator		similar		
He^3	0.1–3 MeV ~ 50 keV res.	0.01%	≤ 1 MHz counting	Energy spectrum
Scintillator (NE213 proton recoil)	0.17 MeV @2.5 MeV 0.6 MeV @14 MeV		≤ 1 MHz	Energy spectrum

Table A2. *Coherent sources: CW.*

Type	Wavelength (μm)	Power (mW)	Mechanical tuning	Comments
Electron tube				
Klystron	$\sim 10^4$	To > 1000		Frequency sweep
	~ 2000	≤ 100	~ 1 GHz	~ 100 MHz
Backward wave				
oscillator	3000	≤ 1000	($\sim 30\%$)	Frequency sweep
	1000	≤ 100		$\sim 30\%$
Gyrotron	≥ 1000	> 1000	$\sim 100\%$	High power (tunable nonproduction)
Solid state microwave				
Gunn diode	$> 10^4$	≥ 100	Not usual	Sweep ~ 100 MHz
oscillators	3000	≥ 10	Not usual	
Impatt diode	10^4	~ 200		Sweep to $\sim 20\%$
oscillator	~ 1500	~ 10		State of art
FIR optically pumped lasers				
CH_3F	1222	≤ 10	~ 5 MHz	Various lines
HCOOH	394	≤ 50	~ 5 MHz	available with
$C_2H_2F_2$	261	≤ 20	~ 5 MHz	different gases
$C_2H_2F_2$	184	≤ 150	~ 5 MHz	
CH_3OH	119	≤ 400	~ 5 MHz	
Discharge gas lasers				
HCN	337/(311)	≤ 100	~ 2 MHz	
DCN	195/190	≤ 300	~ 4 MHz	
H_2O	119	≤ 50	~ 5 MHz	
CO_2	9–10.6	$\leq 10^5$	50 MHz	Various line selection and control techniques
HeNe	0.6328	1 to 10^3	Not usual	Also lines at 1.15 and 3.39 μm
HeCd	0.4416	1 to 40	Not usual	
Argon	0.4880/ 0.5145	To 10^4	Various lines from	0.35 to 0.52 μm

diodes are tunable over appreciable ranges, but suffer from greater levels of frequency and amplitude noise fluctuations.

In the far infrared (FIR) there are many different optically pumped laser possibilities. However, only relatively few lines give reasonable power. The CO_2 laser is used almost universally for pumping. These lasers may be tuned somewhat, within the line width of the transition (~ 5 MHz typically).

Gas discharge lasers are available at a variety of wavelengths from 337 μm to short visible wavelengths.

Table A3. *Coherent sources: pulsed.*

	Wavelength (μm)	Energy	Pulse duration	Comments
FIR (D$_2$O)	385	50 mJ	100 ns	CO$_2$ pumped
CO$_2$	9–11	50 J	100 ns	Various possible high power lines
Neodymium	1.06	1 J	20 ns	Repetitive ($\sim$ 50 Hz)
Ruby	0.6943	10 J	50 ns	Nonrepetitive
Dye	Visible tunable	To 1 J	To 1 μs	Repetitive
Nitrogen	0.337	3 mJ	8 ns	Repetitive
Excimer (e.g., XeCℓ)	0.308	0.5 J	25 ns	Repetitive

In general the CW sources find their major use in interferometers, CW scattering from density fluctuations, and as local oscillators for heterodyne receivers.

Table A3 gives examples of pulsed sources that find use primarily in incoherent (and collective) scattering diagnostics, resonance fluorescence, and similar experiments requiring high peak power. In addition to the parameters listed, other important source characteristics include repetition rate, beam divergence, and band width. These may be obtained from manufacturers or from the references given in Further reading.

Further reading

As a general introduction to optical detectors see, for example:

Budde, W. (1983). *Optical Radiation Measurements: Physical Detectors of Optical Radiation. Vol. 4.* New York: Academic.
Kingston, R. H. (1978). *Detection of Optical and Infrared Radiation.* Berlin: Springer.

Various reviews of far infrared detector technology are available, for example:

Blaney, T. G. (1978). *J. Phys. E: Sci. Instrum.* 11:856.
Putley, E. H. (1973). *Phys. Technol.* 4:202.

The principles of laser operation are described by:

Siegman, A. E. (1971). *An Introduction to Lasers and Masers.* New York: McGraw-Hill.

Available lasers are surveyed in

Weber, M. J., ed. (1982). *Handbook of Laser Science and Technology.* New York: CRC Press.

Microwave techniques are discussed, for example, by

Ginzton, E. L. (1957). *Microwave Measurements.* New York: McGraw-Hill.

For detection of energetic photons and nuclear particles one may consult a book such as

Tsoulfandis, N. (1983). *Measurement and Detection of Radiation*. Washington: Hemisphere.

Detailed information on the performance of detectors, sources, or other radiation components is usually obtained most easily from the manufacturers themselves via catalogues, data sheets, specifications, and application notes. Locating appropriate manufacturers is usually done through trade journals and advertisements. Perhaps the most useful and wide ranging review, which includes some information on product specifications, is the annual "Buyer's Guide" in *Laser Focus*. Littleton, Mass.: Pennwell Publishing.

Appendix 4 Definitions and identities of fundamental parameters

Term	Symbol	Equivalent form
Principal constants		
Speed of light	c	2.998×10^8 ms^{-1}
Proton mass	m_p	1.673×10^{-27} kg
Electron mass	m_e	9.109×10^{-31} kg
Electron charge	e	1.602×10^{-19} C
Planck's constant	$h = 2\pi\hbar$	6.626×10^{-34} J s
Boltzmann's constant	k	1.381×10^{-23} J K^{-1}
		$= 8.617 \times 10^{-5}$ eV K^{-1}
Vacuum permeability	μ_0	$4\pi \times 10^{-7}$ H m^{-1}
Vacuum permittivity	ε_0	8.854×10^{-12} F m^{-1}
Plasma parameters		
Debye length	λ_D	$(\varepsilon_0 T_e / e^2 n_e)^{1/2}$
Plasma frequency	ω_{pe}	$(n_e e^2 / \varepsilon_0 m_e)^{1/2}$
Thermal (electron) speed	v_{te}	$(T_e/m_e)^{1/2} = \lambda_D \omega_{pe}$
Cyclotron frequency	ω	eB/m
Larmor (gyro) radius	ρ	$v/\Omega = mv/eB.$
Ion sound speed ($T_i = 0$)	c_s	$(T_e/m_i)^{1/2} = \lambda_D \omega_{pi}$
Beta	β	$2\mu_0 p/B^2$
Radiation and atoms		
Classical electron radius	r_e	$e^2/4\pi\varepsilon_0 m_e c^2 = 2.818 \times 10^{-15}$ m
		$4\pi n_e r_e = \omega_{pe}^2/c^2$
Thomson cross section	σ_T	$8\pi r_e^2/3 = 6.652 \times 10^{-29}$ m^2
Blackbody (Rayleigh–Jeans) intensity	$B(\omega)$	$\omega^2 T/8\pi^3 c^2$
Rydberg energy	R_y	$(m_e/2)(e^2/4\pi\varepsilon_0 \hbar)^2 =$
		2.180×10^{-18} J $= 13.61$ eV
Bohr radius	a_0	$\hbar 4\pi\varepsilon_0/e^2 m_e = 5.292 \times 10^{-11}$ m
Fine structure constant	α	$e^2/4\pi\varepsilon_0 \hbar c = 1/137.04$
		$(2R_y/m_e)^{1/2} = \alpha c$
		$2R_y a_0 = e^2/4\pi\varepsilon_0 = \alpha\hbar c$

Glossary

For each of the following symbols its meaning (or meanings) and the main or subsection in which it appears with that meaning are indicated.

a	plasma minor radius	2.1.4
	probe (dimensions) radius	3.1.3
a_0	Bohr radius	6.2.5, A3
A	area (generally)	
$\mathbf{A}$	vector potential	5.1.1
A_p	probe area	3.2.2
A_s	sheath area	3.2.1
A_{ij}	spontaneous $i \to j$ transition probability	6.1
b	impact parameter	5.3.1, 6.3.2
$\mathbf{b}$	magnetic field perturbation	2.4.1
b_{90}	90° scattering impact parameter	5.3.1
$\mathbf{B}, B$	magnetic field (generally)	2.1.1
$B(\omega)$	blackbody spectral intensity	5.2.4
B_{ij}	induced $i \to j$ transition probability	6.1
c	speed of light	A4
c_s	sound speed	3.3.3
C	bremsstrahlung coefficient	7.2.4
C_m	mth Fourier cosine coefficient	2.2
d	beam diameter	4.2.4, 7.2.4
D	deuterium	8.3.1
D	plasma dimension	7.2.4
$\mathbf{D}$	electric displacement	7.3.2
$\mathbf{D}$	diffusion tensor	3.3.3
$D_\parallel$	parallel diffusion coefficient	3.3.3
$D_\perp$	perpendicular diffusion coefficient	3.3.3
e	electronic charge (magnitude)	1.2.2
	(as subscript) electron	1.2.2
e	base of natural logarithms	4.1.1
E	particle energy	8.1.3, 8.3.1
$\mathbf{E}, E$	electric field	1.3
f	distribution function	1.2
	volumetric quantity	4.4
f_{ji}	oscillator strength	6.1
F	particle energy spectrum	8.1.3
	chord integral measurement	4.4

	polarization discriminant	4.3.3
F_e	Klimontovich distribution function	7.3.1
g_i	statistical weight	6.1
$\bar{g}$	distribution-averaged Gaunt factor	5.3.3, 6.3.1– 6.3.3
G	Gaunt factor	5.3.1, 6.3.1
h	Planck's constant	6.1, A4
$\hbar$	Planck's constant ($\div 2\pi$)	5.3.2, A4
i	square root of -1 (generally)	
	(as subscript) ion	1.2.2
	(as subscript) incident	7.1
$\hat{\mathbf{i}}$	unit vector in incident direction	7.1
I	electric current	2.1.3, 3.1.1
$I(\omega)$	intensity of radiation	5.2.4
$\mathbf{j}, j$	electric current density	1.2.2, 2.1.4, 4.1.1, etc,
$j(\omega)$	radiant volumetric emissivity	5.2.2
j_m	cyclotron emissivity, frequency-integrated over the mth harmonic	5.2.3
J	total particle current	3.1.1
k	Boltzmann's constant	A4
$\mathbf{k}, k$	propagation wave vector	4.1.1
K	kinetic energy in c.m. frame	8.3.3
	number of electrons	6.2.2
ℓ	mean free path	3.1.3
$\mathbf{l}, l$	path length	2.1.1, 4.1.3, etc,
ℓ_i	internal inductance	2.3
L	length of collection region	3.3.3, 7.2.4
	total path length (difference)	4.2.3
	distance	4.2.6
	number of states	6.2.2
m	mass of particle (generally)	
	harmonic number	2.2, 5.2.1
	angular momentum states	6.1
$\mathbf{M}^k$	kth moment of distribution	1.2.1
M	total mass	8.3.3
n	particle density (generally)	1.2.1
	toroidal harmonic number	2.4.1
	turns per unit length	2.1.3
N	number of turns in coil	2.1.1
	refractive index	4.1.1
N_e	Klimontovich particle density	7.3.1

N_j	number of atoms in state j	6.1
p	pressure tensor	1.2.1
p	scalar pressure	1.2.1
P	power	2.1.4
$dP/d\omega$	spectral power density	5.3.1
$d^2P/d\omega\,d\Omega_s$	spectral power density per unit solid angle	5.2.1
P	electric polarization	7.3.2
q	particle charge	8.4.2
	polarization ratio	4.3.3
q_s	magnetic field safety factor	8.2.4, 8.4.2
Q	heat flux tensor	1.2.1
Q	fusion energy release	8.3.3
	quantum efficiency	7.2.4
r	(minor) radius (generally)	2.2
r	position vector (of charge)	5.1.1
r_e	classical electron radius	5.3.1, 7.1.1, A4
R	(major) radius (generally)	2.2
R	position vector (of field point)	5.1.1
R_y	Rydberg energy (of H ground state)	5.3.2
s	ray path distance	5.2.4
	(as subscript) sheath	3.2.1
	(as subscript) scattered	7.1.1
$\hat{s}$	unit vector in scattering direction	7.1.1
S	source rate	3.3.2, 8.1.3
	(generalized) recombination rate coefficient	6.2.4
S, S	Poynting vector	5.1.1, 7.2.3
$S(\mathbf{k}, \omega)$	scattering form factor	7.3.1
S_{ij}	line strength	6.1
t	time (generally)	
	(as subscript) thermal	A4
t'	retarded time	5.1.1
T	temperature	1.2.1
	time period	7.2.2
T	Maxwell stress tensor	2.2
u_{90}	nondimensionalized frequency	5.3.1
$\mathbf{U}_m$	cyclotron radiation E field vector	5.2.1
v, v	velocity (generally)	
V	electric potential, voltage	2.1.1, 3.1.2
V, V	(mean) c.m. velocity	1.2.1, 8.3.3
w	wall thickness	2.4.1
$w(\xi)$	plasma susceptibility function	7.3.2

W	energy (radiated)	5.1.2, 5.3.1
$\mathbf{x}$	position vector (generally)	
x	coordinate (generally)	
X	plasma wave parameter (ω_p^2/ω^2)	4.1.2
y	coordinate (generally)	
Y	plasma wave parameter (Ω/ω)	4.1.2
z	coordinate (generally)	
Z	ion charge number	1.2.1
Z_{eff}	effective charge number	5.3.6
α	fine structure constant	5.3.1
	absorption coefficient	5.2.4
	attenuation coefficient	8.1.2
	diffraction angle	4.2.4
	ionization coefficient	6.2.4
	$(1/k\lambda_D)$	7.3.2
$\alpha_m(\omega)$	cyclotron absorption coefficient (as a function of frequency)	5.2.4
α_m	cyclotron absorption coefficient (frequency-integrated over mth harmonic)	5.2.4
$\boldsymbol{\beta}, \beta$	velocity in units of c	5.2.1
β	ratio of plasma kinetic to magnetic pressure	2.2.1
γ	relativistic (mass increase) factor	5.2.1
$\boldsymbol{\Gamma}, \Gamma$	particle current density	3.1.1
Γ	Salpeter shape function	7.3.2
δ	(as prefix) change in (generally)	
$\delta(x)$	Dirac delta function	A1
δ_{ij}	Kronecker delta	4.1.1
Δ	(as prefix) change in (generally)	
	plasma displacement	2.2.2
	orbit displacement	8.2.4, 8.4.2
ε	dielectric constant (relative permittivity)	4.1.1
$\boldsymbol{\varepsilon}, \varepsilon_{ij}$	dielectric tensor	4.1.1
ε_0	permittivity of free space	A4
ζ	Reciprocal of Euler's constant, 1.78	5.3.1
η	normalized electric potential	3.3.2
$\eta_\pm$	fraction of power in each mode	5.2.5
η_j	generalized (free state) quantum number	5.3.2
θ	(poloidal) angle	2.2
	ray deviation angle	4.2.4
κ	Doppler shift factor	5.1.1

λ	wavelength (generally)	
λ_D	Debye length	3.1.2
Λ	field asymmetry factor	2.2.2
	argument of Coulomb logarithm	2.1.4
μ	magnetic permeability	2.1.3
$\boldsymbol{\mu}$	magnetic moment	8.4.2
μ_0	permeability of free space	A4
ν	(periodic) frequency (generally)	6.1
ξ	argument ($\omega/kv_t\sqrt{2}$) of dispersion function	7.3.2
π	ratio of circumference to diameter of circle	
Π	polarization tensor in scattering	7.2.2
ρ	Larmor radius	3.3.1
$\rho(\nu)$	energy density of radiation	6.1
σ	conductivity	2.1.3
	cross section (generally)	
	Thomson cross section	7.2.11
$\boldsymbol{\sigma}$	conductivity tensor	4.1.1

Cross section for:

σ_c	charge exchange	8.1.2
σ_d	dielectronic recombination	6.3.4
σ_e	electron ionization	8.1.1
σ_i	(electron) ionization	6.3.2
σ_n	D–D neutron reaction	8.3.1
σ_p	proton ionization	8.1.2
σ_r	(radiative) recombination	6.3.1
σ_{ij}	excitation $i \rightarrow j$	6.3.3
τ	time duration or time constant (generally)	
	optical depth	5.2.4
ϕ	(toroidal) angle	2.1.4
	electric potential	5.1.1
	wave phase	4.2.1
$\hat{\phi}$	unit vector in toroidal direction	2.2.2
Φ	magnetic flux	2.1.3
χ	electric susceptibility	7.3.2
ψ	poloidal magnetic flux function	2.2.2
ω	angular frequency (generally)	
ω_c	(Ω/γ) relativistic cyclotron frequency	5.2.1
ω_p	plasma frequency	4.1.2
Ω	(rest) cyclotron frequency	4.1.2
Ω_s	solid angle	5.1.2

∇	gradient operator	
$\wedge$	vector product (e.g., $\mathbf{A} \wedge \mathbf{B}$)	
$+$	(as subscript) ordinary mode	4.3.2, 5.2.5
$-$	(as subscript) extraordinary mode	4.3.2, 5.2.5
$\parallel$	(as subscript) parallel	5.2.1
$\perp$	(as subscript) perpendicular	5.2.1
$\langle \ \rangle$	mean value	
$[\ ]$	retarded value	
	(in 5.1 and 7.1–7.2.2 only)	
x	convolution operator	A1

References

Allen, J. E., Boyd, R. L. F., and Reynolds, P. (1957). *Proc. Phys. Soc.* 70B:297.

Bekefi, G. (1966). *Radiation Processes in Plasmas.* New York: Wiley.

Bely, O. (1966). *Proc. Phys. Soc.* 88:587.

Berezovskii, E. L., et al. (1985). *Nucl. Fusion* 25:1495.

Bernstein, I. B. and Rabinowitz, I. (1959). *Phys. Fluids* 2:112.

Bienisek, F. M., et al. (1980). *Rev. Sci. Instr.* 51:206.

Bohm, D. (1949). In *Characteristics of Electrical Discharges in Magnetic Fields.* A. Guthrie and R. K. Wakerling, eds. New York: McGraw-Hill.

Bohm, D., Burhop, E. H. S. and Massey, H. S. W. (1949). In *Characteristics of Electrical Discharges in Magnetic Fields.* A. Guthrie and R. K. Waterling, eds. New York: McGraw Hill.

Bornatici, M. (1980). In Electron Cyclotron Emission and Electron Cyclotron Heating, Proc. Joint Workshop. Culham Laboratory Report No. CLM-ECR.

Bornatici, M., Cano, R., DeBarbieri, O., and Engelmann, F. (1983). *Nucl. Fusion* 23:1153.

Boyd, T. J. M. and Sanderson, J. J. (1969). *Plasma Dynamics.* London: Nelson.

Brau, K., et al. (1983). Princeton University Report No. PPPL 2013.

Breene, R. G. (1961). *The Shift and Shape of Spectral Lines.* London: Pergamon.

Bretz, N., et al. (1978). *Appl. Opt.* 17:192.

Brotherton-Ratcliffe, D. and Hutchinson, I. H. (1984). Culham Laboratory Report No. CLM-R246.

Brussard, P. J. and van de Hulst, H. C. (1962). *Rev. Mod. Phys.* 34:507.

Brysk, H. (1973). *Plasma Phys.* 15:611.

Burgess, A. (1965). *Astrophys. J.* 141:1588.

Carolan, P. G. and Evans, D. E. (1972). In *Ionization Phenomena in Gases*, proc. 10th Int. Conf., Oxford. Oxford: Donald Parsons.

Chandler, G. I., et al. (1986). *Appl. Opt.* 25:1770.

Chapman, S. and Cowling, T. G. (1970). *The Mathematical Theory of Non-Uniform Gases.* 3rd ed. New York: Cambridge Univ. Press.

Chen, F. F. (1965). In *Plasma Diagnostic Techniques.* R.H. Huddlestone and S. L. Leonard, eds. New York: Academic.

Chen, F. F. (1984). *Introduction to Plasma Physics and Controlled Fusion.* 2nd ed. New York: Plenum.

Chrien, R. E. and Strachan, J. D. (1983). *Phys. Fluids* 26:1953.

Clemmow, P. C. and Dougherty, J. P. (1969). *Electrodynamics of Particles and Plasmas.* Reading, Mass.: Addison-Wesley.

Cohen, S. A. (1978). *J. Nucl. Mater.* 76/77:68.

Colestock, P. L., Connor, K. A., Hickcock, R. L., and Dandl, R. A. (1978). *Phys. Rev. Lett.* 40:1717.

Cormack, A. M. (1964). *J. Appl. Phys.* 35:2908.

Crandall, D. H. (1983). In *Atomic Physics of Highly Ionized Atoms.* Richard Marrus, ed. New York: Plenum.

DeMarco, F. and Segre, S. E. (1972). *Plasma Phys.* 14:245.

Dirac, P. A. M. (1958). *Principles of Quantum Mechanics*, 4th ed. London: Oxford Univ. Press.

Dodel, G. and Kunze, W. (1978). *Infrared Phys.* 18:773.

Drawin, H. W. (1968). In *Plasma Diagnostics.* W. Lochte-Holtgreven, ed. Amsterdam: North-Holland.

Druyvesteyn, M. J. (1930). *Z. Phys.* 64:790.

Elwert, G. (1939). *Ann. Phys.* 34:178.

Emmert, G. A., Wieland, R. M., Mense, A. T., and Davidson, J. N. (1980). *Phys. Fluids* 23:803.

Eubank, H., et al. (1978). In *Plasma Physics and Controlled Nuclear Fusion Research*, Proc. 7th Int. Conf., Innsbruck. Vol. 1, p. 167. Vienna: IAEA.

Fisher, W. A., Chen, S.-H., Gwinn, D., and Parker, R. R. (1984). *Nucl. Instr. and Methods in Phys. Res.* 219:179.

Foord, M. E., Marmar, E. S., and Terry, J. L. (1982). *Rev. Sci. Instr.* 53:1407.

Foote, J. H. (1979). *Nucl. Fusion* 19:1215

Fonk, R. J., Darrow, D. S., and Jaehnig, K. P. (1984) *Phys. Rev.* A 29:3288.

Forrest, M. G., Carolan, P. G., and Peacock, N. J. (1978). *Nature* 271:718.

Freeman, R. L. and Jones, E. M. (1974). Culham Report No. CLM-R137. (Available from Her Majesty's Stationery Office, London.)

Fried, B. D. and Conte, S. D. (1961). *The Plasma Dispersion Function*. New York: Academic.

Gabriel, A. H. (1972). *Mon. Not. R. Astr. Soc.* 160:99.

Gamow, G. and Critchfield, C. (1949). *Theory of the Atomic Nucleus and Nuclear Energy Sources*. Chapter 10. London: Oxford Univ. Press.

Gandy, R. F., Hutchinson, I. H., and Yates, D. H. (1985). *Phys. Rev. Lett.* 54:800.

Gaunt, J. A. (1930). *Phil. Trans. Roy. Soc.* A 229:163.

Gibson, A. and Reid, G. W. (1964). *Appl. Phys. Lett.* 5:195.

Ginsburg, V. L. (1961). *Propagation of Electromagnetic Waves in Plasma*. New York: Gordon and Breach.

Goldston, R. J. (1978). *Phys. Fluids* 21:2346.

Goldston, R. J. (1982). In *Diagnostics for Fusion Reactor Conditions*, Proc. Int. School of Plasma Phys., Varenna. P. E. Stott et al., eds. Brussels: Commission of E.E.C.

Granetz, R. S. and Camacho, J. F. (1985). *Nucl. Fusion* 25:727.

Granetz, R. S., Hutchinson, I. H., and Overskei, D. O. (1979). *Nucl. Fusion* 19:1587.

Griem, H. R. (1964). *Plasma Spectroscopy*. New York: McGraw-Hill.

Gruebler, W., Schmelzbach, P. A., Konig, V., and Marmier, P. (1970). *Helv. Phys. Acta* 43:254.

Gull, S. F. and Daniell, G. S. (1978). *Nature* 272:686.

Gurney, E. F. and Magee (1957). *J. Chem. Phys.* 26:1237.

Harrison, E. R. and Thompson, W. B. (1959). *Proc. Phys. Soc.* 74:145.

Heald, M. A. and Wharton, C. B. (1965). *Plasma Diagnostics with Microwaves*. New York: Wiley.

Heitler, W. (1944). *The Quantum Theory of Radiation*. 2nd ed. London: Oxford Univ. Press.

Hivley, L. (1977). *Nucl. Fusion* 17:873.

Hosea, J., et al. (1978). *Phys. Rev. Lett.* 40:839.

Hosea, J., et al. (1980). In *Plasma Physics and Controlled Nuclear Fusion Research*, Proc. 8th Int. Conf., Brussels. Vienna: IAEA.

Hsuan, H., et al. (1978). In *Heating in Toroidal Plasmas*, Proc. Joint Varenna–Grenoble Symp. Vol. 2 p. 87. London: Pergamon.

Hughes, M. H. and Post, D. E. (1978). *J. Comp. Phys.* 28:43.

Hulse, R. A., Post, D. E., and Mikkelson, D. R. (1980). *J. Phys. B.* 13:3895.

Hutchinson, I. H. (1976a). *Plasma Phys.* 18:246.

Hutchinson, I. H. (1976b). In *Plasma Physics and Controlled Nuclear Fusion Research*, Proc. 6th Int. Conf., Berchtesgaden. Vol. 2, p. 227. Vienna: IAEA.

Hutchinson, I. H. (1976c). *Phys. Rev. Lett.* 37:338.

Hutchinson, I. H. (1987). Submitted to *Phys. Fluids*.

Hutchinson, I. H. and Kato, K. (1986). *Nucl. Fusion* 26:179.

Hutchinson, I. H. and Komm, D. S. (1977). *Nucl. Fusion* 17:1077.

Jacobson, A. R. (1978). *Rev. Sci. Instr.* 49:318.

Jacobson, A. R. (1982). *Rev. Sci. Instr.* 53:918.

Jahoda, F. C. and Sawyer, G. A. (1971). In *Methods of Experimental Physics*. R. H. Lovberg and H. R. Griem, eds., Vol. 9. New York: Academic.

Jarvis, O. N., et al. (1986). *Rev. Sci. Instrum.* 57:1717.

Johnson, J. L., et al. (1976). In *Plasma Physics and Controlled Nuclear Fusion Research*, Proc. 6th Int. Conf., Berchtesgaden. Vol. 2, p. 395. Vienna IAEA.

Johnson, L. C. and Hinnov, E. (1973). *J. Quant. Spectrosc. Radiat. Transfer* 13:333.

Jones, R. and Wykes, C. (1983). *Holographic and Speckle Interferometry*. New York: Cambridge Univ. Press.

Karzas, W. J. and Latter, R. (1961). *Astrophys. J. Suppl. Series* 6:167.

Kasparek, W. and Holtzhauer, E. (1983). *Phys. Rev.* A 27:1737.

Koch, H. W. and Motz, J. W. (1959). *Rev. Mod. Phys.* 31:920.

Kramers, H. A. (1923). *Phil. Mag.* 46:836.

Laframboise, J. (1966). In *Rarified Gas Dynamics*, Proc. 4th Int. Symp., Toronto. J. H. deLeeuw, ed., Vol. 2, p. 22. New York: Academic. (Also Univ. of Toronto Institute for Aerospace Studies Report No. 100.)

Lam, S. H. (1965). *Phys. Fluids* 8:73.

Landau, L. D. and Lifschitz, E. M. (1951). *The Classical Theory of Fields*. Reading, Mass.: Addison-Wesley.

Lipschultz, B., Prager, S. C., Todd, A. M., and Delucia, J. (1980). *Nucl. Fusion* 20:683.

Lotz, W. (1967). *Z. Phys.* 216:241.

Luhmann, N. C., Jr. (1979). In *Infrared and Millimeter Waves*. K. Button, ed., Vol. 2. New York: Academic.

Marmar, E. S. and Rice, J. E. (1985). Private communication.

McCormick, K., et al. (1977). In *Controlled Fusion and Plasma Physics*, Proc. 8th Europ. Conf., p. 140. Prague: European Physical Society.

McCormick, K., et al. (1986). In *Controlled Fusion and Plasma Physics*, Proc. 13th Europ. Conf., Vol. 2, p. 323. Schliersee: European Physical Society.

McWhirter, R. P. (1965). In *Plasma Diagnostic Techniques*. R. H. Huddlestone and S. L. Leonard, eds., New York: Academic.

Meddens, B. J. H. and Taylor, R. J. (1974). MIT Report No. PRR7411.

Miyamoto, K. (1980). *Plasma Physics for Nuclear Fusion*. Cambridge, Mass.: MIT Press.

Morse, P. M. and Feshbach, H. (1953). *Methods of Theoretical Physics*. New York: McGraw-Hill.

Mukhovatov, V. S. and Shafranov, V. D. (1971). *Nucl. Fusion* 11:605.

Nagle, D. E., Quinn, W. E., Ribe, F. L., and Riesenfield, W. B. (1960). *Phys. Rev.* 119:857.

Nee, S. F., Pechacek, R. E., and Trivelpiece, A. W. (1969). *Phys. Fluids* 12:2651.

Olsen, R. E., et al. (1978). *Phys. Rev. Lett.* 41:168.

Olsen, R. E. (1980). In *Electronic and Atomic Collisions*. N. Oda, ed. Amsterdam: North-Holland.

Park, H., et al. (1982). *Rev. Sci. Instr.* 53:1535.

Peacock, N. J. (1978). In *Diagnostics for Fusion Experiments*, Proc. Int. School of Plasma Phys., Varenna. E. Sindoni and C. B. Wharton, eds. London: Pergamon.

Pechacek, R. E. and Trivelpiece, A. W. (1967). *Phys. Fluids* 10:1688.

Piotrowicz, V. A. and Carolan, P. G. (1983). *Plasma Phys.* 25:1065.

Post, D. E., Jensen, R. V., Tarter, C. B., Grasberger, W.H., and Lokke, W. A. (1977). *At. Data and Nucl. Data Tables* 20:397.

Razdobarin, G., et al. (1979). *Nucl. Fusion* 19:1439.

Rice, J. E., Molvig, K., and Helava, H. I. (1982). *Phys. Rev.* A 25:1645.

Rindler, W. (1960). *Special Relativity*. Edinburgh: Oliver and Boyd.

Rohr, H., et al. (1982). *Nucl. Fusion* 22:1099.

Salpeter, E. E. (1960). *Phys. Rev.* 120:1528.

Scheuer, P. A. G. (1968). *Astrophys. J.* 151:L139.

Schmidt, G. (1979). *Physics of High Temperature Plasmas.* 2nd ed. New York: Academic.

Schott, G. A. (1912). *Electromagnetic Radiation.* New York: Cambridge Univ. Press.

Schott, L. (1968). In *Plasma Diagnostics.* W. Lochte-Holtgreven, ed. Amsterdam: North-Holland.

Seaton, M. (1959). *Mon. Not. R. Astron. Soc.* 119:81.

Seaton, M. (1962). In *Atomic and Molecular Processes.* D. R. Bates, ed. New York: Academic.

Seaton, M. J. and Storey, P. J. (1976). In *Atomic Processes and Applications.* P. G. Burke and B. L. Moiseiwitsch, eds. Amsterdam: North-Holland.

Selden, A. C. (1982). Culham Laboratory Report No. CLM R220. (Available from Her Majesty's Stationery Office, London.)

Semet, A., et al. (1980). *Phys. Rev. Lett.* 45:445.

Shafranov, V. D. (1963). *Plasma Phys.* 5:251.

Sheffield, J. (1972). *Plasma Phys.* 14:385.

Sheffield, J. (1975). *Plasma Scattering of Electromagnetic Radiation.* New York: Academic.

Shkarofsky, I. P., Johnston, T. W., and Bachynski, M. P. (1966). *The Particle Kinetics of Plasmas.* Reading, Mass.: Addison-Wesley.

Slater, J. C. (1968). *Quantum Theory of Matter.* New York: McGraw-Hill.

Soltwisch, H. (1983). In *Controlled Fusion and Plasma Physics.* Proc. 11th Europ. Conf., page 123. Aachen: European Physical Society.

Sommerfield, A. (1939). *Atombau and Spektrallinen.* Braunschewig: Friedrich Vieweg und Sohn.

Spitzer, L. (1962). *The Physics of Fully Ionized Gases.* 2nd ed., p. 136. New York: Wiley Interscience.

Stangeby, P. C. (1982). *J. Phys. D: Appl. Phys.* 15:1007.

Stenzel, R. L., et al. (1983). *Rev. Sci. Instr.* 54:1302.

Strachan, J. D. et al. (1979). *Nature* 279:626.

Strachan, J. D., and Jassby, D. L. (1977). *Trans. Am. Nucl. Soc.* 26:509.

Suckewer et al. (1979). *Phys. Rev. Lett.* 43:207.

Swartz, K., Hutchinson, I. H., and Molvig, K. (1981). *Phys. Fluids* 24:1689.

Texter, S., et al. (1986). *Nucl. Fusion* 26:1279.

Thomson, J. J. (1912). *Phil. Mag.* 23:449.

Thorne, A. P. (1974). *Spectrophysics.* London: Chapman and Hall.

Tonks, L. and Langmuir, I. (1929). *Phys. Rev.* 34:876.

Traving, G. (1968). In *Plasma Diagnostics.* W. Lochte-Holtgreven, ed. Amsterdam: North-Holland.

Trubnikov, B. A. (1958). *Soviet Phys.–Doklady* 3:136.

Unsold, A. (1955). *Physik der Sternatmospheren.* Berlin: Springer.

Veron, D. (1974). *Opt. Comm.* 10:95.

Veron, D. (1979). In *Infrared and Milimeter Waves.* K. J. Button, ed., Vol. 2. New York: Academic.

von Goeler, S. (1978). In *Diagnostics for Fusion Experiments,* Proc. Int. School of Plasma Phys., Varenna. E. Sindoni and C. Wharton, eds. London: Pergamon.

von Goeler, S., et al. (1982). In *Diagnostics for Fusion Reactor Conditions,* Proc. Int. School of Plasma Phys., Varenna. P. E. Stott, et al., eds. Brussels: Commission of E.E.C.

West, W. P. (1986). *Rev. Sci. Instrum.* 57:2006.

Wiese, W. L., Smith, M. W., and Glennon, B. M. (1966). *Atomic Transition Probabilities.* NSRDS-NBS4. Washington: U.S. Government Printing Office.

Wolfe, S. M., Button, K. J., Waldman, J., and Cohn, D. R. (1976). *Appl. Opt.* 15:2645.

Zhuravlev, V. A. and Petrov, G. D. (1979). *Soviet J. Plasma Phys.* 5:3.

Index